BIOINSTRUMENTATION

BIOINSTRUMENTATION

M.H. Fulekar
Bhawana Pandey
School of Environment and Sustainable Development
Central University of Gujarat
Gandhinagar, Gujarat
India

I.K. International Publishing House Pvt. Ltd.

NEW DELHI • BANGALORE

Published by

I.K. International Publishing House Pvt. Ltd.
S-25, Green Park Extension
Uphaar Cinema Market
New Delhi–110 016 (India)
E-mail: info@ikinternational.com
Website: www.ikbooks.com

ISBN 978-93-82332-39-8

10 9 8 7 6 5 4 3 2 1

Published by Krishan Makhijani for I.K. International Publishing House Pvt. Ltd., S-25, Green Park Extension, Uphaar Cinema Market, New Delhi–110 016 and Printed by Rekha Printers Pvt. Ltd., Okhla Industrial Area, Phase II, New Delhi–110 020

In Loving Memory
of
Powara Fulekar
and
Prof. K.N. Pandey

Preface

INSTRUMENTS are a vital part of the methodology of modern science. The development of instruments and literature will aid in updating the content of biology while establishing biology to be a science as quantitative and analytic as chemistry and physics. Biology has evolved from a descriptive cataloging discipline into a fully experimental science. **INSTRUMENTATION** needs in the biological sciences began with the development of the telescope and the microscope by Galileo in the early 1600s. The invention of the compound microscope in 1610 gave birth to microscopic anatomy. In the 19th and early 20th century, microscopic anatomy opened up the areas of biological chemistry. The research in biology has attempted to follow this evolution by including more emphasis on process, quantification, problem solving, and "open-ended" laboratory activity. The measurement of physical, chemical and biological parameters using instruments in any living organisms (including human beings) is what referred to as **"BIOINSTRUMENTATION"**.

The realization of the present-day accomplishments in many disciplines such as physical, chemical and biological sciences has come out through the applications of inventions of modern scientific instruments. Technology includes the development and use of modern scientific instruments and techniques which involve ideas of fresh insights, new concepts, and innovative approaches to scientific problems. The design of bioinstrumentation is the cumulative result of synchronizing the technologies involved in physics, chemistry and biology. The importance of modern instrumentation in the understanding of many biological processes makes it obvious that such instruments should be introduced in graduate/ postgraduate teaching/research syllabus.

The present book deals with modern instrumental knowledge and their applications in biology, with special reference to UV-VIS-FT-RI Spectroscopy, Florescence Spectroscopy, Flame Photometry, Atomic Absorption Spectroscopy, Gas Chromatography, High Pressure Liquid Chromatography, Thin Layer Chromatography, Gas Chromatography—Mass spectroscopy, Nuclear Magnetic Resonance, Bioreactor, Gel Electrophoresis PCR, Biosensor and Microscopy.

"Bioinstrumentation" will be useful as textbook and/or reference book to undergraduate/ postgraduate students, researchers, scientists, academicians, industrial personnel to understand basic principles, theories, components and functions, several other fundamentals and applications in chemistry and biology. The biology teacher, both graduate and postgraduate level equipped with the appropriate selection of instruments can provide to his/her students demonstrations and laboratories which go beyond structural description and illuminate many functional phenomena of the living state. The multidisciplinary research projects in the area of physical, chemical and biological sciences depend upon access to modern scientific instruments.

Prof. R.K. Kale, Vice Chancellor Central University of Gujarat is gratefully acknowledged for his inspiration, motivation, and encouragement for innovative and novel work.

We, the authors of the book, sincerely thank to family members — Dr. Kalpana, Jaya, Jyoti, Vinay Sacchidananad, Dilip, Mrs Gayatri, Santosh, Rachana, Janmejay, Mrityunjay, and Aishwarya.

Authors feel proud to acknowledge by heart to Prof. K. N. Pandey, a distinguished academician who is a source of inspiration for novelty work.

The technical support and constant encouragement received from our colleagues Dr. Anamika Singh, and Jaya Sharma, from the environmental biotechnology laboratory, University of Mumbai are acknowledged with thanks.

We strongly believe that the successful completion of this manuscript is possible because of the blessings of "Lord Buddha".

M. H. Fulekar
Bhawana Pandey

Contents

CHAPTER 1

Instrumentation and their Applications in Biology

Instrumentation in the biological sciences began with the development of the telescope and the microscope by Galileo in the early 1600s. Optical aids were of key importance in observational biology and these devices and their photographic successors led to major advances in descriptive biology, taxonomy and related fields. The invention of the compound microscope in 1610 gave birth to microscopic anatomy. In the 19th and early 20th century, microscopic anatomy opened up the areas of histology, cell biology and biological chemistry. These disciplines, along with the physiological sciences and neurobiology, have matured together; their focus has been concerned with structure and function at the molecular, cellular and organizational level. They have increasingly depended upon instruments to improve power of observation to derive quantitative data and to separate cellular components.

The realization of the present-day accomplishments in many of these disciplines has come about through the applications of inventions of modern scientific instruments. The accomplishment of modern scientific research stands on two pillars, i.e., ideas and technology. Ideas involve fresh insights, new concepts, and innovative approaches to scientific problems. Technology includes the development and use of complex scientific instruments and techniques. Ideas and technology are complementary and to a large extent, interdependent, most research projects in the physical and biological sciences depend upon access to modern complex scientific instruments.

Spectroscopy is a fundamental analytical tool utilized throughout all of the sciences. For chemistry and biology alone, there are thousands of applications. In the past two decades there have been monumental advances in the miniaturization of components used in spectrophotometric systems. The key components include detector arrays, laser diodes and fibre optics. Currently, there are numerous commercially available miniature spectrometer systems as well as discrete components that are used by researchers in designing their own systems. Current literature is reviewed for chemical and biological applications specifically using miniaturized spectrometer systems with the focus being on ultraviolet-visible-near infrared including FT-IR spectrometers. The applications include laboratory applications, environmental sensing, on-site industrial analysis, life sciences and ecological applications.

Fluorescence phenomenon was named by George Stokes in the year 1852. In the field of spectroscopy, the analysis of fluorescence from a sample of material can be done through a special form of electromagnetic spectroscopy. This fluorescence spectroscopy uses a beam of eight which is tailored for the type of fluorescence detection needed in compounds. The usual option is an ultraviolet light that is used to excite electrons in the molecules which will cause them to produce a light emitting phenomenon of lower energy which can be within the visible and non-visible light spectrums. Fluorescence can be used as a practical property of materials in various fields of science. Mineralogy and gemology use fluorescence to determine various qualities in samples. The use of fluorescent chemicals as a sensor or marker is useful in various applications. Special fluorescent labeling is possible with proteins and nucleic acids which can be used in various biological experimentations by attaching fluorophore materials, dyes and biological detectors.

Flame photometry is a branch of atomic spectroscopy in which the species examined in the spectrometer are in the form of atoms. Flame photometers use optical filters to monitor for the selected emissions wavelength produced by the analyte species. Emission techniques measure the intensity of light that is emitted as electrons return to the lower energy levels. Flame photometry is suitable for qualitative and quantitative determination of several cations, especially for metals that are easily excited to higher energy levels at a relatively low flame temperature (mainly Na, K, Rb, Cs, Ca, Ba, Cu).

Atomic absorption spectroscopy is used for determination of the presence and concentration of metal in liquid samples. Metals include Fe, Cu, Al, Pb, Ca, Zn, Cd and many more. Typical concentrations range in the low mg/L (ppm) range. In Atomic Absorption Spectrometry (AAS), a light of a specific wavelength is passed through the atomic vapor of an element of interest, and measurement is made of the alternation of the intensity of the light as a result of absorption. Quantitative analysis by AAS depends on:

(a) accurate measurement of the intensity of the light; and

(b) the assumption that the radiation absorbed in proportional to atomic concentration.

Metals absorb ultraviolet light in their elemental form when they are excited by heat, either by flame or graphite furnaces. Each metal has a characteristic wavelength that is absorbed. The AAS instrument looks for a particular metal by focusing a beam of UV light at a specific wavelength through a flame and into a detector. The sample of interest is aspirated into the flame. If that metal is present in the sample, it will absorb some of the light, thus reducing its intensity. The instrument measures the change in intensity. A computer data system converts the change in intensity into absorbance.

Chromatography is the science which deals with the separation of molecules based on differences in their structure and/or composition. In general, chromatography involves a preparation of the materials to be separated – the "test preparation" over a stationary support. The molecules in the test preparation will have different interactions with the stationary support leading to separation of similar molecules. Test molecules which display lighter interactions with the support will tend to move more slowly through the support than those molecules with weaker interactions. In this way, different types of molecules can be separated from each other as they move over the support material.

- **Gas chromatography** is a non-analytical scientific technique to separate a mixture of vaporizable substance and resolve the mixture into a single component. Gas-liquid chromatography is the most widely used chromatographic technique for environmental analysis. Analysis of organic components is possible for a variety of materials such as water, soils and ambient air. It is often used to test hazardous wastes.
- **High-performance liquid chromatography** (HPLC) or high pressure liquid chromatography is a chromatographic technique to separate, identify and quantify chemical compounds. In biology, HPLC allows for the rapid and sensitive analysis of peptide and protein structures. It has simplifies our understanding of biological processes and in the development of peptide and protein based pharmaceuticals.
- **Thin Layer Chromatography** (TLC) like any other chromatography technique, has a stationary and mobile phase. TLC is a widely used separation technique in life science and chemistry studies. It is easy, cheap and gives rapid results for basic analytical studies of phytochemicals.
- **Gas Chromatography – Mass Spectroscopy** (GC-MS) is one of the so-called hyphenated analytical techniques. As the name implies, it is actually two techniques that are combined to form a single method of analyzing mixtures of chemicals. Gas chromatography separates the components of a mixture and mass spectroscopy characterizes each component individually by combining the two techniques. An analytical chemist can both qualitatively and quantitatively evaluate a solution containing a number of chemicals.

The uses of GC-MS are numerous. They are used extensively in the medical, pharmacological, environmental, etc.

Nuclear Magnetic Resonance Spectroscopy, most commonly known as NMR spectroscopy, is the name given to a technique which exploits the properties of certain nuclei.

Nuclei with non-zero spins, when placed in a strong magnetic field process at specific orientations with respect to the applied magnetic field, when appropriate energy is supplied in the form of radio frequency, these nuclei flip to a higher energy state. The energy absorbed during this transition is a function of nucleus type and its chemical environment in the molecule. The excited nuclei are allowed to process freely and come back to their equilibrium positions. During this process an electric signal is induced in a suitably placed RF coil. This signal which is monitored with respect to time is called free induction decay (FID). The FID which is in time domain gives its equivalent frequency domain signal on Fourier transformation. A plot of the absorption frequency versus the intensity of the absorption constitutes the NMR spectrum.

Major applications of NMR spectroscopy lie in the area of synthetic organic chemistry, inorganic chemistry, bio-organic chemistry, bio-inorganic chemistry, polymer chemistry and organometallic chemistry.

Bioreactor is a vessel in which a chemical process involves organisms or biochemically active substances derived from such organisms. Bioreactors are commonly cylindrical, ranging in size from some liter to cube metre, and are often made of stainless steel. Bioreactor design is quite complex engineering task wherein environmental conditions like gas (i.e., air, oxygen, nitrogen, carbon dioxide), flow rates, temperature, pH and dissolved oxygen levels and agitation speed, are closely monitored and controlled using automatic operational modes. Bioreactor treatment may be performed using microorganisms growing in suspension in the fluid or attached on a solid growth support medium. In suspended growth systems such as fluidized beds or sequencing batch reactors, contaminated water is circulated in an aeration basin where a microbial population aerobically degrades organic matter and produces carbon dioxide, water and biomass.

Gel Electrophoresis, PCR and Biosensors are specialized instruments which find use in microbiology and genetic testing. Most of the procedures used in recombinant DNA technology rely on a researcher's ability to purify a DNA fragment of interest. In an important procedure called agarose gel electrophoresis, DNA fragments are separated by size as they move through a gel matrix and then describe a method called blotting, which identifies the DNA fragments of interest along the length of a gel.

Polymerase Chain Reaction (PCR), invented by Kary B. Mullis, at the Cetus Corporation, who was awarded the 1993 Nobel Prize for Chemistry for PCR, is a technique to exponentially amplify *in vitro* a small quantity of a specific nucleotide sequence in the presence of template sequence, two oligonucleotide primers that hybridize to opposite strands and flank the region of interest in the target DNA, a thermostable (taw) DNA polymerase. The reaction is cycled involving template denaturization, primer annealing, and the extension of the annealed primers by DNA polymerase until enough copies are made for further analysis.

Biosensors use optical and electrical means to detect the disease at the outpatient level. Now biosensors that use embedded processors and photodetectors are available that can detect pathogens and other biomolecules through surface plasmon resonance technique. Another research effort aims to develop a biosensor (using immunoassays and superior transduction techniques such as optical sensing) for detecting bacteria in the environment.

Compound microscope gave birth to microscopic anatomy in early 1610. Light microscopy is a fundamental tool for observing and characterizing living specimens by combination with the full range of biochemical, molecular and cell biology techniques. Stereomicroscopy is often used in the field to immediately observe organisms in water samples. While laboratory based methods use brightfield, fluorescence, confocal, reflected light, DIC and other phase contrast methods can be used to investigate the detailed structure and physiology of living and fixed specimens. Polarizing microscopy can reveal the fine detail in the crystalline structure of some organisms such as diatoms and in thin sections of fossil remains. It was, indeed, fortunate that the electron microscope came to maturity in the later 1940s, a time when federal support for research was moving into an exponential growth phase. This instrument has been the source of the essential morphological data that helps our understanding of life processes.

CHAPTER 2

Basic Terminology and Conversion-Formulas & Equations

2.1 GRAM EQUIVALENT CONCEPT

Equivalent Weight (E)

Equivalent weight

It is defined as the mass of an element/compound/ion which combines or displaces 1 part of hydrogen or 8 parts of oxygen or 35.5 parts of chlorine by mass.

$$\text{Equivalent weight,} \quad E = \frac{\text{Molecular weight}}{\text{Valence factor}} = \frac{M_o}{x}$$

Determination of equivalent weight amounts to determining valence factor, x.

In the case of an element, the equivalent weight is defined as:

$$\text{Equivalent weight,} \quad E = \frac{\text{Atomic weight}}{\text{Valency}} = \frac{A}{x}$$

Note that atomic weight substitutes molecular weight and valency substitutes valence factor in the definition. Valencies of hydrogen, calcium and oxygen are 1, 2 and 2 respectively. Hence, their equivalent weights are $1/1 = 1$, $40/2 = 20$ and $16/2 = 8$ respectively.

Equivalent Weight of an Acid

The valence factor of an acid is equal to its basicity. The basicity of an acid is equal to the furnishable hydrogen ion (proton) in its aqueous solution. Importantly, basicity is not same as the number of hydrogen atoms in acid molecule. Consider acetic acid (CH_3COOH). It contains 4 hydrogen atoms in it, but only 1 furnishable hydrogen ion. As such, basicity of acetic acid is 1. With this background, equivalent weight of an acid is defined as:

$$\text{Equivalent weight,} \quad E = \frac{\text{Molecular weight of acid}}{\text{Basicity}}$$

Basicity of sulfuric acid is 2. Hence, equivalent weight of sulfuric acid (H_2SO_4) is $(2 \times 1 + 32 + 4 \times 16)/2 = 98/2 = 49$. Similarly, basicity of oxalic acid is 2. Hence, equivalent weight of oxalic acid ($H_2C_2O_4$) is $(2 \times 1 + 2 \times 12 + 4 \times 16)/2 = 90/2 = 45$.

Equivalent Weight of a Base

The valence factor of a base is equal to its acidity. The acidity of a base is equal to furnishable hydroxyl ion (OH^-) in its aqueous solution. With this background, equivalent weight of a base is defined as:

$$\text{Equivalent weight,} \qquad E = \frac{\text{Molecular weight of base}}{\text{Acidity}}$$

Acidity of KOH is 1, whereas acidity of $Ca(OH)_2$ is 2. Hence, equivalent weight of KOH is $(39 + 16 + 1)/1 = 56/1 = 56$. Similarly, equivalent weight of $Ca(OH)_2$ is $\{40 + 2 \times (16 + 1)\}/2 = 74/2 = 37$.

Equivalent Weight of a Compound

The valence factor of a compound depends on the manner a compound is involved in a reaction. The compounds of alkali metal salts and alkaline earth metal salts are, however, constant. These compounds are ionic and they dissociate in ionic components in aqueous solution. In this case, valence factor is equal to the number of electronic charge on either cation or anion.

$$\text{Equivalent weight,} \qquad E = \frac{\text{Molecular weight of compound}}{\text{Number of electronic charge on cation or anion}}$$

The number of electronic charge on cation of $NaHCO_3$ is 1. Hence, equivalent weight of $NaHCO_3$ is $(23 + 1 + 12 + 3 \times 16)/1 = 84$.

If the defining ratio of the equivalent weight of a compound (AB) formed of two radicals (say, A and B), then rearrange the ratio as:

$$\text{Equivalent weight, } E = \frac{\text{Molecular weight of Radical } A}{\text{Number of electronic charge}} + \frac{\text{Molecular weight of Radical } B}{\text{Number of electronic charge}}$$

Thus, equivalent weight of AB = Equivalent weight of A + Equivalent weight of B

Equivalent Weight of an Ion

The valence factor of an ion is equal to the number of electronic charge on the ion. Therefore, equivalent weight of an ion is defined as:

$$\text{Equivalent weight,} \qquad E = \frac{\text{Molecular weight of ion}}{\text{Number of electronic charge}}$$

The number of electronic charge on carbonate ion (CO_3^{2-}) is 2. Hence, equivalent weight of carbonate ion is (12 + 3 × 16)/1 = 60/2 = 30. Similarly, equivalent weight of aluminium ion (Al^{3+}) is 27/3 = 9.

Equivalent Weight of an Oxidizing or Reducing Agent

In a redox reaction, one of the reacting entities is oxidizing agent (OA). The other entity is reducing agent (RA). The oxidizer is recipient of electrons, whereas the reducer is the releaser of electrons. The valence factor for either an oxidizing or reducing agent is equal to the number of electrons transferred from one entity to another.

$$\text{Equivalent weight,} \quad E = \frac{\text{Molecular weight of compound}}{\text{Number of electrons transferred in redox reaction}}$$

Alternatively,

$$\text{Equivalent weight,} \quad E = \frac{\text{Molecular weight of compound}}{\text{Change of oxidation number in redox reaction}}$$

Potassium dichromate in acidic medium is a strong oxidizer. It means it gains electrons during redox reaction. Potassium dichromate in acidic solution results in:

$$K_2Cr_2O_7 + 14H^+ + 6e^- \rightarrow 2K^+ + 2Cr^{3+} + 7H_2O$$

$$\text{Equivalent weight of } K_2Cr_2O_7 = \frac{294.2}{6} = 29$$

Study of redox reaction is in itself an exclusive and extensive topic.

Gram Equivalent (geq)

It is equal to the mass in grams numerically equal to equivalent weight. If the mass of a chemical entity is "g" grams, then the given mass contains gram equivalents given by:

$$\text{Gram equivalent (geq)} = \frac{g}{E}$$

This formula is widely used to express grams of substance in terms of gram equivalent and vice versa.

Relation Between Moles and Gram Equivalents (geq)

Gram equivalents is given by: $\text{geq} = \frac{g}{E}$

Substituting for equivalent weight, we have:

$$\text{geq} = \frac{g}{E} = \frac{xg}{M_o}$$

Moles is given by:

$$n = \frac{g}{M_o}$$

Combining expressions, we have:

$$geq = xn$$

gram equivalent = valence factor Xmoles

2.2 NORMAL SOLUTION

The solution that contains one gram equivalent weight of a substance per liter of solution is known as normal solution.

Example 1 N solution of NaOH – gram, molecular weight of NaOH = 40 g

Acidity of NaOH = 1

∴ Equivalent weight of NaOH = 40

When 40 g of NaOH is dissolved in DDW and the final volume of the solution is made up to 1L, this solution is known as 1 N solution of NaOH.

Normality: Normality of a solution is designed as the number of gram equivalents of a substance dissolved per liter of the solution and is represented by the symbol *N*.

Normality $$N = \frac{\text{Gram equivalent of substance}}{\text{Volume of solution, } L}$$

Molar solution: The solution that contains gram molecular weight of a substance dissolved in 1 L of solution is called molar solution.

Example 1 M solution of glucose –

Gm. Molecular weight of glucose, $C_2H_{12}O_6$ – 180 g

When 180 g glucose is dissolved in DDW and the final volume of the solution is made up to 1 L, this solution is known as 1 M solution of glucose.

Molarity: Molarity of a solution is defined as the number of gram moles of the substance dissolved per liter of the solution and is represented by the symbol.

Molarity, $$M = \frac{\text{Gram moles of substance}}{\text{Volume of solution, } L}$$

Relation between normality and molarity of a solution

$$\text{Normality} = \text{Molarity} \times \frac{\text{Molecular mass of substance}}{\text{Equivalent mass of substance}}$$

For acids: normality = molarity × basicity of acid

For bases: normality = molarity × acidity of base

Mass percentage (w/v): Mass percentage is defined as the number of parts by the mass of are substance per hundred parts by volume of solution.

Example 1 g NaOH is dissolved in DDW. The final volume of the solution is brought up to 100 mL. This solution is said to be 1-% (*w*/*v*) NaOH solution.

Volume percentage (v/v): Volume percentage is defined as the number of parts by volume of a substance per hundred parts by volume of solution.

Example If 1 mL HCl is mixed with DDW and the final volume is brought up to 100 mL, the solution is said to be 1% (*v*/*v*) HCl solution.

Relation Between Normality and Molarity

Normality is defined as:

$$\text{Normality } (N) = \frac{\text{geg}}{V_L}$$

Substituting for gram equivalent,

$$\Rightarrow \quad \text{Normality } (N) = \frac{\text{geg}}{V_L} = \frac{xM_o}{V_L}$$

where x is the valance factor. Now, molarity is defined as:

$$M = \frac{M_O}{V_L}$$

Note that we have used the convention that M represents molarity and M_O or subscripted symbol M_{NH_3} represents the molecular weight. Now, combining two equations, we have:

$$\Rightarrow \quad N = xM$$

$$\text{Normality} = \text{Valence factor} \times \text{Molarity}$$

Molecular Mass

The molecular mass of a substance is the relative mass of its molecule as compared with the mass of a ^{12}C atom taken as 12 units. It indicates the number of times one molecule of the substance is heavier than $\frac{1}{12}$ of a ^{12}C atom

Calculation of Molecular Mass

Molecular mass is equal to the sum of the atomic masses of all atoms present in one molecule of the substance. Example: H_2O Mass of H atom = 1 Mass of 2 H atoms = 2 Mass of O atom = 16 Molecular mass = 2 + 16 = 18 g.

Relative Molecular Mass

Relative molecular mass or RMM is the molecular weight of an element or a compound. It is the number of times one molecule of the substance is heavier than 1/12 the mass of an atom of carbon (^{12}C).

$$\text{RMM} = \frac{\text{Mass of one molecule of the substance}}{(1/12)\text{ Mass of the atom of carbon } (C^{12})}$$

Gram Molecular Mass

Molecular mass expressed in grams is numerically equal to gram molecular mass of the substance. Molecular mass of O_2 = 32 gram molecular mass of O_2 is the relative molecular mass expressed in grams. Note that relative atomic mass is a ratio and has no unit while gram molecular mass and gram atomic mass are expressed in grams.

Mole Concept

Since it is not possible to calculate the weight of particles individually, a collection of such particles called mole is taken for all practical purposes. It was discovered that the number of atoms present in 12 g of carbon of ^{12}C isotope is 6.023×10^{23} atoms. This is referred to as Avogadro's number after the discoverer, Avogadro. A mole of a gas is the amount of a substance containing 6.023×10^{23} particles. It is a basic unit of the amount or quantity of a substance. The substance may be atoms, molecules, ions or group of ions.

Avogadro discovered that under standard conditions of temperature and pressure, (1 atm and 273 K) a sample of gas occupies a volume of 22.4 L.

Molar Volume

One mole of any gas at STP will have a volume of 22.4 L. This is called molar volume.

The molar volume [22.4 L at STP] plays a vital role in stoichiometric calculations because it is the link between volume and mass in reactions involving gases.

Relationship Between Gram Molecular Weight and Gram Molecular Volume

Gram molecular weight (GMW) or mole is the relative molecular mass of a substance expressed in grams. It is also called gram molecular weight of that element.

Gram molecular volume (GMV) or molar volume is the volume occupied by one gram molecular weight of a gas at STP (standard temperature and pressure).

All gases of equal volumes contain same number of molecules under the same conditions of temperature and pressure. Equal number of molecules of different gases will occupy equal volumes under the same conditions of temperature and pressure.

One mole of gas = 6.023×10^{23} molecules

$$\text{Gram molecular volume} = \frac{\text{Gram molecular weight}}{\text{Weight/volume of gas at STP}}$$

$$\therefore \text{Molar volume of } O_2 = \frac{32}{1.429 \text{ g/L}} = 22.4 \text{ L}$$

$$\therefore \text{Molar volume of } H_2 = \frac{2.016}{0.09 \text{ g/L}} = 22.4 \text{ L}$$ 1 mole of a gas = 22.4 L at STP

Preparation of Acids and Alkalis of Particular Normality

When a stock solution of acid/alkali with known normality is available, a solution of particular normality can be prepared by using the relationship given below:

$$N_1 \times V_1 = N_2 \times V_2$$

where N_1 = Normality of stock solution

V_1 = Volume of stock solution

N_2 = Normality of particular solution

V_2 = Volume of particular solution

Example If it is desired to make 100 mL of N/5 acid solution from a stock solution of 1 N acid strength, the calculation will be as follows:

$$1 \text{ N} \times \text{V1} - 1/5 \text{ N} \times 100$$

$$V_1 = 20 \text{ mL}$$

Twenty mL of 1N acid should be diluted to 100 mL with DDW and thoroughly mixed to obtain N/5 strength of acid. Thus, solution should be standardized against a weighted sample of a primary standard.

Preparation of Acid and Alkali Solutions

1. ***Redistilled water:*** Prepare by redistilling single-distilled water from an all purex distillation apparatus.
2. ***Detonized distilled water:*** Slowly pass distilled water through a 25 cm column of glass rubbing (1 to 2.5 cm in diameter) that has been charged with 2 parts of volume of a strongly basic anion exchange resin in the hydroxyl form and 1 part by volume of a strongly acidic cation-exchange resin in the hydrogen form. Use ion-exchange resins of a quality suitable for analytical work.

Chemical Reagents

Prepare the following reagents by cautiously adding the required amount of concentrated acid or alkali to the designated volume of the proper type of distilled water : mix thoroughly, and dilute to 1000 mL.

Acid solutions

See Table 1.1 for preparation of HCl. H_2SO_4 and HNO_2 solutions.

Table 2.1 *Preparation of uniform acid solutions**

Desired Component	*Hydrochloric Acid (HCl)*	*Sulfuric Acid (H_2SO_4)*	*Nitric Acid (HNO_3)*
Specific gravity (20/4C) if ACS grade conc. acid	1.174 – 1.189	1.834 – 1.816	1.409 – 1.418
Percent of active ingredient in conc. reagent	16 – 17	96-98	69-70
Normality of conc. reagent	11 -12	16	15 – 16
Volume (mL) of conc. reagent to prepare 1 L of			
18 N solution	–	500 (1 + 1) †	–
6 N solution	500 (1 + 1) †	167 (1 + 5) †	180
1 N solution	83 (1 + 1) †	28	64
0.1 N solution	8.1	2.8	6.4
Volume (mL) of 6 N reagent to prepare 11 of 0.1 N solution	17	17	17
Volume (mL) of 1 N reagent to prepare 11 of 0.02 N solution	20	20	20

* All Values Approximate.

† $a + b$ system of specifying preparatory volumes appears frequently throughout this manual and means that if the volume of the concentrated reagents is diluted with k volumes of diluted water to form the required solution.

(i) Stock sodium hydroxide, NaOH, 15 N (for preparation of 6 N, 1 N, and 0.1 N solutions.) Cautiously dissolve 625 g solid NaOH in 800 mL distilled water to form 1 liter of solution. Remove sodium carbonate precipitate from the solution by keeping it at the boiling point for a few hours in a hot water bath or by letting the particles settle for at least 48 hours in an alkali-resistant container (waxlined or polyethylene) protected from atmospheric carbon dioxide with a atmospheric carbon dioxide with a soda lime tube. Use the supernate for the preparation of the dilute solutions listed in Table 1.2.

Alternatively, prepare the dilute solutions by dissolving the weight of solid NaOH indicated in Table 1.2 in carbon dioxide free distilled water and diluting to 1000 mL.

Table 2.2 *Preparation of uniform sodium hydroxide solutions*

Normality of NaOH Solution	*Required Weight of NaOH to Prepare 1000 mL of solution g*	*Required Volume of 15 N NaOH to Prepare 1000 mL of Solution mL*
6	240	400
1	40	67
0.1	4	6.7

Store the NaOH solutions in polyethylene (rigid, heavy type) bottles with polyethylene screw caps, paraffin-coated bottles with rubber or neoprene stoppers, or pyres bottles with rubber or neoprene stoppers. Check the solutions periodically. Protect them by attaching a tube of carbon dioxide absorbing granular material such as soda lime. Ascarite,* Caroxite, t or equivalent. Use at least 70 cm of rubber tubing to minimize vapor diffusion from the bottle. Replace the absorption tube before it becomes exhausted. Withdraw the solution by a siphon to avoid opening the bottle.

(ii) *Ammonium hydroxide solutions:* NH_4OH: Prepare 5 N, 3 N, and 0.2 N ammonium hydroxide solutions by diluting 333 mL, 200 mL, and 13 mL, respectively, of the concentrated reagent (sp gr 0.90, 20.0%, 15 N) to 1000 mL with distilled water.

2.3 BUFFER SOLUTIONS

A buffer solution is one which resists change in pH when small quantities of an acid or an alkali are added to it.

Acidic Buffer Solutions

An acidic buffer solution is simply one which has a pH less than 7. Acidic buffer solutions are commonly made from a weak acid and one of its salts – often a sodium salt.

A common example would be a mixture of ethanoic acid and sodium ethanoate in solution. In this case, if the solution contained equal molar concentrations of both the acid and the salt, it would have a pH of 4.76. It wouldn't matter what the concentrations were, as long as they were the same.

You can change the pH of the buffer solution by changing the ratio of the acid to salt, or by choosing a different acid and one of its salts.

Alkaline Buffer Solutions

An alkaline buffer solution has a pH greater than 7. Alkaline buffer solutions are commonly made from a weak base and one of its salts.

A frequently used example is a mixture of ammonia solution and ammonium chloride solution. If these were mixed in equal molar proportions, the solution would have a pH of 9.25. Again, it doesn't matter what concentrations you choose as long as they are the same.

2.4 EXPRESSION OF ANALYTICAL RESULTS

Parts Per Million (PPM)

Water analysis involves the detection of minute amounts of a variety of substances. The expression of results in percentage would require the use of cumbersome figures. For this reason, the results of a water analysis are usually expressed in parts per million (ppm) instead of percentage. One part per million equals one ten-thousandth of one percent (0.0001%), or one part (by weight) in a million parts, for example, 1 oz in 1,000,000 oz of water, or 1 lb in

1,000,000 lb of water. It makes no difference what units are used as long as both weights are expressed in the same unit.

When elements are present in minute or trace quantities, the use of parts per million results in small decimal values. Therefore, it is more convenient to use parts per billion (ppb) in these cases. One part per billion is equal to one-thousandth of one part per million (0.001 ppm). For example, in studies of steam purity using a specific ion electrode to measure sodium content, values as low as 0.001 ppm are not uncommon. This is more conveniently reported as 1.0 ppb.

In recent times, the convention for reporting analytical results has been shifting toward the use of milligrams per liter (mg/L) as a replacement for parts per million and micrograms per liter (μg/L) as a replacement for parts per billion.

Test procedures and calculations of results are based on the milliliter (mL) rather than the more common cubic centimeter (cc or cm^3). The distinction between the two terms is very slight. By definition, a milliliter is the volume occupied by 1 g of water at 4°C, whereas a cubic centimeter is the volume enclosed within a cube 1 cm on each edge (1 mL = 1.000028 cm^3).

Miligram Per Liter (mg/L)

The milligrams per liter (mg/L) convention is closely related to parts per million (ppm). This relationship is given by:

$$\text{ppm} \times \text{solution density} = \text{mg/L}$$

Thus, if the solution density is close or equal to 1, then ppm = mg/L. This is normally the case in dilute, aqueous solutions of the type typically found in industrial water systems. Control testing is usually conducted without measurement of a solution's density. For common water samples, this poses no great inaccuracy, because the density of the sample is approximately 1 milligram per liter (mg/L) and parts per million (ppm) begin to diverge as the solution density varies from 1. Examples of this are a dense sludge from a clarifier underflow (density greater than 1) or closed cooling system water with high concentrations of organic compounds (density less than 1).

In water and wastewater analysis, the physical and chemical constituents of the sample can be expressed in the following units.

Milligrams Per Liter: mg/L

Milligrams per liter (mg/L) is a weight-volume relationship, i.e., the weight of physical or chemical constituent in unit volume of water or wastewater sample. This offers a convenient basis for a calculation of suspended and dissolved matter in the sample. Furthermore, mg/L is directly applicable to the metric system.

$$\text{mg/L} = \text{g/m}^3 \text{ and g/L } \times 10^3 = \text{kg/m}^3$$

Sometimes the concentration of pollutants is expressed in lb/million gallon. It is related to mg/L by the following expression:

$$\text{mg/L} \times 8.34 = \text{lb/million gallon}$$

Parts Per Million: ppm

Parts per million is a weight to weight ratio. This is mainly used to measure the gaseous pollutants in air and water.

Relation Between mg/L and ppm

The units mg/L and ppm are related by the following:

$$\text{ppm} = \frac{\text{mg/L}}{\text{Specific gravity of sample}}$$

In the case of water, specific gravity is 1, therefore

$$\text{ppm} = \text{mg/L (in water)}$$

The wastewater collected from domestic resources has specific gravity 1, therefore

$$\text{ppm} = \text{mg/L, (in domestic wastewater)}$$

This relation is not true for industrial wastewater because it differs from domestic wastewater in specific gravity depending on the process and the material produced.

2.5 UNIT CONVERSION AND ABBREVIATIONS

Metric and English units

1 liter (L)	=	1000 milliliter (mL)
1 gallon (gal)	=	3.784 L
1 L	=	0.264 gal
1 quart (qrt)	=	0.9464 L = 946.4 mL
1 L	=	1.057 qt
1 fluid ounce	=	29.6 mL
1 mL	=	1000 microliters (μL)
1 kg	=	1000 grams (g)
1 g	=	1000 milligrams (mg)
1 mg	=	1000 micrograms (μg)
1 μg	=	1000 nanograms (ng)
1 ng	=	1000 picogram (pg)
1 kg	=	2.205 pounds (lb)
1 lb	=	453.6 g
1 ounce (oz)	=	28.35 g

1 cubic meter (m^3)	=	1,000,000 cubic centimeter (cc) or cm^3
1 cc	=	1 mL
1 m^3	=	1000 L

Conversions

- Concentration of a solution as weight percent (w/w) = $\frac{\text{Mass of solute (g)}}{\text{100 g solution}} \times 100\%$
- Concentration of a solution as volume percent (v/v) = $\frac{\text{mL substance}}{\text{100 mL solution}} \times 100\%$
- Concentration of a solution as weight per volume percent (v/v)
  $= \frac{\text{Mass of substance (g)}}{\text{100 mL solution}} \times 100\%$
- Concentration as part million (ppm) = mg substance/L solution: that is, ppm = mg/L or μg/mL or ng/μL; part per billion (ppb) = μg/L or ng/mL
- 1 Atmosphere (atm) = 760 torr
  = 760 mm mercury
  = 14.6 pounds per square inch (psi)
  = 101.306 kPa
- $^oF = (1.8 \times {^oC}) + 32$
- $^oC = \frac{(^oF - 32)}{1.8}$
- K (Kelvin) = °C + 273
- Density = mass (The units commonly used for density are g/mL, for volume liquids, gm/cm^3 for solids and g/L for gases).
- 1 Mole (mol) = molecular or formula weight in grams (i.e., 1 mol NaOH = 40 g NaOH)
- Molarity (M) = $\frac{\text{Mol of substances}}{\text{Liter solution}}$
- Normality (N) = $\frac{\text{Gram equivalent weight of substance}}{\text{Liter solution}}$ or Milligram equivalent/mL, solution
- STP = Standard Temperature and Pressure, which is 0°C and 1 atm
- NTP = Normal Temperature and Pressure, which is 25°C and in atm (At STP volume of 1 mol of any gas = 22.4 L, while at NTP 1 mol of any gas would occupy 24.45 L)
- Ideal gas equation: $PV = nRT$

where P = pressure
V = volume
T = temperature (K)
R = 0.082 L atm/mol K (when pressure and volume are expressed in atm and L respectively)

- Vapor pressure is the pressure of a vapor in equilibrium with its liquid or solid form. It is temperature dependent, and expressed in mm Hg or torr. It is a characteristic of the volatility of a substance. The higher the vapor pressure of a substance, the more volatile it is. Vapor pressure data in this text are presented at the temperature 20°C.
- Density of a gas at STP = $\dfrac{\text{Gram molecular weight (mole)}}{\text{22.4 L}}$
- Density of a gas at NTP = $\dfrac{\text{Gram molecular weight (mole)}}{\text{24.45 L}}$
- To determine how heavy a gas is with respect to air, divide the molecular weight of the gas by 29 $\left(\text{i.e., gas } x = \dfrac{\text{mol. wt. of } x}{\text{air}}\right) 29$

2.6 PRECISION AND ACCURACY

Measurements are an important part of experiments. Two different and often confused concepts pertaining to measurements are **accuracy** and **precision**. Precision and accuracy are terms used to describe systems and methods that measure, estimate, or predict. In all these cases, there is some parameter you wish to know the value of. This is called the true value, or simply, truth. The method provides a measured value, that you want to be as close to the true value as possible. Precision and accuracy are ways of describing the error that can exist between these two values.

Unfortunately, precision and accuracy are used interchangeably in non-technical settings. In fact, dictionaries define them by referring to each other! In spite of this, science and engineering have very specific definitions for each.

Accuracy is how close to "true" measurements are. Measuring devices or techniques can easily be inaccurate and lead to false measurements, and no matter how accurate a device may be, there is still a tolerance for error. No measurement is perfect. Accuracy must be accounted for in your results.

Precision is how consistent your results are for the same phenomena over several measurements, or how repeatable a device's (like a spring's) performance can be made. Precision as a measure of variation, must be accounted for in your calculations and results.

Standard Deviation (σ)

Standard deviation is calculated from several individual results of a sample concentration taken at different times under similar experimental conditions represented as $x_1, x_2, x_3 .... x_n$, the number of readings taken n times, and means of the results. These terms are related as follows:

$$\bar{x} = (x_1 + x_2 + x_3 + \ldots x_n)/n = \Sigma x_n/n$$

if the deviation of each result of a sample from the mean is represented by d, then the standard deviation (σ) can be calculated as follows:

$$d_1 = x_1 - d_2 = x_2 - \bar{x}$$

$$d_3 = x_3 - d_n = x_n - \bar{x}$$

Then

$$\sigma^2 = (d_1^2 + d_2^2 + d_3^2 + \ldots d_n^2)/n$$

In the above equation, σ^2 is known as variance. The square root of variance gives the standard deviation and is denoted by σ.

$$\sigma = \sqrt{d_1^2 + d_2^2 + d_3^2 + \ldots d_n^2)/n}$$

It is tedious to calculate deviation for an individual sample from the mean value. The standard deviation can be readily calculated from the sum of the values of x, i.e., (Σx) and the sum of the values of x^2 that means Σx^2 using the following expresssion.

$$\sigma^2 = \Sigma x^2 - (\Sigma x)^2/n + (n - 1)$$

$$\sigma = \sqrt{\Sigma x^2 - (\Sigma x)^2/n + (n - 1)}$$

Relation can calculate the Standard Error of Mean (SEM).

$$\text{SEM} = \sigma/\sqrt{n} = 2\sqrt{S^2/n}$$

Where

$$S^2 = \Sigma x^2 - (\Sigma x)^2/\sigma$$

Example

The total petroleum hydrocarbon (TPM) in an effluent sample in six replicate analyses was found to be as follows: 5.3, 4.9, 5.1, 5.5, 4.7 and 5.0 mg/L.

Determine the standard deviation

x	x^2
5.3	28.09
4.9	24.01
5.1	26.01
5.5	30.25

4.7	22.09
5.0	25.00
30.5	155.45

$$\Sigma x = 30.5; \Sigma x^2 = 155.45$$

$$(\Sigma x)^2 = (30.5)^5 = 930.25; n = 6$$

$$x = \sqrt{\frac{155.45 - \frac{930.25}{6}}{6-1}} = \sqrt{\frac{0.41}{5}} = 0.29 \text{ mg}$$

CHAPTER 3

UV-VIS Spectroscopy

3.1 INTRODUCTION

A spectrophotometer can be either single beam or double beam. In a single-beam instrument (such as the Spectronic 20), all of the light passes through the sample cell. I_0 must be measured by removing the sample. This was the earliest design, but is still in common use in both teaching and industrial labs.

In a double-beam instrument, the light is split into two beams before it reaches the sample. One beam is used as the reference; the other beam passes through the sample. Some double-beam instruments have two detectors (photodiodes), and the sample and reference beam are measured at the same time. In other instruments, the two beams pass through a beam chopper which blocks one beam at a time. The detector alternates between measuring the sample beam and the reference beam.

Samples for UV/VIS spectrophotometry are most often liquids, although the absorbance of gases and even of solids can also be measured. Samples are typically placed in a transparent cell, known as a cuvette. Cuvettes are typically rectangular in shape, commonly with an internal width of 1 cm. (This width becomes the path length, L, in the Beer-Lambert law.) Test tubes can also be used as cuvettes in some instruments. The best cuvettes are made of high quality quartz, although glass or plastic cuvettes are common. Glass and most plastics absorb in the UV, which limits their usefulness to visible wavelengths.

An ultraviolet-visible spectrum is essentially a graph light absorbance versus wavelength in a range of ultraviolet or visible regions. Such a spectrum can often be produced directly by a more sophisticated spectrophotometer, or the data can be collected one wavelength at a time by simpler instruments. Wavelength is often represented by the symbol λ. Similarly, for a given substance, a standard graph of the extinction coefficient (ε) vs. wavelength (λ) may be made or used if one is already available. Such a standard graph would be effectively "concentration-corrected" and thus independent of concentration. For the given substance, the wavelength at which maximum absorbance in the spectrum occurs is called λ_{max}, pronounced "Lambda-max".

The Woodward-Fieser rules are a set of empirical observations which can be used to predict λ_{max}, the wavelength of the most intense UV/VIS absorption, for conjugated organic compounds. The wavelengths of absorption peaks can be correlated with the types of bonds in a given molecule and are valuable in determining the functional groups within a molecule. UV/VIS absorption is not, however, a specific test for any given compound. The nature of the solvent, the pH of the solution, temperature, high electrolyte concentrations, and the presence of interfering substances can influence the absorption spectra of compounds as can variations in slit width (effective bandwidth) in the spectrophotometer. UV-VIS spectroscopy has wide application in Chemistry, Physics and Biology.

3.2 CONCEPTS

An obvious difference between certain compounds is their color. Thus, quinone is yellow; chlorophyll is green; the 2,4-dinitrophenylhydrazone derivatives of aldehydes and ketones range in color from bright yellow to deep red, depending on double bond conjugation; and aspirin is colorless. In this respect the human eye is functioning as a spectrometer analyzing the light reflected from the surface of a solid or passing through a liquid. Although we see sunlight (or white light) as uniform or homogeneous in color, it is actually composed of a broad range of radiation wavelengths in the ultraviolet (UV), visible and infrared (IR) portions of the spectrum. As shown on the right, the component colors of the visible portion can be separated by passing sunlight through a prism, which acts to bend the light in differing degrees according to, wavelength. Electromagnetic radiation (Fig. 3.1a) such as visible light is commonly treated as a wave phenomenon, characterized by a wavelength or frequency. Wavelength is defined on the left below, as the distance between adjacent peaks (or troughs), and may be designated in meters, centimeters or nanometers (10^{-9} meters). Frequency is the number of wave cycles that travel past a fixed point per unit of time, and is usually given in cycles per second, or hertz (Hz). Visible wavelengths cover a range from approximately 400 to 800 nm (Fig. 3.1b). The longest visible wavelength is red and the shortest is violet. Other common colors of the spectrum, in order of decreasing wavelength, may be remembered by the mnemonic: ROY G BIV. The wavelengths of what we perceive as particular colors in the visible portion of the spectrum are displayed and listed below. In horizontal diagrams, such as the one on the bottom left, wavelength will increase on moving from left to right.

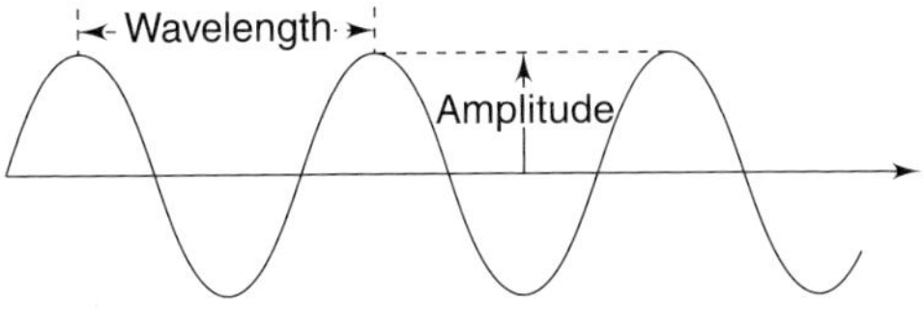

Fig. 3.1 (a) *Electromagnetic radiation.*

When white light passes through or is reflected by a colored substance, a characteristic portion of the mixed wavelengths is absorbed. The remaining light will then assume the complementary color to the wavelength(s) absorbed. This relationship is demonstrated by the

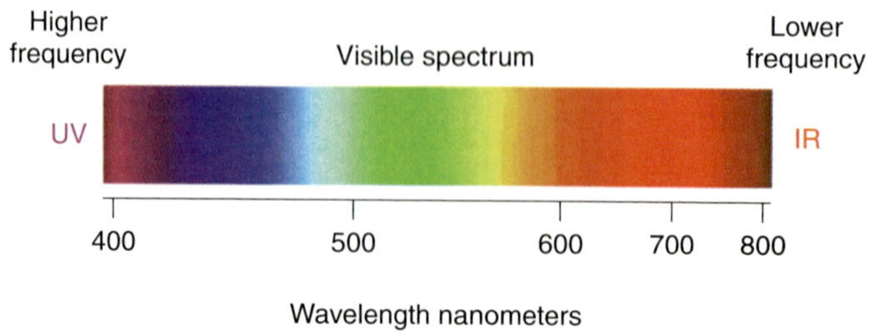

- Violet : 400 - 420 nm
- Indigo : 420 - 440 nm
- Blue : 440 - 490 nm
- Green : 490 - 570 nm
- Orange : 585 - 620 nm
- Yellow : 570 - 585 nm
- Red : 620 - 780 nm

Fig. 3.1 (b) *Visible wavelengths cover a range 400 to 800 nm.*

color wheel shown on the right. Here, complementary colors are diametrically opposite to each other. Thus, absorption of 420-430 nm light renders a substance yellow, and absorption of 500-520 nm light makes it red. Green is unique in that it can be created by absorption close to 400 nm as well as absorption near 800 nm.

Early humans valued colored pigments, and used them for decorative purposes. Many of these were inorganic minerals, but several important organic dyes were also known. These included the crimson pigment, kermesic acid, the blue dye, indigo, and the yellow saffron pigment, crocetin. A rare dibromo-indigo derivative, punicin, was used to color the robes of the royal and wealthy. The deep orange hydrocarbon carotene is widely distributed in plants, but is not sufficiently stable to be used as permanent pigment, other than for food coloring. A common feature of all these colored compounds, displayed below, is a system of extensively conjugated pi-electrons.

3.2.1 Definitions and Units

Radiation is a form of energy and we are constantly reminded of its presence via our sense of sight and ability to feel radiant heat. It may be considered in terms of a wave motion where the wavelength, λ, is the distance between two successive peaks (Fig. 3.2). The frequency, ν, is the number of peaks passing a given point per second. These terms are related so that:

$$c = \nu\lambda$$

where c is the velocity of light in a vacuum.

The full electromagnetic radiation spectrum is continuous and each region merges slowly into the next. For spectroscopy purposes, we choose to characterize light in the ultraviolet and visible regions in terms of wavelength expressed in nanometers. Other units which may

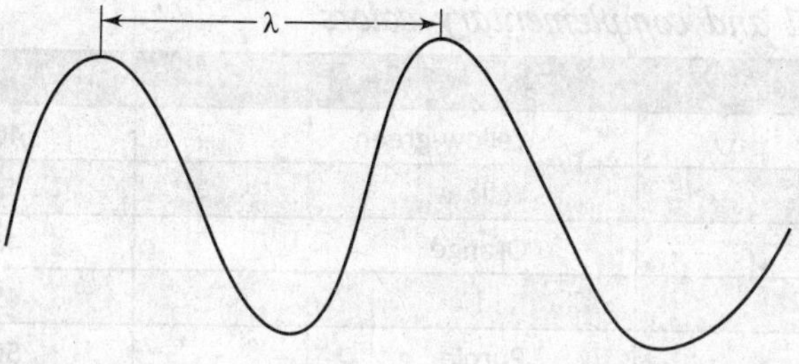

Fig. 3.2 *The wavelength λ of electromagnetic radiation.*

be encountered, but whose use is now discouraged, are the Angstrom (Å) and the millimicron (mμ).

$$1 \text{ nm} = 1 \text{ m}\mu = 10\text{Å} = 10^{-9} \text{ meters}$$

3.3 ELECTROMAGNETIC SPECTRUM

The electromagnetic spectrum of energy extends from the gamma rays emitted by radioactive elements with wavelengths of less than 0.1 nanometer to radio waves with a wavelength greater than 250 millimeters. However, this will deal only with a very small section of this spectrum, namely the ultraviolet (185 to 380 nanometers). A schematic diagram is shown in Fig. 3.3.

The terms used in spectrophotometry are gradually changing, and unless one is familiar with the old and new forms, confusion may result. The new and old terms are given in Table 3.1

Table 3.1 *Terms in common usage in spectrophotometry*

New	*Old*	*Value*
	Angstrom, Å	10^{-10} meter
Nanometer	Millimicron	10^{-9} meter
Micrometer	Micron	10^{-6} meter
Millimeter	Millimeter	10^{-3} meter
Nanogram		10^{-9} meter
Microgram		10^{-6} meter
Milligram		10^{-3} meter

Visible light can be divided into six principal colors as shown in Table 3.2. These are the principal colors seen when white light is diffracted into its primary colors by a prism or diffraction grating. Various shades and tones of these colors are possible. When visible light is absorbed by a compound, there are resulting energy changes involving the valence electrons.

Table 3.2 *Principal and complementary colors*

Color Absorbed	*Color Observed*	*Absorbed Radiation (nm)*
Violet	Yellow-green	400-435
Blue	Yellow	435-480
Green-blue	Orange	480-490
Blue-green	Red	490-500
Green	Purple	500-560
Yellow-green	Violet	560-580
Yellow	Blue	580-595
Orange	Green-blue	595-605
Red	Blue-green	605-750

The ultraviolet portion of the spectrum is that portion of the sun's energy which causes sunburn and similar skin damage. The absorption of ultraviolet energy changes involving the ionization of atoms and molecules.

Infrared radiation is the portion of the spectrum associated with the generation of the heat. Absorption of infrared energy results in molecular vibrations, such as bending or stretching of the inter-atomic bonds, and molecular rotation. The type of vibration depends on the wavelength of the incident radiation.

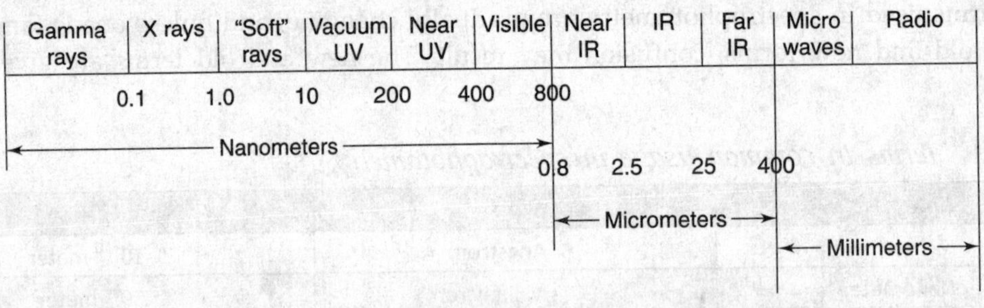

Fig. 3.3 *Schematic diagram of electromagnetic spectrum.*

When certain types of molecules absorb ultraviolet energy, energy in the ultraviolet or visible regions is emitted as the excess energy is released. This is called fluorescence and is a valuable tool for the chemist in identifying and quantitating such compounds.

3.4 QUANTUM THEORY

To gain an understanding of the origins of practical absorption spectrometry, a short diversion into quantum theory is necessary. For this purpose, it is best to think of radiation as a stream of particles known as photons instead of the waves considered earlier. Atoms and molecules exist in a number of defined energy states or levels and a change of level requires the absorption or emission of an integral number of a unit of energy called a quantum, or in our context, a photon.

The energy of a photon absorbed or emitted during a transition from one molecular energy level to another is given by the equation

$$e = h\nu$$

where h is known as Planck's constant and ν is the frequency of the photon. We have already seen that $c = \nu\lambda$

therefore, $$E = hc/\lambda$$

Thus, the shorter the wavelength, the greater the energy of the photon and vice versa.

A molecule of any substance has an internal energy which can be considered the sum of the energy of its electrons, the energy of vibration between its constituent atoms and the energy associated with rotation of the molecule.

The electronic energy levels of simple molecules are widely separated and usually only the absorption of a high energy photon, that is one of very short wavelength, can excite a molecule from one level to another (Fig. 3.4).

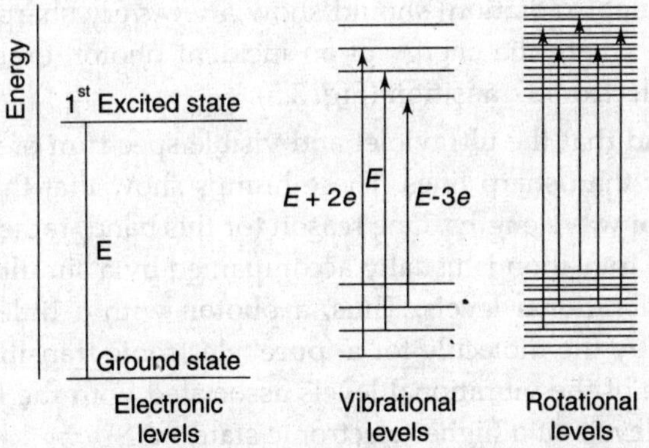

Fig. 3.4 *Energy levels of a molecule.*

In complex molecules the energy levels are more closely spaced and photons of near ultraviolet and visible light can effect the transition. These substances, therefore, will absorb light in some areas of the near ultraviolet and visible regions.

The vibrational energy states of the various parts of a molecule are much closer together than the electronic energy levels and thus protons of lower energy (longer wavelength) are sufficient to bring about vibrational changes. Light absorption due to only to vibrational changes occurs in the infrared region. The rotational energy states of molecules are so closely spaced that light in the far infrared and microwave regions of the electromagnetic spectrum has enough energy to cause these small changes.

For ultraviolet and visible wavelengths, one should expect from this discussion that the absorption spectrum of a molecule (i.e., a plot of its degree of absorption against the

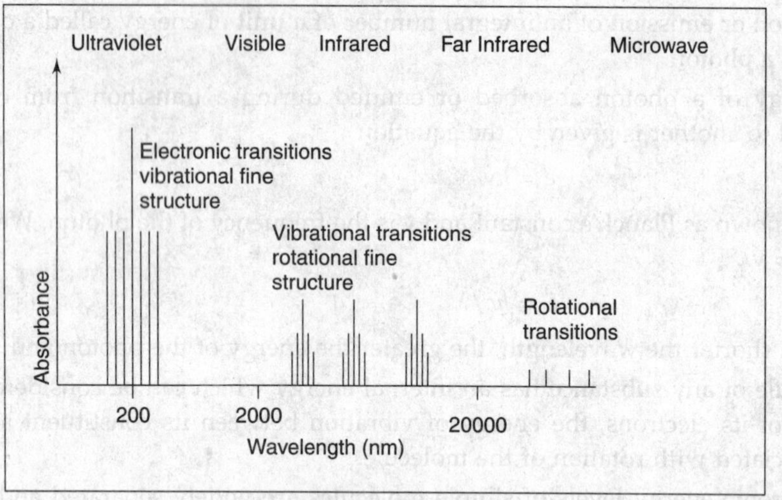

Fig. 3.5 *Idealized absorption spectrum.*

wavelength of the incident radiation) should show a few very sharp lines. Each line should occur at a wavelength where the energy of an incident photon exactly matches the energy required to excite an electronic transition (Fig. 3.5).

In practice it is found that the ultraviolet and visible spectrum of most molecules consists of a few humps rather than sharp lines. These humps show that the molecule is absorbing radiation over a band of wavelengths. One reason for this band, rather than line absorption is that an electronic level transition is usually accompanied by a simultaneous change between the more numerous vibrational levels. Thus, a photon with a little too much or too little energy to be accepted by the molecule for a 'pure' electronic transition can be utilized for a transition between one of the vibrational levels associated with the lower electronic state to one of the vibrational levels of a higher electronic state.

If the difference in electronic energy is E and the difference in vibrational energy is e, then photons with energies of E, $E+e$, $E+2e$, $E-e$, $E-2e$, etc. will be absorbed.

Furthermore, each of the many vibrational levels associated with the electronic states also has a large number of rotational levels associated with it. Thus, a transition can consist of a large electronic component, a smaller vibrational element and an even smaller rotational change. The rotational contribution to the transition has the effect of filling in the gaps in the vibrational fine structure.

In addition, when molecules are closely packed together as they normally are in solutions, they exert influences on each other which slightly disturb the already numerous, and almost infinite energy levels and blur the sharp spectral lines into bands. These effects can be seen in the spectra of benzene as a vapor and in solution. In the vapor, the transitions between the vibration levels are visible as bands superimposed on the main electronic transition bands.

In solution they merge together and at high temperature or pressure even the electronic bands can blur to produce single wide band.

3.5 BEER-LAMBERT LAW

The Beer-Lambert law states that the concentration of a substance in solution is directly proportional to the 'absorbance', A, of the solution.

Absorbance A = constant × concentration × cell length.

The law is only true for monochromatic light, that is, light of a single wavelength or narrowband of wavelengths, and provided that the physical or chemical state of the substance does not change with concentration.

When monochromatic radiation passes through a homogeneous solution in a cell, the intensity of the emitted radiation depends upon the thickness (l) and the concentration (C) of the solution.

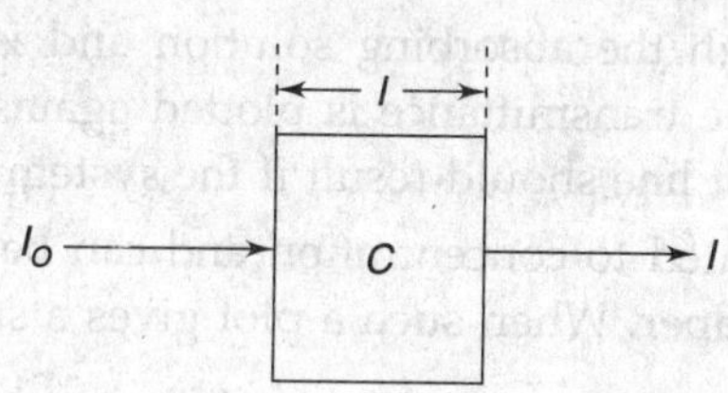

I_o is the intensity of the incident radiation and I is the intensity of the transmitted radiation. The ratio i/I_0 is called transmittance. This is sometimes expressed as a percentage and referred to as % transmittance.

The absorption of energy by solutions follows two basic laws. The Bouguer (1729), or Lambert (1760), law states that when a beam of plane-parallel monochromatic light enters an absorbing medium at right angles to the plane surfaces of the medium, the rate of decrease in intensity with the length of the light path through the absorbing medium is proportional to the intensity of the beam. Mathematically, this can be expressed as:

$$I = I_0 e^{-Kb}$$

where

I = unabsorbed intensity
I_o = incident intensity
K = constant
b = cell thickness (light path length)

Bernard's (1852), or Beer's (1852), law states that the intensity of the energy decreases exponentially with the increase in concentration. Mathematically, this can be written as

$$2.303 \log = \frac{K^{l}_{o} C}{I}$$

where C is the concentration of the absorbing material.

Combination of these two laws forms the basic law of spectrophotometry. It takes the form of:

$$\log = a\ b\ c\ \frac{I_o}{I}$$

where a is the absorptivity, a constant dependent upon the wavelength of the radiation and the nature of the absorbing material, whose concentration c is expressed in grams per liter. The product of the absorptivity and the molecular weight of the absorbing substance is called the "molar absorptivity" (r).

Absorbance A is the product of the absorptivity, the optical path length and the concentration, i.e:

$$A = a\ b\ c$$

The absorbance of a 1 cm layer of a solution containing 1 percent by weight of the absorbing solute is represented by the term A . The term "transmittance %" is the percent of the incident light passing through the absorbing solution and is related to the concentration exponentially. Thus, when the transmittance is plotted against concentration on semi-logarithmic graph paper a straight line should result if the system follows Beer's law.

Absorbance is directly related to concentration and can be plotted against concentration on linear coordinate graph paper. When such a plot gives a straight line the system follows Beer's law (Fig. 3.6).

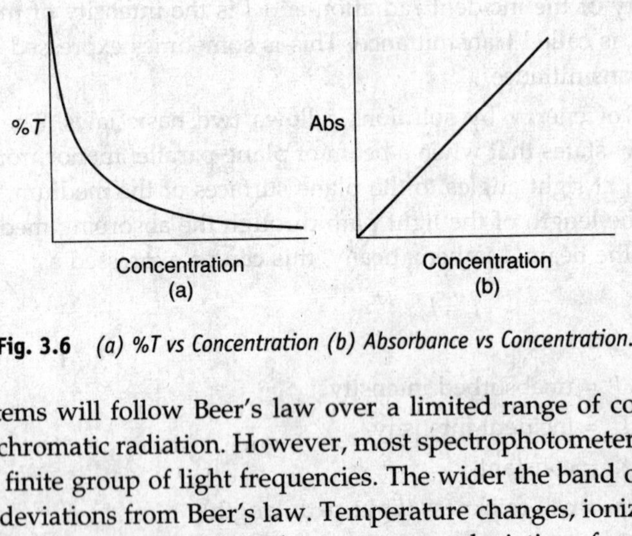

Fig. 3.6 *(a) %T vs Concentration (b) Absorbance vs Concentration.*

Most dilute systems will follow Beer's law over a limited range of concentration. Beer's law requires monochromatic radiation. However, most spectrophotometers and all filter photometers employ a finite group of light frequencies. The wider the band of the radiation, the greater will be the deviations from Beer's law. Temperature changes, ionization of the solute, stray light and changes in the pH of the solute may cause deviations from Beer's law.

While it is desirable to have the analytical system follow Beer's law, it is not essential if a good reproducible calibration curve can be prepared.

3.6 VISIBLE LIGHT SPECTROPHOTOMETRY

3.6.1 Introduction

Analytical methods utilizing the visible section of the electromagnetic spectrum are of great importance. Most methods for the determination of metals in trace concentrations involve the production of a colored complex with some organic reagent. To be of value for analytical purposes the color producing reaction should have the following characteristics:

1. Reagent and the color complex should be stable.
2. The reaction should be stoichiometric.
3. The color development should be rapid and color should resist fading.
4. The reaction should be specific for the element to be determined.
5. The reaction should show no more than minor variation with pH, temperature and other factors.
6. The color complex should be soluble in a solvent which is transparent in the area of spectral absorbance of the complex.
7. The color complex should have a sharp absorbance band.

Methods of color development generally fall into the following categories:

1. Redox methods
2. Complex formation
3. Diazo and coupling reactions
4. Condensations and addition
5. Salt formation
6. Chromophoric changes in valence
7. Substitution.

3.6.2 Instrumentation

The initial techniques of spectrophotometry involved the direct comparisons of colors produced in unknown solutions with those of standards prepared under similar conditions. Observations were made with the naked eye using a common light source. Such techniques can still be used to obtain a rough estimate of the concentration. From this beginning, Nessler tubes developed. The color reaction is carried out in both — a series of standards and unknowns. The solutions are placed in a series of long flat-bottom tubes and diluted so that the column of solution is either 10 or 20 centimeters in depth. After mixing, the color of the unknowns is matched against the standards. The unknown solutions can be bracketed between two standards and a rough approximation of the concentration can be made.

The Duboscq colorimeter was developed from this technique. Light illumination from a common light source is passed up through the bottom of a pair of matched cups, through the solution and through a matched set of glass plungers. A prism system brings the light beams to a common axis. Light from each cup illuminates one-half of the viewed field. The intensities of the two halves of the viewed field are matched visually by raising or lowering the plunger in one cup.

The depth is measured on a scale. From previously prepared calibration curves it is possible to estimate the concentration. The ability of the eye to match intensities varies with wavelength and intensity. The eye is best at about 500 nanometers and under the best conditions can be accurate to within 1 or 2 percent.

Another version of this technique is the wedge comparator. The light beam is split into two segments. One passes through the standard solution and one through the unknown. A neutral wedge of glass is moved into the exit beam of the standard solution to attenuate the light intensity until it matches the unknown.

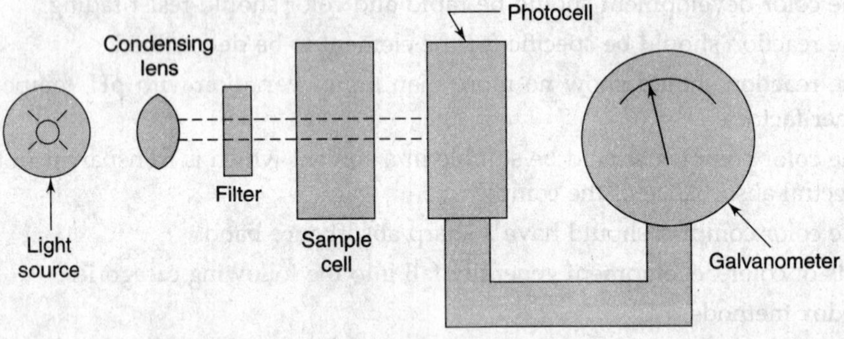

Fig. 3.7 *Schematic diagram of filter photometer.*

The instrumental development was the single-beam photometer. The basic design of this type of instrument is shown in Fig. 3.7. The light from the source passes through a filter, through the solution, and strikes a photocell. With the solvent in the light path and the proper filter in position the light intensity is adjusted to give a reading of 100 on the scale. Next the standards are inserted and the scale readings are recorded. Then the unknown solutions are inserted and read. The intensity of the light source can be adjusted either by a rheostat in series with the light source or a diaphragm in the light path.

To eliminate errors due to variations in the light source with time, the double-beam photometer was developed. In this instrument the filtered light is separated into two beams. One beam is reflected to a reference photocell. The remaining beam passes through the solution to be measured and strikes a second photocell. The net output of the two photocells, connected in opposition, is balanced by a variable resistor to give a zero reading on a galvanometer for the blank solution. The standards and unknowns are inserted into the beam and the deflections of the galvanometer are read; from the calibration curve the concentrations of the unknown can be determined. Filters available for these instruments were generally wide band-pass filters and lacked the narrow spectral band width required by Beer's law.

The next development was the prism or grating single-beam spectrophotometer. The basic instrumental design is shown in Fig. 3.8. In this instrument light from the source is refracted by a prism or diffracted by a grating into its spectrum. A series of adjustable exit slits limits the wavelengths striking the sample.

This spectrophotometer has a narrow band-pass which improves the conformity to Beer's law. The position of the grating or the prism is set to the proper wavelength and the blank is inserted in the beam (Fig. 3.8). After balancing the instrument with the shutter closed, the shutter is opened and the meter set to 100 percent transmittance or 0 absorbance

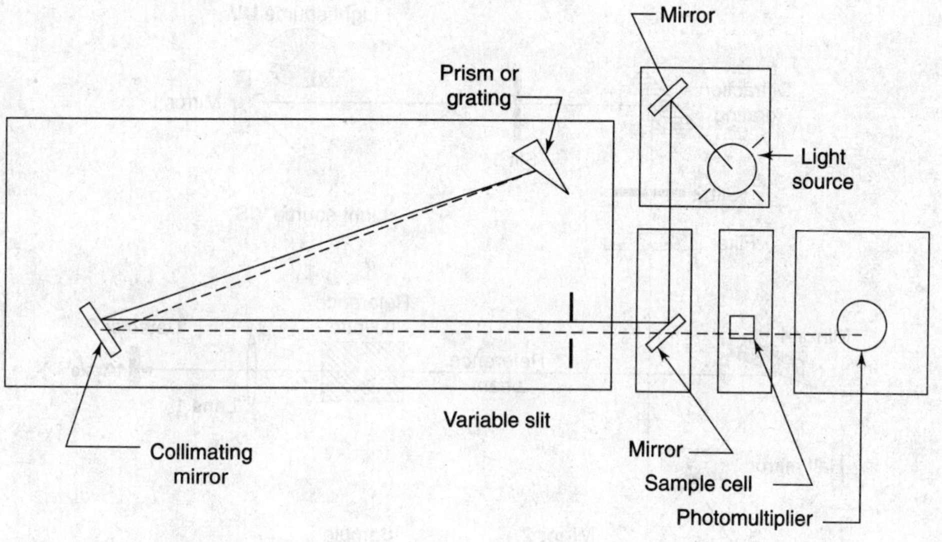

Fig. 3.8 *Schematic diagram of prism or grating spectrophotometer.*

by adjustment of the slit width and the sensitivity control. The standards are inserted in the beam and the absorbance determined for each concentration. The absorbance of the unknowns can be measured and the concentration determined from the calibration curve.

The light-measuring devices used in this type of instrument are either photocells or photomultiplier tubes. Generally, two photocells are available to cover the entire spectral range. These instruments are more expensive than photometers but give greater accuracy and reproducibility as well as monochromatic character of the light used in the analysis.

3.6.3 UV-Visible Spectroscopy

Components of a typical spectrometer are shown in Figs. 3.9 and 3.10 a&b. The functioning of this instrument is relatively straightforward. A beam of light from a visible and/or UV light source (colored red) is separated into its component wavelengths by a prism or diffraction grating. Each monochromatic (single wavelength) beam in turn is split into two equal intensity beams by a half-mirrored device. One beam, the sample beam (colored magenta), passes through a small transparent container (cuvette) containing a solution of the compound being studied in a transparent solvent. The other beam, the reference (colored blue), passes through an identical cuvette containing only the solvent. The intensities of these light beams are then measured by electronic detectors and compared. The intensity of the reference beam, which should have suffered little or no light absorption, is defined as I_0. The intensity of the sample beam is defined as I. Over a short period of time, the spectrometer automatically scans all the component wavelengths in the manner described. The ultraviolet (UV) region scanned is normally from 200 to 400 nm, and the visible portion is from 400 to 800 nm.

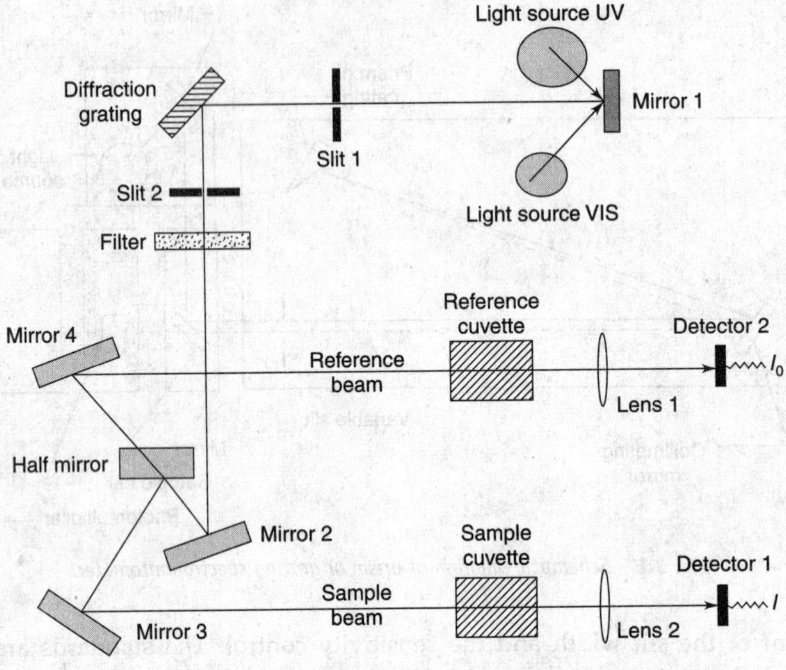

Fig. 3.9 *A schematic diagram of double-beam spectrophotometer.*

If the sample compound does not absorb light of of a given wavelength, **I** = $\mathbf{I_0}$. However, if the sample compound absorbs light then **I** is less than $\mathbf{I_0}$, and this difference may be plotted on a graph versus wavelength, as shown on the right. Absorption may be presented as transmittance ($T = I/I_0$) or absorbance (A= log $\mathbf{I_0/I}$). If no absorption has occurred, T = 1.0 and A = 0. Most spectrometers display absorbance on the vertical axis, and the commonly observed range is from 0 (100% transmittance) to 2 (1% transmittance). The wavelength of maximum absorbance is a characteristic value, designated as λ_{max}.

Different compounds may have very different absorption maxima and absorbances. Intensely absorbing compounds must be examined in dilute solution, so that significant light energy is received by the detector, and this requires the use of completely transparent (non-absorbing) solvents. The most commonly used solvents are water, ethanol, hexane and cyclohexane. Solvents having double or triple bonds, or heavy atoms (e.g., S, Br & I) are generally avoided. Because the absorbance of a sample will be proportional to its molar concentration in the sample cuvette, a corrected absorption value known as the molar absorptivity is used when comparing the spectra of different compounds. This is defined as:

Molar Absorptivity, $\varepsilon = A/\ \mathbf{c}\ \mathbf{l}$ (where A = absorbance, **c** = sample concentration in moles/liter & **l** = length of light path through the cuvette in cm.)

For the spectrum on the right, a solution of 0.249 mg of the unsaturated aldehyde in 95% ethanol ($1.42 \bullet 10^{-5}$ M) was placed in a 1 cm cuvette for measurement. Using the above formula, ε = 36,600 for the 395 nm peak, and 14,000 for the 255 nm peak. Note that the absorption extends into the visible region of the spectrum, so it is not surprising that this compound is orange colored.

Molar absoptivities may be very large for strongly absorbing compounds (ε >10,000) and very small if absorption is weak (ε = 10 to 100).

Absorption Intensity

Molar absoptivities may be very large for strongly absorbing chromophores (>10,000) and very small if absorption is weak (10 to 100). The magnitude of ε reflects both the size of the chromophore and the probability that light of a given wavelength will be absorbed when it strikes the chromophore. A general equation stating this relationship may be written as follows:

$\varepsilon = 0.87 \bullet 10^{20}$ **P** • **a** (where **P** is the transition probability (0 to 1) & **a** is the chromophore area in cm^2)

What magnitudes of energies are involved?

1. **ITV** (ultraviolet) radiation: l = 190 - 400 nm Visible: l = 400 - 800 nm

$$E = h\nu = \frac{hc}{\lambda} = \frac{[(1.58 \times 10^{-37}\ \text{kcal-s})\ (3 \times 1017)]}{300\ \text{nm}} = 1.58 \times 10^{-22}\ \text{kcal/photon.}$$

2. This corresponds to approximately 95 kcal/mole of photons, the magnitude of energy in a covalent bond.
3. Energies of these magnitudes correspond to electronic transitions ⇒ promotion of an electron from one molecular orbital to an unoccupied molecular orbital of higher energy. Absorption of a photon of light has a high probability of occurring when the orientation of the electric vector of the light wave, **E**, is oriented in a direction which will induce a movement of an electron from one molecular orbital to another oribital of higher energy. Another way of stating this is to say that **E** is oriented along a transition dipole of the molecule.

3.6.4 Measurements

Instrumentation

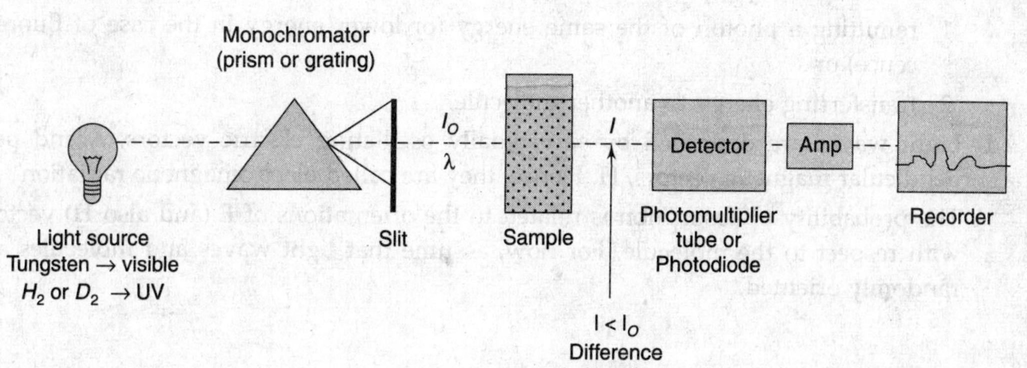

Fig. 3.10(a) *Single-beam spectrophotometer.*

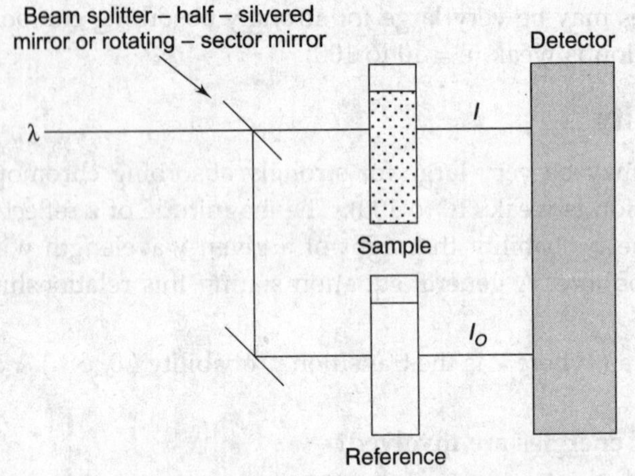

Fig. 3.10 (b) Split beam spectrophotometer.

Interaction of Electromagnetic Radiation (Light) with Molecules

(a) Molecules have different states (configurations) characterized by differences in energy.

(b) Light has energy determined by its wavelength, or frequency.
Thus, there is an inverse relationship between wavelength and energy. In the diagram below, gamma rays have the largest energy and radio waves the smallest.

(c) If light striking a molecule in State 1 has an energy equal to the difference in energy between molecular State 1 and State 2, there is a probability that a molecule will absorb a photon and jump to the higher energy state, State 2.

State 2 _____		State 2 ______X _____
	+hv ----->	
State 1 _____X_____		State 1 _______

The molecule can jump back to state 1 by:

1. remitting a photon of the same energy (or lower energy in the case of fluorescence) or
2. transferring energy to another molecule

(d) Light waves are described by sinusoidally oscillating electric vectors, *e*, and perpendicular magnetic vectors, H. Hence, they are called electromagnetic radiation
The probability of absorption is related to the orientations of **E** (and also **H**) vectors with respect to the molecule. For now, assume that light waves and molecules are randomly oriented.

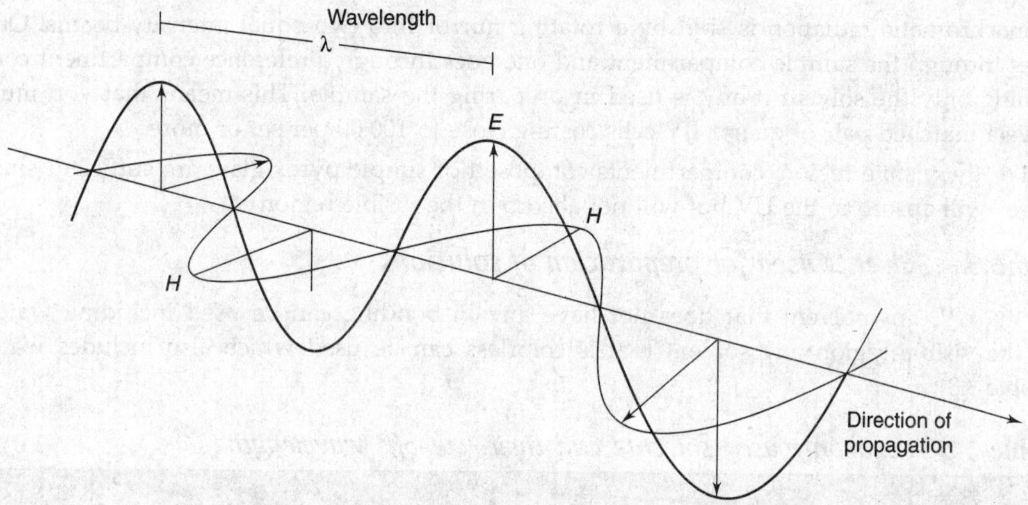

3.6.5 The UV/Visible Spectrophotometer

There are instruments that generate only UV spectra and those that deal only with visible absorption. However, some instruments can generate both UV and visible spectra. Two differences that must be made are the radiation source and the sample compartments.

3.6.5.1 *Radiation source*

For the UV region, the radiation source is a hydrogen discharge lamp that emits polychromatic UV radiation which can then be filtered into monochromatic UV radiation. The instrument scans a wavelength region from 150 nm to 350 nm allowing a monochromatic UV radiation to pass through the sample. When there is sufficient amount of energy from the radiation source then energy will be absorbed and the instrument compares the intensity before and after passing through the sample. The absorbance is recorded on the *y*-axis and the wavelength scan on the *x*-axis of a graphic display.

For the visible region the radiation source must be changed to a tungsten filament similar to the ones found in a common incandescent light bulb. The glowing of the filament produces a polychromatic visible radiation source (i.e., white light). The instrument filters out all but one wavelength and passes it through the sample. The wavelengths are scanned from 400 nm (violet) to 750 nm (red). Those samples whose lamda max is closer to the 400 nm side will appear redder. Those samples whose lamda max is nearer the 750 nm end will appear more blue. The way the absorbance is registered is the same as in the UV region (See above paragraph).

3.6.5.2 *Sample compartment*

For UV spectra the sample must be placed in sample compartments made of quartz since quartz will not absorb in the UV region except at very short wavelengths (vacuum UV). This makes the sample compartments very expensive. If one uses a double-beam instrument, the monochromatic radiation is split by a rotating mirror into two equal intensity beams. One goes through the sample compartment and one goes through a reference compartment containing only the solvent if any is used in preparing the sample. This means that you must have a matched pair of quartz IJV cells costing close to 100.00 per set or more.

For the visible region, compartments composed of simple pyrex glass are sufficient since pyrex will absorb in the UV but will not absorb in the visible region.

3.6.5.3 Solvents used for preparation of solutions

In the UV, any solvent that does not have any Pi bonding can be used including water. In the visible region any solvent that is colorless can be used which also includes water (Table 3.3).

Table 3.3 *Commonly used solvents and their 'cut-off' wavelengths*

Solvent	Cut-off (nm)
Iso-octane	202
Ethyl alcohol	205
Cyclohexane	200
Acetone	325
Tetrachloroethylene	290
Benzene	280
Carbon tetrachloride	265
Chloroform	245
Ethyl ether	220
Isopropyl alcohol	210
Methyl alcohol	210

3.6.5.4 Filters and monochromators

A source is generally emitting a continuous spectra. Therefore, a device is required to select a narrow band from wavelengths of the continuous spectra. For this selection, filters or monochromators or both are used.

(a) **Filters:** A light filter is a device that allows light of the required wavelength to pass but absorbs light of other wavelengths wholly or partially. Thus, a suitable filter can select a desired wavelength band. It means that a particular filter may be used for a specific analysis is carried out for several species, a large number of filters have to be used and interchanged. This method is very useful for routine analysis.

Filters are of two types: Absorption and interference filters.

Absorption filters work by selective absorption of unwanted wavelengths.

An absorption filter is a solid sheet of glass that has been coloured by a pigment which is dissolved or dispersed in the glass. Dyed gelatin or similar materials can also be used as absorption filters.

Absorption filters are classified as either cut-off or band-pass filters.

Table 3.4 briefly describes the absorption characteristics of common filters.

Table 3.4 *Absorption charactersitics of common filters*

Colour of Filters	Approximate Range of Absorption Band in μ
Yellow	450
Orange	500
Red	575
Purple	450-650
Blue	480
Green	400-475
	575-700

(b) **Monochromators:** A monochromator successfully isolates band of wavelengths usually much more than a narrower filter. The essential elements of a monochromator are an entrance slit, a dispersing element (a prism or grating) and an exit slit. The function of a prism or a grating is to disperse the heterochromatic radiation into its component wavelengths.

3.6.5.5 Slits

There are two slits: entrance and exit slit. The main function of the entrance slit is to provide a narrow source of light so that there should be no overlapping of monochromatic images. From this, the exit slit selects a narrowband of dispersed spectrum for observation by the detector. The usual practice is that both entrance and exit slits have equal width. Under such conditions the width of a monochromatic image of the entrance slit is such that it will be just passed by the exit slit. Most of the instruments are provided with variable slit widths to give the instrument greater flexibility and the entrance and exit slits are adjusted simultaneously to the same width.

3.6.5.6 Cells

The cell holding the sample (usually a solution) should be transparent to the wavelength region being recorded. For visible region they are generally made of color-corrected fused glass. The thickness of the cell is generally 1 cm (internal distance). Cells may be rectangular or cylindrical in shape or cylindrical width flat ends.

3.6.5.7 Detection of radiation

In order to detect radiations, three types of photosensitive devices are photovoltaic cells (also called barrier-layer cells), phototubes (also called photoemissive tubes) and photomultiplier tubes.

3.7 APPLICATIONS OF UV/VISIBLE SPECTROPHOTOMETRY

UV spectra and visible spectra can be used to identify an unknown compound by comparative analysis. One can compare the UV or visible spectra of the unknown with the spectra of known suspects. Those that match are evidence that they could be one and the same. However, using a match on UV or visible is not conclusive. Usually, IR and Proton NMR spectra must nail down the exact identification. Such companies as Sadler publish catalogs containing IR, UV, Visible, Proton NMR, C-13 NMR, and mass spectra of known compounds for comparison. One can subscribe to the catalogs and receive periodic updates and additions to be inserted into the catalog volumes.

UV and Visible spectra can also be used to determine the concentration of a chromaphore compound in a mixture using the Beer-Lambert law. Usually, 3-5 standard samples of the chromaphore are prepared, and the absorbance of each standard is measured. The absorbance can be plotted against the concentration and a standard curve can be generated. By measuring the absorbance of the unknown mixture, one can locate the absorbance on the *y*-axis of the standard curve, draw a perpendicular over till it intersects the standard curve and then a perpendicular down until it intersects the *x*-axis. At the point of intersection on the "*x*" axis, the concentration of the chromaphore can be determined. Of course, only one chromaphore may absorb at the wavelength used to measure the standards.

One can also use a proportional method where the absorbance of a known concentration sample is measured and the absorbance of the unknown can be measured. Then at lamda max:

$$A_x = bc_x \text{ (molar extinction coefficient)}$$

$A_{std}\ bc_{std}$ (molar extinction coefficient)

Since $b_{2,}$ the path length and the extinction coeff will be constant:

$$A_x/C_x = b \text{ (extinction coefficient)}$$

$$A_{std}\ C_{std} = b \text{ (extinction coefficient)}$$

$$A_x/C_x = A_{std}/C_{std}$$

$$(A_x)\ (C)/A_{std} = C_X$$

3.7.1 Quantitative Analysis

For quantitative analysis, it is normally chosen to use radiation to use radiation of a wavelength, at which k, the extinction coefficient, is a maximum, i.e., at the peak of the absorption band, for the following reasons:

- The change in absorbance for a given concentration change is greater, leading to greater sensitivity and accuracy in measurement.
- The relative effect of other substances or impurities is smaller.
- The rate of change of absorbance with wavelength is smaller, and the measurement is not so severely affected by small errors in the wavelength setting.

The extinction coefficient must have units relating to the units used for concentration and path length of the sample.

It is usually written as:

$E^{1\%}_{1\,cm}$	The value of *k* for a 1% sample concentration of 1 cm thickness.
Or	
ε	The molar absorptivity previously described as the molar extinction coefficient; the value of *k* for a sample concentration of 1 gram molecule per liter and a 1 cm thickness.

To carry out an analysis of a known material it should be possible in theory to measure the absorption of a sample at its absorption maximum, measure the thickness, obtain values of *e* from tables and calculate the concentration. In practice, values of *E* depend on instrumental characteristics, so published values from tables are not accurate enough, and a reliable analyst will usually plot the absorbance values of a series of standards of known concentration; the concentrations of actual samples can then be read directly from the calibration graph. Using today's modern software driven instruments, the analyst will usually only require one 'top' standard which can be used to calibrate the instrument so that it displays the concentration of subsequent samples directly in the required units. If the calibration curve is non-linear due to either instrumental or chemistry considerations, then some instruments will automatically construct a 'best fit' calibration when presented with a number of standards.

3.7.2 Analysis of Mixtures

It is relatively rare to find a practical problem in which one has a mixture to be analysed with only one component which absorbs radiation. When there are several such components which absorb at the same wavelength their absorbances add together, and it is no longer true that the absorbance of the sample is proportional to the concentration of one component (Fig. 3.11).

In these cases, several approaches can be adopted with the most important being chemical reaction and multi-wavelength measurements.

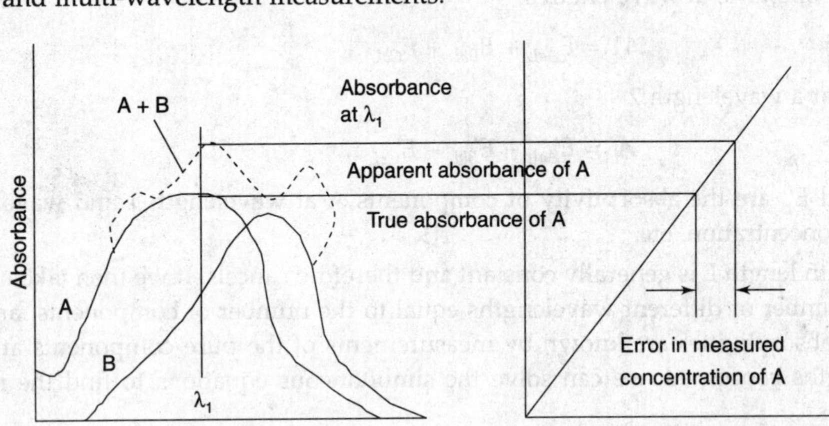

Fig. 3.11 *Error caused by superimposed absorption in a mixture.*

3.7.3 Chemical Reaction

A common method of analysis is to change the required component by adding a chemical reagent which reacts with it specifically to form a highly absorbing compound. An example of this is shown in Fig. 3.12. A quantity of the reagent added to the mixture reacts only with one component and both increase its absorption and change the wavelength of the absorption maximum so that there is no longer interference between the components. The analysis is then reduced to a simple case and its sensitivity is improved. Many hundreds of such specific reagents are now available for all sorts of analyses and sample matrices and are thoroughly detailed in the literature.

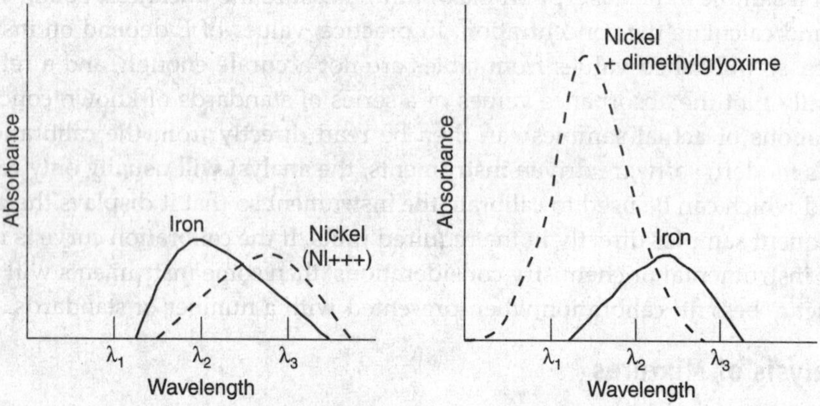

Fig. 3.12 *Improving sensitivity by adding a chemical reagent.*

3.7.4 Multiwavelength Measurements

In a mixture of components, the observed absorption at any wavelength is the sum of the individual absorption spectra of the components thus:

Measure absorbance at wavelength 1

$$A1 = E_{AaL} + E_{BbL} + E_{CcL} \ldots$$

Similarly, at a wavelength 2

$$A' = E'_{AaL} + E'_{BbL} + E'_{CcL} \cdots$$

where E_A and E'_A are the absorptivity of components A at wavelength 1 and wavelength 2 and A is its concentration, etc.

The cell path length L is generally constant and therefore cancels. If we then take measurements at a number of different wavelengths equal to the number of components, and if the value of the absorptivities are known by measurements of the pure components at each of the wavelengths concerned, we can solve the simultaneous equations to find the required concentrations.

The multiwavelength or multicomponents analysis technique has seen a resurgence of interest over the last few years. This has been due to data processing techniques. A variety of algorithms is available and the analyst is generally required to input the number of components, measurement wavelengths and concentration values of the standards.

3.7.5 UV Spectrophotometric Analysis of DNA and RNA

The concentration of an RNA or DNA sample can be checked by the use of UV spectrophotometry. Both RNA and DNA absorb UV light very efficiently making it possible to detect and quantify either at concentrations as low as 2.5 ng/μL. The nitrogenous bases in nucleotides have an absorption maximum at about 260 nm. Using a 1 cm light path, the extinction coefficient for nucleotides at this wavelength is 20. Based on this extinction coefficient, the absorbance at 260 nm in a 1 em quartz cuvette of a 50 μg/mL solution of double-stranded DNA or a 40 μg/mL solution of single-stranded RNA is equal to 1. You can calculate the concentration (c) the DNA or RNA in your sample as follows:

DNA concentration (μg/mL) = (OD_{260}) × (dilution factor) × (50 μg DNA/mL)/(l OD_{260} unit).

RNA concentration (μg/mL) = (OD_{260}) × (dilution factor) × (40 μg RNA/mL)/(l OD_{260} unit).

In contrast to nucleic acids, proteins have a UV absorption maximum of 280 nm, due mostly to the tryptophan residues. The absorbance of a DNA sample at 280 nm gives an estimate of the protein contamination of the sample. The ratio of the absorbance at 260 nm/ absorbance at 280 nm is a measure of the purity of a DNA sample; it should be between 1.65 and 1.85.

Phenol has an absorbance maximum of 270 but the absorbance spectrum over considerably with that of nucleic acids. If there is phenol contamination in your sample, the absorbance at 260 nm will be high, giving a false measure of DNA concentration. These procedures are specific to the Beckman DU C 640B spectrophotometer.

Procedure for Using the Instrument to Measure OD_{260}, and OD_{280}

1. Turn the machine on using the switch in the back of the machine (lower right-hand side as you face the machine, turn on the monitor screen and printer. Wait for the machine to go through its start-up routine.
2. Open the lid on the top of the machine. Choose the small cuvette holder and place it in the holder in front of the light source. The word FRONT should be towards the front of the machine.
3. Click on the UV light key to turn the UV on. It takes about a minute for the UV lamp to warm up. Quit the diagnostic screen by clicking on the word QUIT in the upper right-hand corner. The Main Menu will appear. Choose Nucleic Acid from the menu. When the nucleic acids analysis menu appears, make sure that the measured absorbances are 260 and 280. Do not use the calculation functions.

4. Use lens paper to clean the surfaces of the cuvette. Rinse the cuvette chamber with 70% ETON. Be sure to remove all of the ETON after the wash. Place 100 µg sample of your blank (nH_20, TE, whatever your DNA or RNA sample dissolved in) in the cuvette chamber. Place the cuvette in the holder and place the lid on the holder.
5. Shut the lid of the machine.
6. Click READ BLANK in the bottom left corner of the screen.
7. Prepare a 1:100 dilution of the sample you want to read.
8. After the machine has read the blank, remove the cuvette, remove the blanking solution from the chamber, rinse and dry the chamber and place your diluted sample in it.
9. Click on READ SAMPLES in the upper left hand of the screen.
10. After the machine has read your sample, the data will appear on the screen. You do not need to print this each time you measure a sample. The machine will collect data for you.
11. Between samples clean the cuvette as described above. You do not need to blank each time unless your samples are dissolved in different solutions. If you use a different cuvette, you must run a blank.
12. When you are finished reading samples, remove and clean the cuvette and put it away. Print your results by clicking on PRINT at the top right of the screen. QUIT the data screen. This will take you back to the main menu. Turn off the UV. You can turn off the machine while the Main Menu screen is active. Turn off the machine, the monitor and the printer.
13. Calculate the concentration of your RNA or DNA sample and the OD_{260}/OD_{280}.

3.7.6 Rate Measurements

Rate studies involve the measurement of the change in concentration of a participant in the reaction as a function of time. Whereas we have previously considered the measurement of a sample whose absorption characteristics are constant, spectrometric rate measurement involves the measurement of a fall or rise in absorption at a fixed wavelength. This measurement provides information on the change in concentration of either the reactants or products. The most widely encountered application of spectrometric rate measurements is for the study of enzymes (proteins that are present in all living tissues). Enzymes cannot be measured directly but their catalytic properties allow their estimation from the speed of the reactions which they catalyse. Enzymes have many uses as reagents or as labels that can be attached to other molecules to permit their indirect detection and measurement. However, their widest use is in the field of clinical diagnostics as an indicator of tissue damage. When cells are damaged by disease, enzymes leak into the bloodstream and the amount present indicates the severity of the tissue damage. The relative proportions of different enzymes can be used to diagnose disease, say of the liver, pancreas or other organs which otherwise exhibit similar symptoms. The reaction most used in these clinical assays is the reduction of nicotinamide adenine dinucleotide (NAD) to $NADH_2$. The spectra of NAD and $NADH_2$ are shown in figure and it can be seen that if the absorbance is monitored at the $NADH_2$ peak at 340 nm the reading will increase as the reaction progresses towards $NADH_2$ (Fig. 3.13).

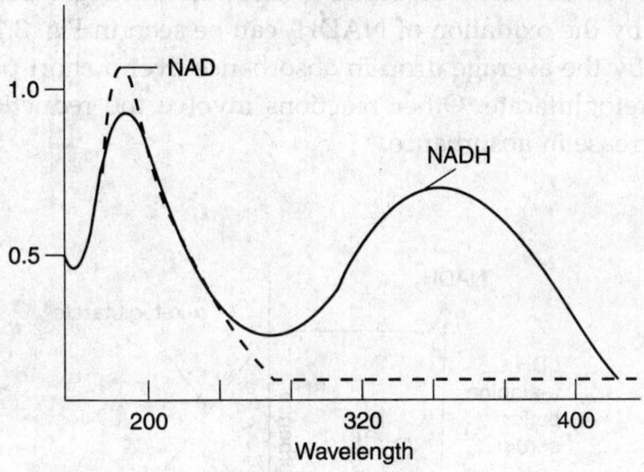

Fig. 3.13 *Absorption spectra of NAD and NADH.*

Other reactions can be monitored by coupling them together so that the product of one reaction feeds the next which can then be optically monitored.

The general equation for an enzyme reaction is:

$$A \xrightarrow{\text{Enzyme } X} B$$

In this reaction enzyme *X* catalyses the conversion of compound *A* to *B*. A coupled reaction utilizes NAD in a secondary reaction, the completion of which depends on the concentration of a product of the primary reaction, e.g.,

$$A \xrightarrow[\text{(Primary reaction)}]{} B \quad B + NADH_2\,C \xrightarrow[\text{(Indicator or secondary reaction)}]{\text{Indicator enzyme}} + NAD$$

This equation, however, will only hold true when the reaction between B and $NADH_2$ is virtually instantaneous. Consider the following coupled reaction:

$$\text{L-alanine} + \alpha - \text{ketoglutarate} \xrightarrow{\text{ALT}} \text{pyruvate}$$

$$\text{Pyruvate} + \text{NADN2} \xrightarrow{\text{LDH}} \text{lactate} + \text{NAD}$$

Reaction 1

The initial reaction is the conversion of L-alanine and α-ketoglutarate to pyruvate in the presence of alanine transaminase. The indicator reaction is the conversion of pyruvate (and $NADH_2$) to lactate which is catalysed by the enzyme lactate dehydrogenase.

The effect of this reaction on absorbance is illustrated in Fig. 3.14. The actual decrease in absorbance caused by the oxidation of $NADH_2$ can be seen in Fig. 3.15. The rate of the reaction is determined by the average drop in absorbance over a short period of time following the addition of α-ketoglutarate. Other reactions involve the reduction of NAD to $NADH_2$ with a resultant increase in absorbance.

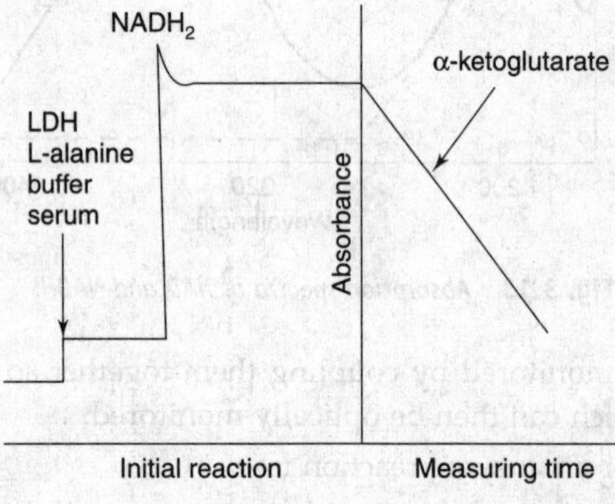

Fig. 3.14 *Enzyme reaction based on oxidation of NADH to NAD.*

The conversion of glucose-6-phosphate to 6-phosphogluconalactone (a reaction catalysed by glucose-6-phosphate dehydrogenase in the presence of $NADH_2$) is an example of this type of reaction.

The equation for this reaction would be as follows:

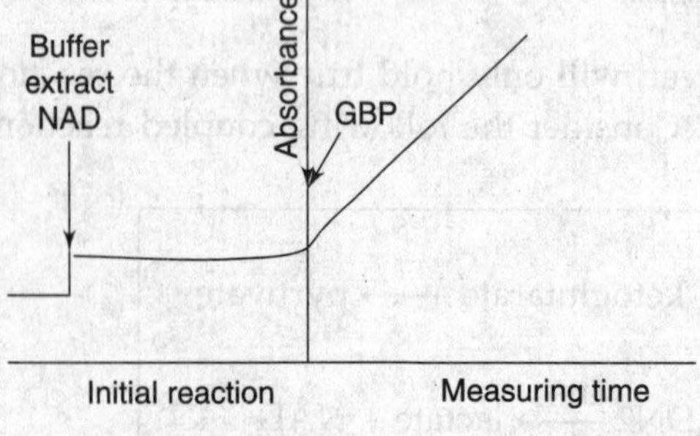

Fig. 3.15 *Enzyme reaction based on reduction of NAD to NADH.*

The change on absorbance is shown in Fig. 3.11. In case of reaction 1, i.e., alanine transaminase, the rate of reaction is calculated by measuring the average decrease in absorbance over a period of time and expressing it as an average drop in absorbance per minute. With reaction 2, the increase in absorbance is measured.

A variety of parameters will influence the rate of reaction. The majority of these, i.e., substrate concentration, enzyme concentration and pH have to be accounted for in chemistry. However, one of the most important is temperature and it is important to precisely control the temperature of the sample and reagents immediately prior to and during measurement. For this requirement, a whole range of thermostatted cell holders and temperature control systems is available.

3.7.7 Spectroprotometric Analysis of the Blood Plasma for Different Mammals

Blood samples from healthy animals at the Faculty of Veterinary Medicine of the University of Agricultural Science and Veterinary Medicine in Lasi were collected in test tubes containing EDTA as an anticoagulant. About 20 mL of blood from sheep and dogs and about 100 mL from cows and from horses are sampled for every test. We collected only 20 mL from pig, sheep and dog because it is hard to collect much blood from these animals. The sampled blood is centrifuged to separate blood plasma (4000 g during 5 minutes). To analyze the blood plasma spectrum for different mammals, we prepared solutions putting 10 μL of blood plasma from different mammals in 5 mL of distilled water.

Solutions of different other concentration from horse blood plasma in distilled water were prepared in order to study the relationship between maximum absorption spectrum and the protein content.

Spectrophotometric analysis in the UV domain of the blood plasma was performed with a spectrophotometer Specord M-42. Measurements of plasma refractive index were performed with an Abbe refractometer.

The protein absorbance band in the UV domain (about 280 nm) determines the characteristic spectrum of the blood plasma. Figure 3.16 presents the characteristic spectrum of the blood plasma for different mammals in the UV domain.

From these spectra one can see that our measurements have shown some differences between the spectra from the different studied species. The highest value of the maximum absorption spectrum (protein content) was obtained for horse and the lowest for dog.

It is known that there is a linear dependence between the protein concentration and the refractive index The results obtained from sprectrophotometric measurements are in good accordance with the protein content obtained by measurements of the refractive index for non-diluted blood plasma of studied mammals, which are presented in Fig. 3.16, by comparison with human blood plasma.

The value of the absorption band strongly depends on the blood plasma concentration. This dependence is presented in Fig. 3.17.

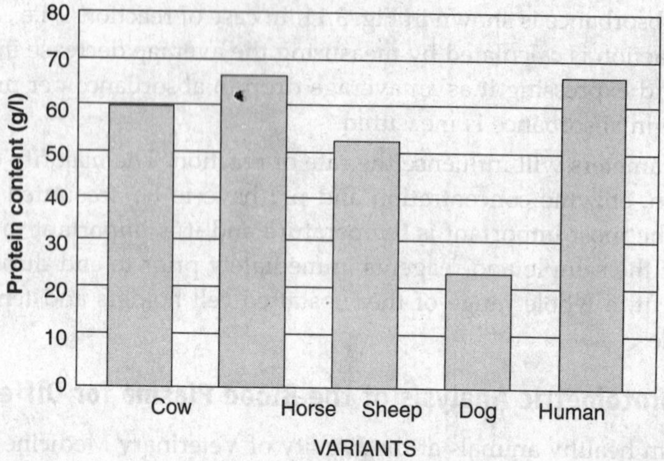

Fig. 3.16 *Protein content of blood plasma from human and different species of mammals.*

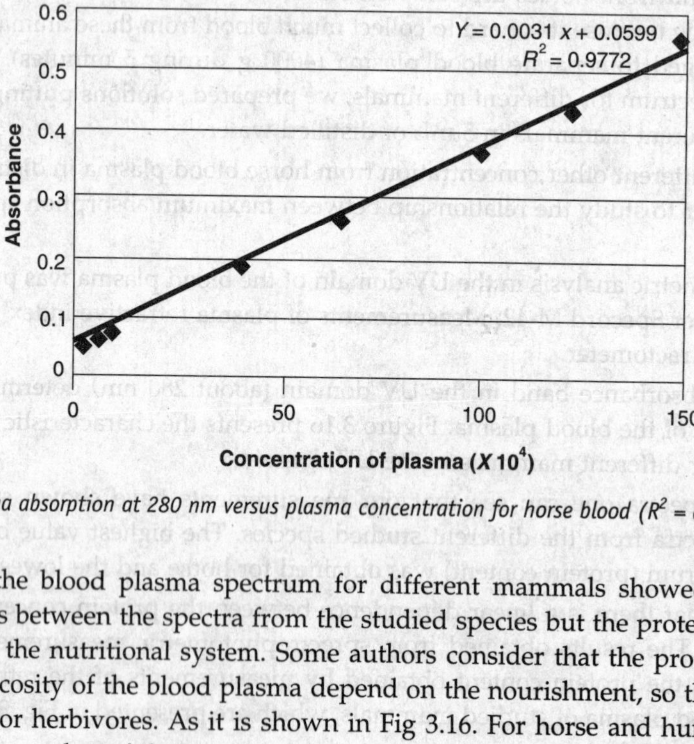

Fig. 3.17 *Plasma absorption at 280 nm versus plasma concentration for horse blood (R^2 = correlation factor).*

Analysis of the blood plasma spectrum for different mammals showed that there are some differences between the spectra from the studied species but the protein content is not in relation with the nutritional system. Some authors consider that the protein content and the dynamic viscosity of the blood plasma depend on the nourishment, so they should have smaller values for herbivores. As it is shown in Fig 3.16. For horse and human sample, we obtained the same value of the protein content and for dog a smaller value than for horse, cow and sheep.

A linear relationship between the protein content and the maximum absorption spectrum in the UV region has been established.

The present study suggests that the spectrophotometric analysis of the blood plasma is a useful tool for determination of plasma protein content from different samples of mammal blood and also it is a less difficult method than the other ones. Due to the fact that the protein absorbance band at 280 nm determines the characteristic spectrum of the blood plasma and the absorption maximum strongly depends on the blood plasma protein concentration, this method permits the determination of a small content of protein from the plasma blood. This means we can use a very small quantity of plasma, which is diluted in distilled water.

3.8 REFERENCES

1. Allen, D.J. Blackmore, Relationship between Paired Plasma and Serum Viscosity and Plasma Proteins in the Horse, *Research in Veterinary Science*, 1984, 36, 360-363.
2. Benishin, C.G., R.Z. Lewanczuk, P.K. Pang, Purification of Parathyroid Hypertensive Factor from Plasma of Spontaneously Hypertensive Rats, Proc NatI. Acad. Sci. USA, 1991, 88(14), 6372-6376.
3. Mattley, Y., G. Leparc, R. Po Cier, L. Garcia-Rubio Light Scattering and Absorption Model for the Quantitative Interpretation of Human Blood Platelet Spectral Data, Photochemjstry and Photobiology 2000, 71(5), 610-619.
4. Narayahan, S., L. Galloway, A. Nonoyama G.F. Leparc, L-H. Garciarubjo R.L. Potter, UV-Visible Spectrophotometric Approach To Blood Typing Li-Phenotyping of Subtype A2 and Weak D and Whole Blood Analysis, Transfusion, 2002, 42(5), 619-626
5. Pe Ribois, C., G. Cazorla A. Cassaigne G. Deleris, Plasma Protein Contents Determined By Fourier Transform Infrared Spectrometry Clinical Chemistry, 2001, 47, 730-738.
6. Raghu, P., Ravinder, P., Sivakumar B., A New Method for Purification of Human Plasma Retino-Bindjng Protein and Transthyretin Biotechnology and Applied Biochemistry, 2003, 38, 19-24.
7. Ueno, I., T. Sakai, M. Yamaoka R. Yoshida A. Tsugita, Analysis of Blood Plasma Proteins in Patients with Alzheimer's Disease by Two-Dimensional Electrophoresis Sequence Homology and Immunodetection Electrophoresis 2000, 21(9), 1832-1845.
8. Windberger, U., A. Bartholovitsch R. Plasenzofti K.J. Korak G. Hejnze, Whole Blood Viscosity, Plasma Viscosity and Erythrocyte Aggregation in Nine Mammalian Species: Reference Values and Comparison of Data, Exp. Physiol. 2003, 88 (3), 431-440.
9. Zahao, B., S.Y. Tham, J. Lu, L. Lee, S. Moochhala, Simultaneous Determination of Vitamins C, E and l^3-Carotene in Human Plasma by High-Performance Liquid Liquid Chromatography with Photodiode Array Detection, J. Pharmaceut. Sci., 2004, 7(2), 200-204.
10. OH. Lowry, N.J. Rosenbrough, A.L Fan and R.J. Randall. J. Biol. Chem- 193 (1951), p. 265.
11. G.L. Peterson. Anal. Biochem. 100(1979), p. 201.
12. F. Capitan, R. Checa, R. Avidad and L.F. Capitan-Vallvey. Talanta 42 (1995), p. 711.

13. L.F. Capitan-Vallvey, R. Avidad and M.D. Fernandez-Ramos. Recent Dev. Pure Appi. Anal. Chem. 1 (1998), p. 149
14. R. Peterson, R. Mountfort, G.P.A.D. Hollom, Guia de campo de las aves de Espana y de Europa, Omega, Barcelona, Spain, 1980.
15. GEP. Box, W. Hunter, J.S. Hunter, Estadistica para investigadores, Editorial Reverte, Barcelona, Spain, 1993.
16. W. Bates and D. McAllister. Anal. Biochem. 59 (1974), p. 190.

CHAPTER 4

Fourier Transform Infrared

4.1 INTRODUCTION

Fourier transform spectroscopy is a measurement technique whereby spectra are collected based on measurements of the coherence of a radiative source, using time-domain or space-domain measurements of the electromagnetic radiation or other type of radiation. It can be applied to a variety of types of spectroscopy including optical spectroscopy, infrared spectroscopy (FT IR, FT-NIRS), Fourier transform (FT) nuclear magnetic resonance, mass spectrometry and electron spin resonance spectroscopy. There are several methods for measuring the temporal coherence of the light, including the continuous wave Michelson or Fourier transform spectrometer and the pulsed Fourier transform spectrograph (which is more sensitive and has a much shorter sampling time than conventional spectroscopic techniques, but is only applicable in a laboratory environment).

Fourier transform infrared (FTIR) spectroscopy is a measurement technique that allows to record infrared spectra. Infrared light is guided through an interferometer and further across the sample (or vice versa). The recorded signal is an interferogram which includes two contributions; one from the instrument itself, and another due to variations in "optical path difference" introduced by the sample. Performing a Fourier transform on these two sets of data (one with sample inserted, one without) results in spectra comparable to those obtained from a conventional (dispersive) infrared spectrometer. The effect due to the sample can be isolated by calculating the ratio or the difference of the two spectra. FTIR spectrometers are cheaper than conventional spectrometers because building an interferometer is easier than the fabrication of a monochromator. In addition, measurement of a single spectrum is faster for the FTIR technique because the information at all frequencies is collected simultaneously. This allows multiple samples to be collected and averaged together resulting in an improvement in sensitivity. Virtually all modern infrared spectrometers are FTIR instruments.

4.2 PRINCIPAL FOURIER TRANSFORM INFRARED (FTIR)

FTIR stands for Fourier Transform Infrared, the preferred method of infrared spectroscopy. In infrared spectroscopy, IR radiation is passed through a sample. Some of the infrared radiation is absorbed by the sample and some of it is transmitted. The resulting spectrum represents the molecular absorption and transmission, creating a molecular fingerprint of the sample. Like a fingerprint no two unique molecular structures produce the same infrared spectrums. This makes infrared spectroscopy useful for several types of analysis.

4.2.1 Physical Principles

Molecular bonds vibrate at various frequencies depending on the elements and the type of bonds. For any given bond, there are several specific frequencies at which it can vibrate. According to quantum mechanics, these frequencies correspond to the ground state (lowest frequency) and several excited states (higher frequencies). One way to cause the frequency of a molecular vibration to increase is to excite the bond by having it absorb light energy. For any given transition between two states the light energy (determined by the wavelength) must exactly equal the difference in the energy between the two states [usually ground state (E_0) and the first excited state (E_1)].

The energy corresponding to these transitions between molecular vibrational states is generally 1-10 kilocalories/mole which corresponds to the infrared portion of the electromagnetic spectrum.

Difference in energy states = Energy of light absorbed

$$E_1 - E_0 = hc/l$$

where h = Planck's constant

c = speed of light, and

l = the wavelength of light.

Fourier Transform Infrared Spectroscopy provides specific information about chemical bonding and molecular structures, making it useful for analyzing organic materials and certain inorganic materials. Chemical bonds vibrate at characteristic frequencies, and when exposed to infrared radiation, they absorb the radiation at frequencies that match their vibration modes. Measuring the radiation absorption as a function of frequency produces a spectrum that can be used to identify functional groups and compounds.

FTIR provides:

- ✦ It can identify unknown materials.
- ✦ It can determine the quality or consistency of a sample.
- ✦ It can determine the amount of components in a mixture.

4.3 THEORY

Fourier Transform Infrared spectrometry was developed in order to overcome the limitations encountered with dispersive instruments. The main difficulty was the slow scanning process. A method for measuring all the infrared frequencies simultaneously, rather than individually, was needed. A solution was developed which employed a very simple optical device called an interferometer. The interferometer produces a unique type of signal which has all the infrared frequencies "encoded" into it. The signal can be measured very quickly, usually on the order of one second or so. Thus, the time element per sample is reduced to a matter of a few seconds rather than several minutes.

Most interferometers employ a beam splitter which takes the incoming infrared beam and divides it into two optical beams. One beam reflects off of a flat mirror which is fixed in place. The other beam reflects off of a flat mirror which is on a mechanism which allows this mirror to move a very short distance (typically a few millimeters) away from the beam splitter. The two beams reflect off of their respective mirrors and are recombined when they meet back at the beam splitter. Because the path that one beam travels is a fixed length and the other is constantly changing as its mirror moves, the signal which exits the interferometer is the result of these two beams "interfering" with each other. The resulting signal is called an interferogram which has the unique property that every data point (a function of the moving mirror position) which makes up the signal has information about every infrared frequency which comes from the source. This means that as the interferogram is measured; all frequencies are being measured simultaneously. Thus, the use of the interferometer results in extremely fast measurements. Because the analyst requires a frequency spectrum (a plot of the intensity at each individual frequency) in order to make identification, the measured interferogram signal cannot be interpreted directly. A means of "decoding" the individual frequencies is required. This can be accomplished via a well-known mathematical technique called the Fourier transformation. This transformation is performed by the computer which then presents the user with the desired spectral information for analysis.

4.4 INSTRUMENTATION

The normal instrumental process (Fig. 4.1) is as follows:

1. **Source:** Infrared energy is emitted from a glowing black-body source. This beam passes through an aperture which controls the amount of energy presented to the sample (and, ultimately, to the detector).
2. **Interferometer:** The beam enters the interferometer where the "spectral encoding" takes place. The resulting interferogram signal then exits the interferometer.
3. **Sample:** The beam enters the sample compartment where it is transmitted through or reflected off of the surface of the sample, depending on the type of analysis being accomplished. This is where specific frequencies of energy, with unique characteristic of the sample, are absorbed.
4. **Detector:** The beam finally passes to the detector for final measurement. The detectors used are specially designed to measure the special interferogram signal.
5. **Computer:** The measured signal is digitized and sent to the computer where the Fourier transformation takes place. The final infrared spectrum is then presented to the user for interpretation and any further manipulation. Because there needs to be a relative scale for the absorption intensity, a background spectrum must also be measured. This is normally a

measurement with no sample in the beam. This can be compared to the measurement with the sample in the beam to determine the "percent transmittance."

This technique results in a spectrum which has all the instrumental characteristics removed. Thus, all spectral features present are strictly due to the sample. A single background measurement can be used for many sample measurements because this spectrum is characteristic of the instrument itself.

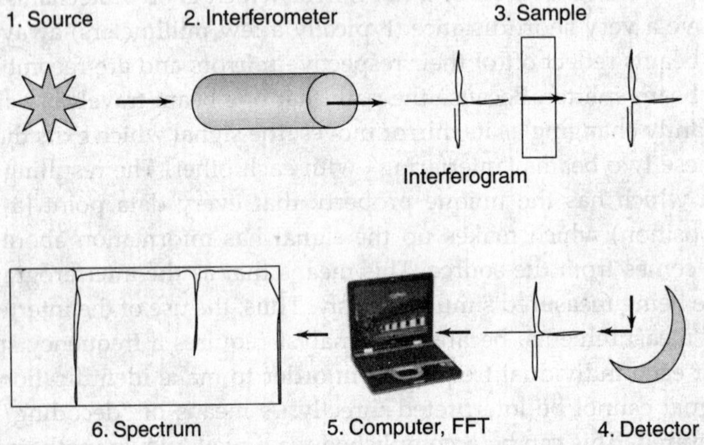

Fig 4.1 *Diagrammatic presentation of the FTIR components.*

4.5 ADVANTAGES OF FTIR

Some of the major advantages of FT-IR over the dispersive technique include:

- **Speed:** Because all frequencies are measured simultaneously, most measurements by FTIR are made in a matter of seconds rather than several minutes. This is sometimes referred to as the Felgett Advantage.
- **Sensitivity:** Sensitivity is dramatically improved with FTIR for many reasons. The detectors employed are much more sensitive, the optical throughput is much higher (referred to as the Jacquinot Advantage) which results in much lower noise levels, and the fast scans enable the co-addition of several scans in order to reduce the random measurement noise to any desired level (referred to as signal averaging).
- **Mechanical Simplicity:** The moving mirror in the interferometer is the only continuously moving part in the instrument. Thus, there is very little possibility of mechanical breakdown.
- **Internally Calibrated:** These instruments employ a He-Ne laser as an internal wavelength calibration standard (referred to as the Connes Advantage). These instruments are self-calibrating and never need to be calibrated by the user.

These advantages, along with several others, make measurements made by FTIR extremely accurate and reproducible. Thus, it is a very reliable technique for positive identification of virtually any sample. The sensitivity benefits enable identification of even the smallest of contaminants. This makes FTIR an invaluable tool for quality control or quality assurance applications whether it is batch-to-batch comparisons to quality standards or analysis of an unknown contaminant. In addition, the sensitivity and accuracy of FTIR detectors, along with a wide variety of software algorithms, have dramatically increased the practical use of infrared for quantitative analysis. Quantitative methods can be easily developed and calibrated and can be incorporated into simple procedures for routine analysis.

The Fourier Transform Infrared (FTIR) technique has brought significant practical advantages to infrared spectroscopy, it has made possible the development of many new sampling techniques which were designed to tackle challenging problems which were impossible by older technology.

4.6 FTIR-SENSITIVITY AND PERFORMANCE

- Highest sensitivity in its class of mid-IR FTIR spectrophotometers with signal-to-noise ratio of 30,000 : 1 or better to meet most critical R&D and QC applications.
- High spectral resolution of 0.5 cm^{-1} makes it most versatile to meet varied applications.
- Assured interferometer protection from high humidity through in-built automatic solid high-polymer electrolytic dehumidifier for long instrument life.
- Fast instrument stabilization and reliable measurements with dynamic alignment mechanism ensures stable interferometer with short warm up times.
- Meets all regulatory requirements with built-in validation program complying with European Pharmacopoeia and ASTM.
- Advanced diagnostics functions with on screen display of instrument status and condition of critical instrument components like beam splitter, laser, source, etc.
- Powerful Windows based 32-bit IR Solution complying with **FDA 21 CRF, Part 11.**
- Comprehensive data processing capability including advanced quantitation protocols.
- New in-built "Pharma Report Program" makes pass/fail judgment on samples with respect to standards based on peak positions and strength ratio.
- New "contaminant analysis program" with built-in algorithms to detect common inorganic, organic and polymers contaminants with high degree of accuracy.
- Advanced automation with automatic accessories recognition function for smart accessories.
- Saves valuable laboratory space with compact design and small footprint.
- Wide range of accessories – DRS, single/multi-reflection ATRs, infrared microscope, etc. to make even complex analysis seem easy.

It has made the use of infrared analysis virtually limitless.

4.7 SAMPLE PREPARATION

Samples for FTIR can be prepared in a number of ways. For liquid samples, the easiest is to place one drop of sample between two plates of sodium chloride (salt). Salt is transparent to infrared light. The drop forms a thin film between the plates. Solid samples can be milled with potassium bromide (KBr) to form a very fine powder. This powder is then compressed into a thin pellet which can be analyzed. KBr is also transparent in the IR. Alternatively, solid samples can be dissolved in a solvent such as methylene chloride, and the solution placed onto a single salt plate. The solvent is then evaporated, leaving a thin film of the original material on the plate. This is called a cast film, and is frequently used for polymer identification.

Solutions can also be analyzed in a liquid cell. This is a small container made from NaCl (or other IR-transparent material) which can be filled with liquid. This creates a longer path length for the sample, which leads to increased sensitivity. Sampling methods include making a mull of a powder with hydrocarbon oil (Nujol) or pyrolyzing insoluble polymers and using the distilled pyrolyzate to cast a film. Materials can be placed in an attenuated total reflectance (ATR) cell and gases in gas cells.

4.7.1 Sample Requirements

Sample requirements vary depending on the sample form and instrument. Samples may be in liquid, solid or gaseous form. When using a microscope attachment on the spectrometer, the analysis area can be as small as 10 µm. Thin organic films on a reflective surface (e.g., gold) can be analyzed *in situ* using the microscope's reflectance mode. The outer 1-10 µm of a material can be analyzed using attenuated total reflectance (ATR).

For small samples or samples with surface contamination visible to the naked eye, a microscope attachment can be used to obtain reflectance or transmittance spectra on particles as small as 20 microns.

4.8 FTIR-QUALITATIVE ANALYSIS AND QUANTITATIVE ANALYSIS

FTIR is most useful for identifying chemicals that are either organic or inorganic. It can be utilized to quantitate some components of an unknown mixture. It can be applied to the analysis of solids, liquids, and gases. The term Fourier Transform Infrared Spectroscopy refers to a fairly recent development in the manner in which the data is collected and converted from an interference pattern to a spectrum. Today's FTIR instruments are computerized which makes them faster and more sensitive than the older dispersive instruments.

4.8.1 Qualitative Analysis

FTIR can be used to identify chemicals from spills, paints, polymers, coatings, drugs, and contaminants. FTIR is perhaps the most powerful tool for identifying types of chemical bonds (functional groups). The wavelength of light absorbed is characteristic of the chemical bond as can be seen in this annotated spectrum (Fig. 4.2).

By interpreting the infrared absorption spectrum, the chemical bonds in a molecule can be determined. FTIR spectra of pure compounds are generally so unique that they are like a molecular "fingerprint". While organic compounds have very rich, detailed spectra, inorganic compounds are usually much simpler. For most common materials, the spectrum of an unknown compound can be identified by comparison to a library of known compounds. We have several infrared spectral libraries including on-line computer libraries. To identify less common materials, IR will need to be combined with nuclear magnetic resonance, mass spectrometry, emission spectroscopy, X-ray diffraction, and/or other techniques.

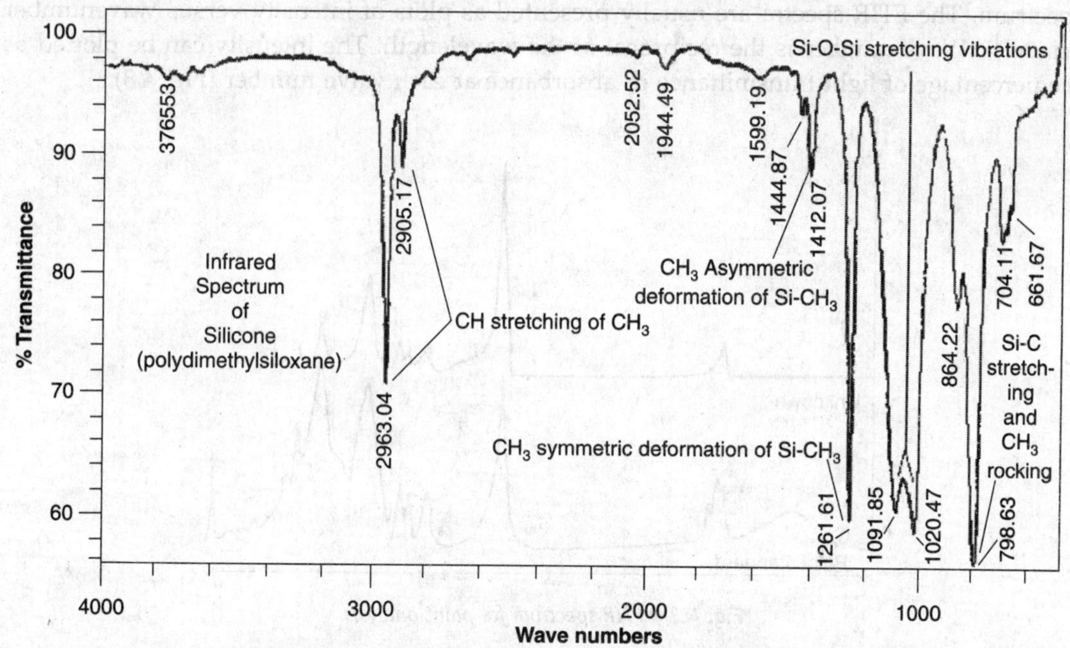

Fig. 4.2 *Infrared spectrum of silicon (Polydimethylsiloxane).*

4.8.2 Quantitative Analysis

Because the strength of the absorption is proportional to the concentration, FTIR can be used for some quantitative analyses. Usually, these are rather simple types of tests in the concentration range of a few ppm up to the percent level.

4.9 DESCRIPTION OF TECHNIQUE

Fourier Transform Infrared Spectroscopy is an analytical technique used to identify organic (and in some cases inorganic) materials. This technique measures the absorption of infrared radiation by the sample material versus wavelength. The infrared absorption bands identify molecular components and structures.

When a material is irradiated with infrared radiation, absorbed IR radiation usually excites molecules into a higher vibrational state. The wavelength of the light absorbed by a particular molecule is a function of the energy difference between the at-rest and excited vibrational states. The wavelengths that are absorbed by the sample are characteristic of its molecular structure.

The FTIR spectrometer uses an interferometer to modulate the wavelength from a broadband infrared source. A detector measures the intensity of transmitted or reflected light as a function of its wavelength. The signal obtained from the detector is an interferogram, which must be analyzed with a computer using Fourier transforms to obtain a single-beam infrared spectrum. The FTIR spectra are usually presented as plots of intensity versus wavenumber (in cm^{-1}). Wave number is the reciprocal of the wavelength. The intensity can be plotted as the percentage of light transmittance or absorbance at each wave number (Fig. 4.3).

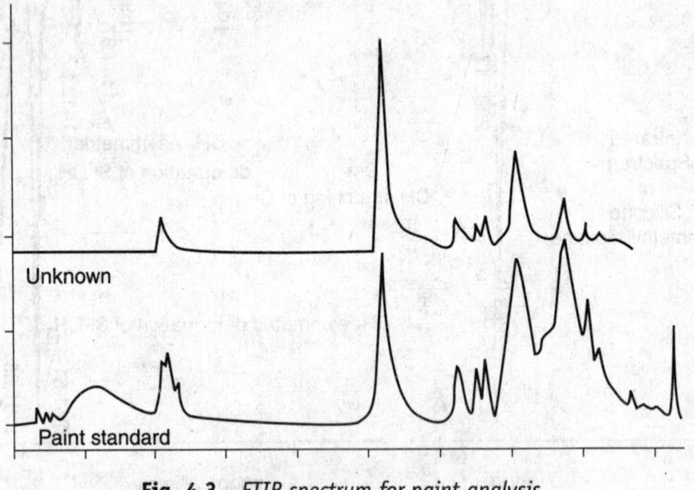

Fig. 4.3 *FTIR spectrum for paint analysis.*

4.9.1 Qualitative Material Identification

To identify the material being analyzed, the unknown IR absorption spectrum is compared with standard spectra in computer databases or with a spectrum obtained from a known material. Spectrum matches identify the polymer or other constituent(s) in the sample. Absorption bands in the range of 4000–1500 wave numbers are typically due to functional groups (e.g., –OH, C=O, N-H, CH_3, etc.). The region from 1500–400 wave numbers is referred to as the fingerprint region. Absorption bands in this region are generally due to intramolecular phenomena and are highly specific to each material. The specificity of these bands allows computerized data searches within reference libraries to identify a material.

4.9.2 Quantitation

Quantitative concentration of a compound can be determined from the area under the curve in characteristic regions of the IR spectrum. Concentration calibration is obtained by establishing a standard curve from spectra for known concentrations.

4.10 APPLICATIONS OF FTIR-GENERAL

1. **Materials Evaluation**
   Identification of most solid; liquid and gas-phase organic compounds,
   Identification of many liquid and gas-phase inorganic compounds,
   Identification of many crystalline and amorphous solid inorganic,
   Identification of polymers, polymer blends and multilayer laminates.
2. **Failure Analysis**
   Qualitative composition verification,
   Quantitative composition verification,
   Coating composition and thickness measurement,
   Verification of parts cleanliness,
   Solvent purity evaluation,
   Optical filter performance verification,
3. **Quality Control Screening**
   Identification of organic deposits on microelectronic packages and devices identification of stains,
   Analysis of inclusions in polymers,
   Analysis of delamination problems in laminates and coatings,
   Analysis of surface degradation due to heat, aging, or chemical attack,
   Analysis of process material degradation or contamination,
   Analysis of lubricants, coolants, power transfer fluids and greases for degradation or contamination.

4.11 RECENT BIOLOGICAL APPLICATIONS

Fourier-Transform Infrared spectroscopy has been used to analyze biological materials since the 1980s. There are several applications of the technique, such as, e.g., the characterization of tissue samples and the differentiation of microorganisms. Some of the advantages of the FTIR technique are that it is easy to use, rapid and cost effective.

1. Application of FTTR for assessment of the biological stability of landfilled municipal solid waste (MSW) during *in situ* aeration.
2. Tracking chemical changes in a live cell by FTIR.
3. Application of FTIR spectroscopy for control of the medium composition during the biodegradation of nitroaromatic compounds
4. Classification and differentiation of food-relevant microorganisms, e.g., differentiation of lactic acid bacteria at different taxonomic levels (e.g., species and strain level).
5. Detection of specific bacterial cell components such as capsular polysaccharides in *Campylobacter jejuni*.

6. Use of FTIR imaging as a new technique to assess the quality of meat. In a pilot study, FTIR imaging was used to monitor the denaturation process during four different heating temperatures (raw, 45°C, 60°C and 70°C) spatially resolved in beef cryosections from two different muscles (longissimus dorsi and semimembranosus). The infrared images were recorded from regions in the tissue. From Fig. 4.3 it is evident, that the infrared images of longissimus dorsi allow visualizing the denaturation process. The scales on the right side of each infrared image are increasing from raw to 60°C indicating an increasing denaturation level. These results demonstrate that infrared imaging can be used to monitor the denaturation process in both tissues. Interestingly, the connective tissue spectra showed a higher β-sheet content indicating a more extensive denaturation in this tissue.
7. Investigation of hydrolysed by-products from the fish industry.
8. Monitoring of fermentation processes under varying process conditions.
9. Application of FTIR spectroscopy in ultra-structural studies of bacteria individual human cell responses to low doses of chemicals studied by synchrotron infrared spectromicroscopy.
10. Application of advanced synchrotron radiation-based Fourier transform infrared (SR-FTIR) microspectroscopy to animal nutrition and feed science: a novel approach.
11. The deposition of thin films of p(FIENIA) has been verified by FTIR spectroscopy.
12. Optimal methods for processing mineralized tissues for Fourier transform infrared microspectroscopy.
13. Effects of transforming growth factor deficiency on bone development: A Fourier transform-infrared imaging analysis.
14. A physical, chemical, and mechanical study of lumbar vertebrae from normal, ovariectomized, and nandrolone decanoate-treated cynomolgus monkey.
15. Osteopontin deficiency increases mineral content and mineral crystallinity in mouse bone.
16. FTIR microscopic imaging of collagen and proteoglycan in bovine cartilage.
17. FTIR imaging spectroscopy of genetically modified bovine chondrocytes.
18. Studies of benign prostate and prostate cancer.
19. Monitoring of denaturation processes in aged beef loin.
20. Characterization of bioactive and resorbable polylactide/ Bioglass(R) composites.
21. High-Fidelity FTIR spectroscopic imaging of primate brain tissue.
22. Imaging of biomineralized tissues using a mercury-cadmium-telluride focal-plane array detector.
23. IR microscopic imaging of pathological states and fracture healing of bone.
24. Distribution of collagen cross-links in normal human trabecular bone.

25. Imaging data of human breast cells by cluster analysis and artificial neural networks.
26. Imaging by Fourier transform infrared microspectroscopy for objective and quantifiable assessment.
27. Investigation into an animal model exhibiting glioblastoma multiforme.
28. Towards a practical Fourier transform infrared chemical imaging protocol for cancer histopathology.
29. Imaging of wheat endosperm cell walls *in situ* reveals compositional and architectural heterogeneity related to grain hardness.
30. Imaging to composite wood materials.
31. Fourier transform infrared bacteria identification with the use of a focal-plane-array detector and microarray printing.
32. The use of FTIR imaging as an analytical tool for the characterization of drug, delivery systems.
33. Visualization of the heterogeneous water sorption in a pharmaceutical formulation under controlled humidity.
34. FTIR imaging of the interface in multicomponent systems using optical effects induced by differences in refractive index.
35. Polarized IR. spectroscopic imaging on molecular orientation of ferroelectric liquid crystals.
36. Studies of peroxide curing of polybutadiene/zinc diacrylate blends.
37. The application of infrared chemical imaging to the detection and enhancement of latent fingerprints: Method optimization and further findings.

4.12 REFERENCES

1. Laurence M. Harwood, Christopher J. Moody. *Experimental organic chemistry: Principles and Practice* (Illustrated edition). pp. 292.
2. Luypaert, J.; Zhang, M.H.; Massart, D.L. (2003), "Feasibility study for the use of near infrared spectroscopy in the qualitative and quantitative analysis of green tea, *Camellia sinensis* (L.)", *Analytica Chimica Acta*, **478(2)**, Elsevier, pp. 303–312
3. Lau, W.S. (1999). *Infrared characterization for microelectronics*. World Scientific.
4. Isotope Dependence of the Lifetime of the 1136-cm[sup -1] Vibration of Oxygen in Silicon K.K. Kohli, Gordon Davies, N.Q. Vinh, D. West, S.K. Estreicher, T. Gregorkiewicz, I. Izeddin, and K.M. Itoh, Phys. Rev. Lett. 96, 225503 (2006), DOI:10.1103/Phys. Rev. Lett. 96, 225503
5. P. Hamm, M.H. Lim, R.M. Hochstrasser (1998). "Structure of the amide I band of peptides measured by femtosecond nonlinear-infrared spectroscopy". *J. Phys. Chem. B* **102**: 6123. doi:10.1021/jp9813286.

6. S. Mukamel (2000). "Multidimensional Femtosecond Correlation Spectroscopies of Electronic and Vibrational Excitations". *Annual Review of Physics and Chemistry* **51**: 691. doi:10.1146/annurev.physchem.51.1.691.
7. N. Demirdöven, C.M. Cheatum, H.S. Chung, M. Khalil, J. Knoester, A. Tokmakoff (2004). "Two-dimensional infrared spectroscopy of antiparallel beta-sheet secondary structure". *Journal of the American Chemical Society* **126**: 7981. doi:10.1021/ja049811j.

CHAPTER

5

Fluorescence Spectrophotometry

5.1 INTRODUCTION

Fluorescence spectroscopy is as one of the most sensitive analytical tools. In fluorescence, the intensity of the emission of the sample is used to measure concentrations as low as 10^{-8} to 10^{-10} grams per milliliter. Few colorimetric procedures are of value at concentrations below 10^{-7} grams per milliliter. The reason for the high sensitivity of fluorescence technique is that the emission signal is measured above a low background level. This is inherently more sensitive than comparing two relatively large signals as in absorption spectroscopy. The sensitivity of fluorescence is as much as 1000 times more sensitive than absorption spectroscopy. For analytical measurements, sensitivity determines the detection limit of a material. For research in life science, it may be important to work at concentration levels that are very low. A more subtle requirement is the need to measure change in fluorescence during the course of an experiment.

Fluorescence is the result of a three-stage process, viz., excitation, excited state and fluorescence emission that occurs in certain molecules (generally, polyaromatic hydrocarbons or heterocyclic) called fluorophores or fluorescent dyes. A fluorescent probe is a fluorophore designed to respond to a specific stimulus or to localize within a specific region of a biological specimen.

5.2 THEORY

When the external energy supply is by means of the absorption of infrared, visible or ultraviolet light, the emitted light is called photoluminescence and this is the process that takes place in any fluorimetric analysis. To account adequately for the processes of absorption and emission of light, it is necessary to assume that radiant energy can only be absorbed in definite units, or quanta. The energy E, carried by any one quantum is proportional to its frequency of oscillation, this is

$$E = hv = \frac{hc}{\lambda} \text{ ergs}$$

where v is the frequency, λ the related wavelength and h = Planck's constant ($6.6 = 4 \times 10^{-27}$ ergs/second).

The energy of a single quantum is too small for convenience and it is usual to talk of the energy associated with N quanta (where $N = 6.023 \times 10^{23}$ the number of single molecules in a gram molecule), which is called an einstein. Thus, if in a photochemical reaction one molecule reacts for each quantum absorbed, then the absorption of one einstein is sufficient energy for the reaction of one gram mole. Since the amount of energy per einstein is proportional to the frequency of the radiation, it varies enormously over the range of the electromagnetic spectrum.

The ultraviolet and visible regions of the spectrum are of most interest in fluorimeter. Absorption in these regional causes the excitation of the outermost electrons of the molecule. The energy associated with radiation of this frequency is quite high, around 100 kilogram calories per einstein, and it is sometimes sufficient to break down the absorbing molecules, as for instance, with the fading of dyes by the action of sunlight.

The absorption of light results in the formation of excited molecules which can in turn dissipate their energy by decomposition, reaction, or re-emission. The efficiency with which these processes take place is called the quantum efficiency and in the case of photoluminescence can be defined as:

$$\Phi e = \frac{\text{einsteins emitted}}{\text{einsteins absorbed}}$$

or

$$\frac{\text{No. of quanta emitted}}{\text{No. of quanta absorbed}}$$

5.2.1 Quantitation

The fraction of a parallel beam of light absorbed by a sample is independent of the intensity of the incident beam and is related to the concentration of the absorbing species by the familiar Beer-Lambert law:

$$\frac{I_o}{I} = e^{-Ecl}$$

Normally written as:

$$\log_{10} \frac{I_o}{I} = Ecl$$

where

I = intensity of transmitted light

I_o = intensity of incident light

E = molecular extinction coefficient

c = concentration in gm moles/L-1

l = path length of sample

and the quantity $\log_{10}$. I_o is known as the absorbance or optical density of the sample.

Absorption spectra are simply a plot of absorbance against wavelength for any sample.

The intensity of fluorescence emission of weakly absorbing, dilute samples is related to the molecular extinction coefficient by the relation:

$$FI = I_o (2.303\ Ecl)\ \Phi_f$$

and so the intensity of fluorescence emission is directly proportional to the intensity of the incident radiation. These two equations explain the difference in sensitivity between absorption and fluorimetric procedures. Sensitivity in absorption measurements is limited by the ability of the instrument to discriminate between the two nearly equal signals due to I and I_o, and the detection limit for even the most favorable cases rarely exceeds 10^{-8} moles.

Conversely, fluorimetric instruments are limited only by the intensity of exciting light and the ability to detect low light levels, so under ideal conditions concentrations of 10^{-12} moles can be measured.

The above equation only holds for dilute samples and it is possible to calculate the concentration level (C_{max}) above which non-linearity is observed from:

$$C_{max} = \frac{0.05}{El}$$

It is most important to establish the concentration at which a plot of fluorescence emission against concentration becomes non-linear either by calculation, or by measurement of a suitable range of standards. Direct interpolation of concentrations outside this limit will give inaccurate results.

5.2.2 Fluorescence and Phosphorescence

Fluorescence is essentially an electronic phenomenon and is primarily concerned with light of wavelengths in the region of 200 to 800 nanometers. When light in this region strikes some compounds they absorb energy at specific wavelengths which are characteristics of the compound. This is called the absorption spectrum of the compound. As a result of this absorption of energy, some of the molecules are raised from the ground state to a higher energy level called singlet or excited state. Since this excited state is unstable, the molecule tends to return to the ground state by emitting the absorbed energy as fluorescence. As some of the released energy is lost by other means, the energy released is lost by other means, the energy released as fluorescence is always less than the absorbed energy. Therefore, the wavelength of the fluorescence is longer than that of the absorbed energy. The absorption and release of energy as fluorescence takes place in 10^{-8} seconds.

In some instances, the absorbed energy may be released in two steps. First, a small amount of energy is lost, allowing the molecule to reach the triplet or metastable state. It then returns to the ground state by releasing energy slowly. The energy lost from the triplet to the ground state is called phosphorescence. As in fluorescence, the energy lost by phosphorescence is less than the total absorbed energy and thus, the wavelength of the phosphorescence is longer than that of the absorbed energy. Since the energy release is slower, phosphorescence is more persistent than fluorescence.

5.3 INSTRUMENTATION

5.3.1 General

All fluorescence instruments contain three basic items: a source of light, a sample holder and a detector. In addition, to be of analytical use, the wavelength of incident radiation needs to be selectable and the detector signal capable of precise manipulation and presentation (Fig. 5.1). In simple filter fluorimeters, the wavelengths of excited and emitted light are selected by filters which allow measurements to be made at any pair of fixed wavelengths.

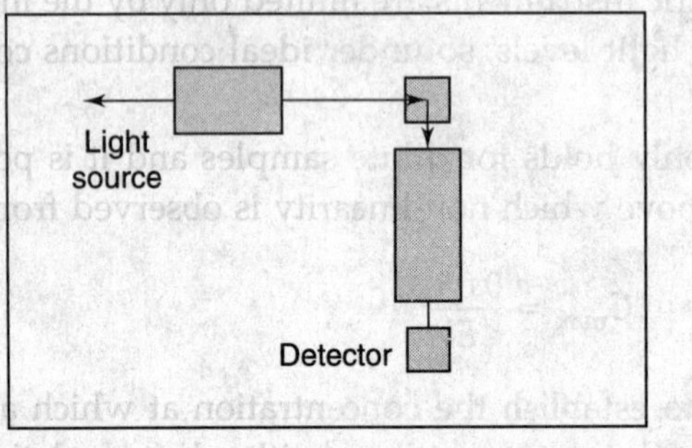

Fig. 5.1 *Essential components of a fluorescence spectrometer.*

Simple fluorescence spectrometers have a means of analyzing the spectral distribution of the light emitted from the sample, the fluorescence emission spectrum, which may be by means of either a continuously variable interference filter or a monochromator. In more sophisticated instruments, monochromators are provided for both the selection of exciting light and the analysis of sample emission. Such instruments are also capable of measuring the variation of emission intensity with exciting wavelength, the fluorescence excitation spectrum.

In principle, the greatest sensitivity can be achieved by the use of filters, which allow the total range of wavelengths emitted by the sample to be collected, together with the highest intensity source possible. In practice, to realize the full potential of the technique, only a small band of emitted wavelengths is examined and the incident light intensity is not made excessive, to minimize the possible photodecomposition of the sample.

5.3.2 Light Sources

Commonly employed sources in fluorescence spectrometry have spectral outputs either as a continuum of energy over a wide range or as a series of discrete lines. An example of the first type is the tungsten-halogen lamp and of the latter, a mercury lamp. Mercury lamps are the most commonly employed sources and have the property that their spectral output depends upon the pressure of the filler gas. The output from a low-pressure mercury lamp is concentrated in the UV range, whereas the most commonly employed lamps, of medium and high pressure have an output covering the whole UV-visible spectrum.

Although in many cases the output from a line source will be adequate, it is rare that an available line will exactly coincide with the optimum excitation wavelength of the sample.

It is therefore advantageous to employ a source whose output is a continuum and the most commonly employed type is the xenon arc. Xenon arc sources can be operated either on a continuous DC basis or stroboscopically; the latter method offers advantages in the size and cost of lamps. The output is essentially a continuum on which are superimposed a number of sharp lines, allowing any wavelength throughout the UV-visible region of the spectrum to be selected (Fig. 5.2).

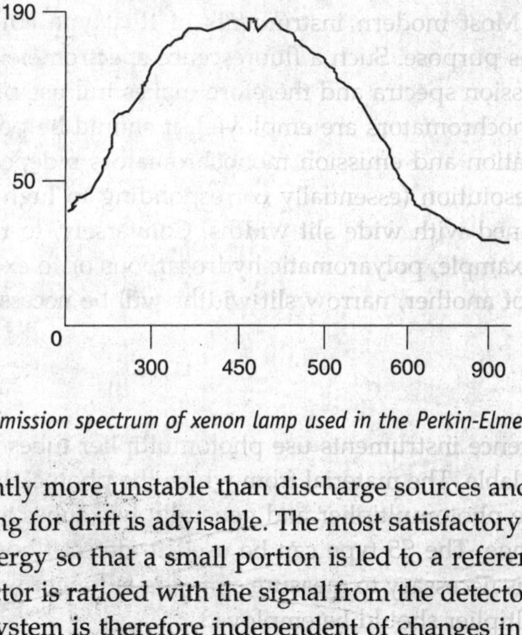

Fig. 5.2 *Emission spectrum of xenon lamp used in the Perkin-Elmer LS series.*

Arc lamps are inherently more unstable than discharge sources and for long-term stability a method of compensating for drift is advisable. The most satisfactory method of doing this is to split the excitation energy so that a small portion is led to a reference detector. The signai from this reference detector is ratioed with the signal from the detector observing the sample. Such a ratio-recording system is therefore independent of changes in the source intensity.

All sources of UV radiation will produce ozone from atmospheric oxygen, which should be dispersed, since it is not only toxic, but also absorbs strongly in the region below 300 nm. For this reason, most lamps are operated in a current of air and, if the supply fan fails, the lamp should be extinguished immediately. Lamps must be handled with great care since fingermarks will seriously decrease the UV output.

5.3.3 Wavelength Selection

The simplest filter fluorimeters use fixed filters to isolate both the excited and emitted wavelengths. To isolate one particular wavelength from a source emitting a line spectrum, a pair of cut-off filters is all that is required. These may be either glass filters or solutions in cuvettes. The emission filter must be chosen so that the Rayleigh-Tyndall scattered light is obscured and the light emitted by the sample transmitted. To avoid high blanks it may also be necessary to filter out any Raman scatter.

Interference filters having high transmission (>> 40%) of a narrow range (10-15 nm) of wavelengths are available and it is possible to purchase filters with a maximum transmission at any desired wavelength. UV filters of this type, however, are expensive and of limited range. A simple filter system is acceptable for much quantitative work, particularly where sufficient chemistry has been carried out to eliminate interfering compounds.

However, it is useful to be able to scan the emission from the sample to check for impurities and optimize conditions. A convenient method is to make use of a continuous interference filter so that an emission spectrum can be recorded, at least over the visible region of the spectrum.

A further refinement would be to use monochromators to select both the excitation and emission wavelengths. Most modern instruments of this type employ diffraction grating monochromators for this purpose. Such a fluorescence spectrometer is capable of recording both excitation and emission spectra and therefore makes full use of the analytical potential of the technique. If monochromators are employed, it should be possible to change the slit width of both the excitation and emission monochromators independently. Many analyses will not require high resolution (essentially corresponding to high selectivity) and greater sensitivity will be obtained with wide slit widths. Conversely, to record the fine structure in the emission of, for example, polyaromatic hydrocarbons or to excite selectively one compound in the presence of another, narrow slit widths will be necessary, and sensitivity will be sacrificed.

5.3.4 Detectors

All commercial fluorescence instruments use photomultiplier tubes as detectors and a wide variety of types are available. The material from which the photocathode is made determines the spectral range of the photomultiplier and generally two tubes are required to cover the complete UV-visible range. The S5 type can be used to detect fluorescence out to approximately 650 nm, but if it is necessary to measure emission at longer wavelengths, a special red sensitive, S20, photomultiplier should be employed.

The limit of sensitivity of a photomultiplier is normally governed by the level of dark current (which is the signal derived from the tube with no light falling on it). The dark current is caused by thermal activation and can usually be reduced by cooling the photomultiplier. Another method of minimizing dark current is to use a stroboscopic source since the ratio of dark current to fluorescence will be very small during each high intensity flash. During the periods between flashes when the dark current is relatively high, the photomultiplier output can be disconnected. The overall result is that the dark current no longer becomes the limitation to sensitivity.

The spectral response of all photomultipliers varies with wavelength, but it is sometimes necessary to determine the actual quantum intensity of the incident radiation and a detector insensitive to changes in wavelength is required. A suitable quantum counter can be made from a concentrated solution of Rhodamine 101 in ethylene glycol which has the property of emitting the same number of quanta of light as it absorbs, but over a very wide wavelength range. Thus, by measuring the output of the quantum counter at one wavelength, the number of incident quanta over a wide wavelength range can be measured.

5.3.5 Read-Out Devices

The output from the detector is amplified and displayed on a PC. It should be possible to change the sensitivity of the amplifier in a series of discrete steps so that samples of widely differing concentration can be compared. A continuous sensitivity adjustment is also useful so that the display can be made to read directly in concentration units.

Digital displays are most legible and free from misinterpretation. Improvement in precision is obtained by the use of integration techniques where the average value over a period of a few seconds is displayed as an unchanging signal. Microprocessor electronics provides outputs directly compatible with printer systems and computers, eliminating any possibility of operator error in transferring data.

Schematic diagram of the integrated spectrofluorometer is demonstrated in Fig. 5.3.

5.3.6 Sample Holders

The majority of fluorescence assays are carried out in solution, the final measurement being made upon the sample contained in a cuvette or in a flow cell. Cuvettes may be circular, square or rectangular (the latter being uncommon), and must be constructed of a material that will transmit both the incident and emitted light. Square cuvettes, or cells will be found to be most precise since the parameters of path length and parallelism are easier to maintain during manufacture. However, round cuvettes are suitable for many more routine applications and have the advantage of being less expensive.

The cuvette is placed normal to the incident beam. The resulting fluorescence is given off equally in all directions, and may be collected from either the front surface of the cell, at right angles to the incident beam, or inline with the incident beam.

Some instruments will provide the option of choosing which collecting method should be employed, a choice based upon the characteristics of the sample. A very dilute solution will produce fluorescence equally from any point along the path taken by the incident beam through the sample. Under these conditions, the right-angled collection method should be used since it has the benefit of minimizing the effect of light scattering by the solution and cell. This is the usual measuring condition in analytical procedures.

5.4 SENSITIVITY

The sensitivity of fluorescence is dependent on both the fluorophore and the instrument. The response of a fluorophore depends on the molar absorptivity and the quantum yield. These factors are, in general, beyond the control of the analyst.

The sensitivity of a spectrofluorometer depends on a number of factors. Instrumental contributions to sensitivity are described below.

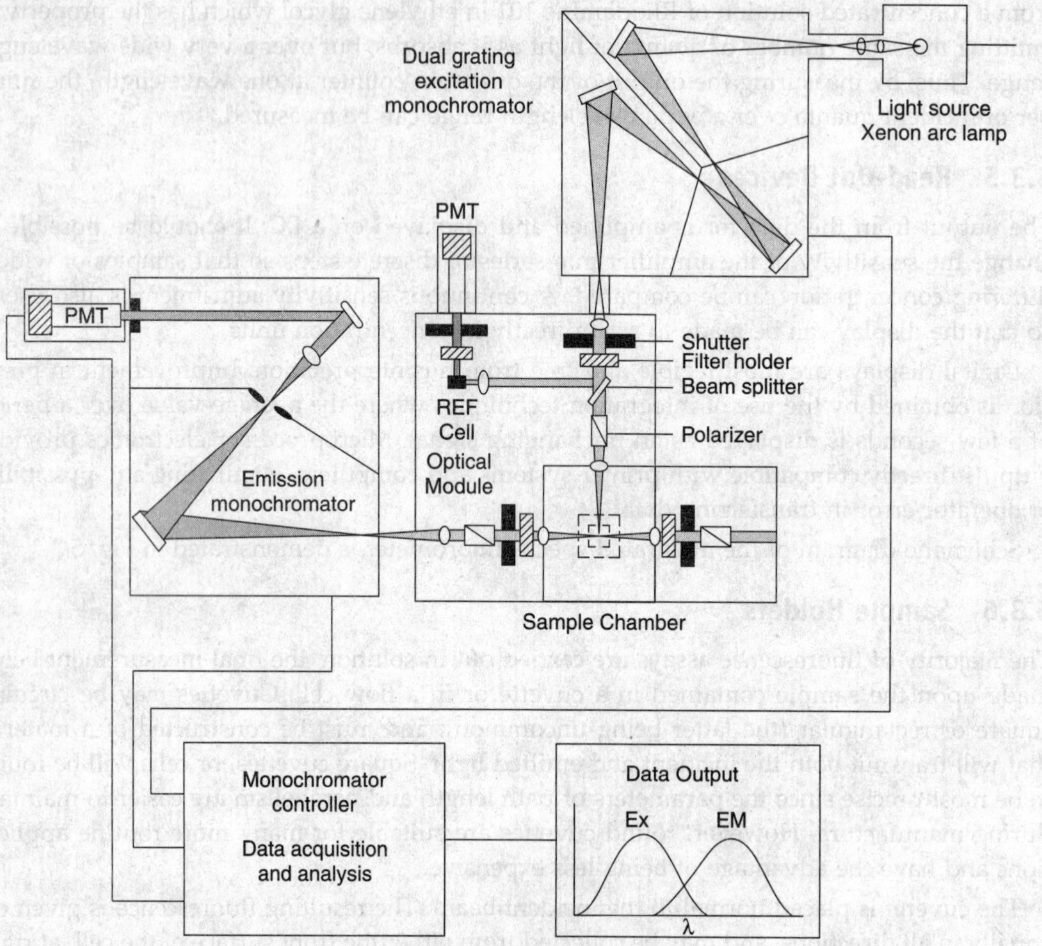

Fig. 5.3 *A schematic diagram of spectrofluorometer.*

1. **Source intensity**. In general, a brighter excitation source will result in brighter emission. The source for most fluorometers is a xenon arc, which has a high intensity between 200-900 nm, the spectral region where most fluorescence experiments are performed. While a high power arc lamp is good for highest intensity, a more important criteria is the brightness of the lamp. The brightness is a function of both the power and the size of the arc. Arc lamps for commercial fluorescence systems range from 75-450 watts.
2. **Efficiency of the optical system**. The light collection efficiency is a function of two factors. A high efficiency optical design is characterized by a low f number. For example, f/2 is more efficient, or faster, than f/4.5. It is imperative that the various optical components (e.g., source, monochromators, sample cuvette position, and detectors) are optically matched. A high speed design, however, is prone to high stray light. Also, the gratings, mirrors, and lenses that are incorporated into the spectrofluorometer

all have associated optical losses and will decrease the intensity of the light passing through the system. The number of optical elements, the thickness of lenses, and the coatings on mirrors and lenses affect the throughput.

3. **Spectral bandpass of the monochromators.** The bandpass of commercial fluorometers may be varied between 0.5-30 nm. Doubling the bandpass of a monochromator will increase the throughput of light by a factor of four. Resolution, however, is worse at high bandpass. The bandpass may be adjusted by the analyst to balance sensitivity and resolution.
4. **Efficiency of the detector.** Both analog and photon counting methods of detection are used in commercial instruments. In general, photon counting is considered to be slightly more sensitive. Both these methods are subject to degradation because of noise in the electronics.

It is the combination of the effects from all of these contributions that determine the sensitivity of the fluorescence spectrometer. It follows, then, that the most appropriate way to quantify the sensitivity of a spectrofluorometer is to measure a standard sample using the complete instrument system.

5.5 FACTORS AFFECTING FLUORESCENCE

In general, the compounds which fluoresce are aromatic or contain conjugated double bonds (i.e., alternating single and double bonds). Those compounds containing electrons which undergo energy transformations readily should fluoresce. Any radical, which when added to the molecule increases the freedom of these electrons, will enhance the fluorescence. Conversely, any radical which tends to restrict the electrons' ability to absorb energy will decrease the fluorescence.

Two different spectra are generally shown for compounds showing fluorescence. The excitation spectrum is obtained by measuring the variation in intensity of a strong emission wavelength as the wavelength of the excitation energy is changed. Conversely, when the wavelength and intensity are measured over the emission range using a strong excitation wavelength, an emission spectrum is obtained.

The effects of substitution upon fluorescence can be illustrated with benzene, aniline and nitrobenzene. In dilute solutions, aniline is 40 to 50 times more fluorescent than benzene whereas nitrobenzene does not fluoresce. The - NH group increases the freedom of the electrons while the - NO group tends to decrease the freedom.

Many factors may affect the intensity of fluorescence. Among the more important are instrumental parameters, concentration, solvent, pH temperature and the stability of the compound to light. Instrumental slit widths and light intensity can affect the intensity and all instrumental parameters must be kept constant during any set of determination.

(i) **Concentration effects:** Concentration plays a very important role in the intensity of emitted light. Generally, the fluorescence is viewed at right angles to the incident light. The fluorescence emitted must pass through the cell and some is reabsorbed by the solution. The higher the concentration of the compound, the greater the energy which is reabsorbed and lost. Consequently, a linear relationship between concentration and fluorescence exists only in a very dilute solution.

(ii) **Solvent effects:** Solvents used in fluorescence measurements may affect the results radically. Many solvents may contain impurities which will fluoresce and need extensive purification to make them usable. Solvents such as water, simple alcohols, ether or hexane can be used. The fluorescence wavelength may shift rather widely as the solvent is varied.

(iii) **pH effects:** Relatively small changes in pH will sometimes radically affect the intensity and spectral characteristics of fluorescence. Accurate pH control is essential and, when particular buffer solutions are recommended in an assay procedure, they should not be changed without investigation. Most phenols are fluorescent in neutral or acidic media, but the presence of a base leads to the formation of non-fluorescent phenate ions. 5 hydoxyindoles, for example, serotonin, show a shift in fluorescence emission maximum from 330 nm at neutral pH to 550 nm in strong acid without any change in the absorption spectrum.

Ionization of the fluorescence compound may change or eliminate fluorescence. Thus, pH may become an important factor in any measurement. Aqueous buffer solutions are frequently used to control pH.

(iv) **Temperature effects:** Changes in temperature affect the viscosity of the medium and hence the number of collisions of the molecules of the fluorophore with solvent molecules. Fluorescence intensity is sensitive to such changes and the fluorescence of many certain fluorophores shows temperature dependence. In such cases the use of thermostatted cell holders is to be recommended. Normally, it is sufficient to work at room temperature with the proviso that any sample procedure involving heating or cooling must also allow sufficient time for the final solution to reach ambient temperature before measurement.

Fluorescence intensity tends to increase as the temperature is lowered and decreases as the temperature is raised. The fluorescence may change by as much as 5 percent per degree of temperature change.

(v) **Inner-filter effects:** Fluorescence intensity will be reduced by the presence of any compound which is capable of absorbing a portion of either the excitation or emission energy. At high concentrations this can be caused by absorption due to the fluorophore itself. More commonly, particularly when working with tissue or urine extracts, it is the presence of relatively large quantities of other absorbing species that is troublesome. The purpose of extraction procedures is usually to eliminate such species so that the final measurement is made upon a solution essentially similar to the standard.

Some compounds tend to decompose under the influence of ultraviolet light. Thus, as the concentration of the solute decreases the intensity of fluorescence decreases, except in those cases where the decomposition products may fluoresce.

(vi) **Quenching:** Quenching is the term applied to the loss of fluorescence. As mentioned earlier, the compound itself may cause concentration quenching. Some compounds can reduce or eliminate fluorescence. Quenching can be caused by inner-filter effects, energy degradation, chemical change, absorption and/or intersystem transfer. Decrease of fluorescence intensity by interaction of the excited state of the fluorophore with its surroundings is known as quenching and is fortunately relatively rare. Quenching is not random. Each example is indicative of a specific chemical interaction, and the common instances are well known. Quinine fluorescence is quenched by the presence of halide ion despite the fact that the absorption spectrum and extinction coefficient of quinine is identical in 0.5 M H_2SO_4 and 0.5 M HCl.

It should be pointed out that all glassware must be kept very clean. Many detergents fluoresce and should not be used to clean cuvettes (sample cells). Chromic acid absorbs ultraviolet light and should not be used for cleaning glassware for fluorescent procedures. Concentrated nitric acid is frequently used to clean cuvettes.

5.6 APPLICATIONS

The use of fluorescence in environmental and biological specimens has developed in the last few years. Its use will probably increase rapidly as new organic compounds are added to the hazardous chemicals list.

The analysis of airborne particulate material for beryllium is one example of fluorescence analysis. Under proper conditions beryllium can be reacted with morin to produce a compound with strongly fluorescent properties. This reaction has been the basis for many published methods for the determination of microgram quantities of beryllium.

As interest in the polynuclear aromatic hydrocarbon content of the effluent from combustion processes has increased, the use of the spectrophotofluorometer has increased rapidly. Most of these compounds have specific excitation and emission spectra, and very minute quantities of these compounds can be determined in complex mixtures, after separation by celium chromatography, gas chromatography or thin layer chromatography.

The extended use of these fluorescent technologies is expected to lead to a greater number of analytical methods in this phase of spectrophotometry.

5.7 REFERENCES

1. The luminescence of biological systems, Johnson, F.H., Amer. Assoc. Adv. Sci., Washington, D.C. (1955).
2. Fluorescence and phosphorescence analysts, Hercules, D.M., Ed. Wiley- Interscience Publishers, New York, London, Sydney (1965).
3. Fluorometric analysis, Konstantinova- Schlezinger, M.A., Ed., Davis Publishing Co., New York (1965).
4. Handbook of fluorescence spectra in aromatic molecules, Berlman, LB., Academic Press, New York, Vol. 1 (1965), Vol. II (1971).

5. Fluorometric techniques in clinical pathology, Phillips, R., et al., Progress in Clinical Pathology, Grune and Stratton, New York (1965).
6. Biochemical fluorescence: concepts, Vol. I, Chen, R.F., et al., 408 (1965).
7. Fluorescence, theory, instrumentation and practice, Edited by Guilbault, G.G., published by Marcel Dekker, Inc., New York (1967).
8. Guide to fluorescence literature, Passwater, R.A., Plenum Press, New York, Vol. I (1967), Vol. II (1969).
9. Fluorescence and phosphorescence of proteins and nucleic acids, Koney, S.U., Plenum Press, New York (1967).
10. Luminescence of organic substances, Schmillen, A., et al., Hellwege Verlag, Berlin (1967).
11. Photoluminescence of solutions, Parker, C.A., Elsevier Publishing Co., New York (1968).
12. Fluorescence assay in biology and medicine, Udenfriend, S., Academic Press, Vol. I (1962), New York Vol. II (1969).
13. Theory and interpretation of fluorescence and phosphorescence, Becker R., Wiley-Interscience (1969).
14. Instrumental methods of chemical analysis, 3rd edition, Ewing, G.W., McGraw-Hill Book Co. (1969).
15. Fluorimetric assay of enzymes, Roth, M., Methods of Biochemical Analyses Vol. 17, David Glick, Editor Wiley Interscience, New York (1969).
16. Luminescence and biophysics of cells, Baremboim, G.M., et al., Plenum Press, New York (1969).
17. Fluorescence analysis: a practical approach, While; C.E., et al., Marcel Dekker (1970).
18. Fluorescence spectroscopy: an introduction for biology and medicine, Pesce, A.J., et al., Marcel Dekker (1971).
19. Fluorometric analysis, Weissler, A., et al., Standard methods of chemical analysis – Welcher, Editor D.Van Nostrand Co., Princeton, N.J. (1970).
20. Biochemical fluorescence: concepts, Vol. II, Chen, R.F., et al., M. Dekker, New York (1970) 535 pp.
21. Luminescence spectrometry in analytical chemistry, Winefordner, J.D., et al., Wiley-Interscience Publishers, New York (1972).
22. Methods In pharmacology, Vol. II, physical techniques, Chignell, C.F., Appleton-Crofts (1972).
23. Practical fluorescence: theory, methods and techniques, Guilbault, G.G., Marcel Dekker (1973).
24. Fluorometric techniques in clinical chemistry, Elevitch, F.R., Little Brown & Co. (1973).
25. Modern fluorescence spectroscopy, Wehry E.L., Ed; Plenum Press, New York (1976), Vol. I 238 pp.

26. Excited states of biological molecules, Birks, J.B., Ed; John Wiley, London, (1976) 652 pp.
27. Modern fluorescence spectroscopy, Wehry, E.L., Ed. Plenum Press, New York (1976), Vol. 11459 pp.
28. Physical Biochemistry-applications to biochemistry and molecular biology, Freifelder, D., W. H. Freeman & Co. (1976).
29. Fluorescence and phosphorescence spectroscopy: physiochemical principles and practice, Schulman, S.G., Pergamon, Elmsford, New York (1977).
30. Molecular fluorescence spectroscopy, Guilbault, G.G., Cooper Anal. Chem. 8, 71-205, (1977), Svehlag. Ed; Elsevier.
31. Developments in instrumentation and techniques, West, M.A., from Photochemistry Vol. Chem. Soc. London.
32. The handbook of derivatives for chromatography, Blau, K., King, G., Heyden & Son Ltd. (1977).
33. Standardization & fluorescence spectrometry, Miller, G.N., Ed, Chapman and Hall, 1981.
34. Fluorescence detection in liquid chromatography, Rhys-Williams, A.T., PerkinElmer Ltd.
35. Fluoroimmunoassay – a review of present methods, Rhys-Williams, A.T., PerkinElmer Ltd.
36. An introduction to phosphorescence spectroscopy, Rhys-Williams, A.T., PerkinElmer Ltd.
37. Nomenclature, symbols, units and their usage in spectrochemical analysis – VI molecular luminescence spectroscopy, Melhuish, W.H., Zander, M., Pure App. Chem. 53, 1953, 1981.
38. Froehlich, P. 1989. Understanding the sensitivity specification of spectrofluorometers. International Laboratory. 19:42-45.
39. 1993 Annual Book of ASTM Standards, Volume 14.01. 1993. TSM Standard E 579-84, Test Method for Limit of Detection of Fluorescence of Quinine Sulfate, p. 142-143. American Society for Testing and Materials, Philadelphia, PA.
40. 1993 Annual Book of ASTM Standards, Volume 14.01. 1993. TSM Standard E 386-90, Standard Practice for Data Presentation Relating to High-Resolution Nuclear Magnetic Resonance (NMR) Spectroscopy, p. 93-101. American Society for Testing and Materials, Philadelphia, PA.
41. Dereniak, E.L. and D.G. Crowe. 1984. *Optical Radiation Detectors*. John Wiley & Sons, New York.
42. Jacobs, S.F. 1978. *Handbook of Optics*. McGraw-Hill Book Co., New York.
43. For example: (a) SigmaPlot, Jandel Scientific. (b) LabCalc or GRAMS 386, Galactic Software. (c) PSIPlot, Poly Software Intl. Other data analysis products are also available.

CHAPTER 6

pH Electrodes & Meter and pH Measurement

6.1 INTRODUCTION

The term pH is derived from a combination of "*p*" for the word power and "*H*" for the symbol of the element hydrogen. Together the measuring is the power of hydrogen. pH serves as a convenient way to compare the relative acidity or alkalinity of a solution at a given temperature. The pH value states the relative quantity of hydrogen ions (H^+) contained in a solution. Greater the concentration of H^+, the more acidic the solution and the lower the pH. In this relationship, pH is defined as the negative logarithm of hydrogen activity. pH measurement is used in a wide variety of applications: agriculture, wastewater treatment, industrial processes, environmental monitoring and in research and development.

6.2 DEFINITION OF pH

The pH value of a solution is the negative log of its hydrogen ion activity (α), which is the product of hydrogen ion concentration [H^+] and the activity coefficient of hydrogen ($\gamma_H{}^+$) at that concentration.

$$\mathbf{pH = -\log \alpha = -\log \alpha_H{}^+[H^+]}$$

In pure water and in dilute solutions, the H^+ activity can be considered the same as the H^+ concentration.

$$\mathbf{pH = -\log gH^+[H^+] = -\log [H^+]}$$

The pH of a solution measures the degree of acidity or alkalinity relative to the ionization of water. Pure water dissociates to yield 10^{-7} M of [H^+] and [OH^-] at 25°C; thus, the pH of water is 7 minus the point of neutrality.

$$\mathbf{pH_{water} = -\log [H^+] = -\log 10^{-7} = 7}$$

Most pH readings range from 0 to 14. Solutions with a higher [H^+] than water (pH less than 7) are acidic; solutions with a lower [H^+] than water (pH greater than 7) are basic or alkaline (Fig. 6.1).

Acids ⟵⟶ (pH 0–5) Bases ⟵⟶ (pH 9–13)

0	1	2	3	4	5	6	7	8	9	10	11	12	13	14
+414	+355	+296	+237	+177	+118	+59	00	–59	–118	–177	–237	–296	–355	–414

Fig. 6.1 *Demonstrate acids and bases - pH range.*

6.3 pH THEORY

A pH of 7 describes a neutral solution because the activities of hydrogen and hydroxide ions are equal. When the pH is below 7, activity of hydrogen ion is greater than that of hydroxide ion. A solution is more acidic as the hydrogen ion activity increases and the pH value decreases. Conversely, when the pH is above 7, the solution is described as basic (or alkaline) because the activity of hydrogen ion is greater than that of hydrogen ion. pH of some of the compounds are given in Fig. 6.2.

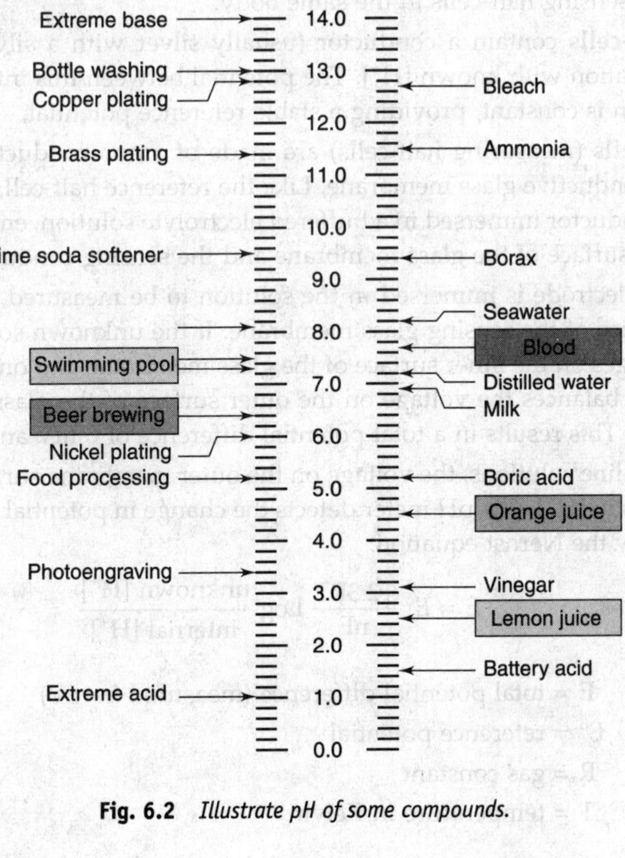

Fig. 6.2 *Illustrate pH of some compounds.*

pH electrode measurements are made by comparing the readings in a sample with the readings on standards whose pH has been defined (buffers). When a pH sensing electrode comes in contact with a sample, a potential develops across the sensing membrane surface and that membrane potential varies with pH. A reference electrode provides a second, unvarying potential to quantitatively compare the changes of the sensing membrane potential. Combination of pH electrodes are composed of a sensing electrode with the reference electrode built into the same electrode body. Combination electrodes provide the same selectivity and response as a half-cell system but often the convenience of working and maintaining only one electrode. A meter serves as the read-out device and calculates the difference between the reference electrode and sensing electrode potentials in millivolts. The millivolts are then converted to pH unit and shown on the meter display. The sample or standard solution is then the final component of the system.

Electrode behavior is described by the Nernst equation.

6.3.1 pH Measurement

Measuring pH involves comparing the potential of solutions with unknown $[H^+]$ to a known reference potential. pH meters convert the voltage ratio between a reference half-cell and a sensing half-cell to pH values. Today, most electrodes are combination electrodes with both the reference and sensing half-cells in the same body.

Reference half-cells contain a conductor (usually silver with a silver chloride coating) immersed in a solution with known $[H^+]$. The potential between this internal conductor and the known solution is constant, providing a stable reference potential.

Sensing half-cells (measuring half-cells) are made of a non-conducting glass (or epoxy) tube sealed to a conductive glass membrane. Like the reference half-cell, the sensing half-cell also contains a conductor immersed in a buffered electrolyte solution, ensuring constant voltages on the inner surface of the glass membrane and the sensing conductor.

When the pH electrode is immersed in the solution to be measured, a potential is established on the surface of the sensing glass membrane. If the unknown solution is neutral, the sum of fixed voltages on the inner surface of the glass membrane and on the sensing conductor approximately balances the voltage on the outer surface of the glass membrane and the reference half-cell. This results in a total potential difference of 0 mV and a pH value of 7.

In acidic or alkaline solutions, the voltage on the outer membrane surface changes proportionally to changes in $[H^+]$. The pH meter detects the change in potential and determines $[H^+]$ of the unknown by the Nernst equation:

$$E = E^\circ \frac{+2.3RT}{nF} \text{Log} \frac{\text{unknown } [H^+]}{\text{internal } [H^+]}$$

where

E = total potential difference (measured in mV)

E° = reference potential

R = gas constant

T = temperature in Kelvin

n = number of electrons

F = Faraday's constant

$[H^+]$ = hydrogen ion concentration

6.3.2 Nernst Equation - Electrode Potential as Function of pH

Nernst equation describes potential of electrochemical cell as a function of concentrations of ions taking part in the reaction:

$$E = E_0 - \frac{RT}{nF} \ln (Q)$$

where Q is a reaction quotient and n is the number of electrons exchanged. For constant temperature expression RT/F has constant value. To simplify calculations it is often combined with conversion factor between natural logarithm (denoted here by ln) and decimal logarithm (denoted here by log) to form value of 0.0591 (for 25°C). Thus, for example for half reaction of MnO_4^- in acidic media:

$$MnO_4^- + 8H^+ + 5e^- \rightarrow Mn^{2+} + 4H_2O$$

potential is given by

$$E = 1.51 - \frac{0.0591}{5} \log \left(\frac{[Mn^{2+}]}{[MnO_4^-]\,[H^+]^8} \right)$$

To be precise we should not use concentrations, but thermodynamic activities of the ions present in the solution.

When applied to the full cell formed from half-cells present on inside and outside of glass pH electrode, Nernst equation takes form:

$$E = E_0 + 0.0591 \log \left(\frac{[H^+_{inside}]}{[H^+_{outside}]} \right)$$

or (using pH definition)

$$E = E_0 + 0.0591\ pH_{outside}$$

pH inside of the electrode has constant value, thus it can be included in the potential part of the above equation.

This form of the equation describes behavior of the glass electrode used for pH measurements.

It is necessary to add two remarks here. First, the above equation assumes ideal response from the glass electrode. Second, 0.0591 value was calculated for 25°C temperature. In most cases electrode doesn't behave as ideal and the temperature is not exactly 25°C, thus slope of the electrode response is not 0.0591. To overcome these problems the electrode should be calibrated before each use or series of uses.

6.3.3 Reaction Quotient

In chemistry, **reaction quotient (Q_r)** is a function of the extent of reaction: ξ, the relative proportion of products and reactants present in the reaction mixture at some instant of time. The reaction quotient, Q_r, relating to a chemical reaction equation, has the same expression as the equilibrium constant for that reaction. However, the reaction quotient is computed using the current concentrations or pressures, not the equilibrium concentrations.

For a chemical mixture with certain initial concentrations of reactants and products, it is useful to know if the reaction will shift to the right/in the forward direction (increasing the concentrations of the products) or if it will shift to the left/in the reverse direction (increasing the concentrations of the reactants). Given a general equilibrium expression such as

$$kA + mB \ldots \rightleftharpoons nC + pD \ldots$$

where A, B, C, and D are chemical species involved in this reaction and k, m, n, and p are the stoichiometric coefficients for the reaction, the reaction quotient, Q_r, is defined as:

$$Q_r = \frac{\{C_t\}^n \{D_t\}^p \ldots}{\{A_t\}^k \{B_t\}^m \ldots}$$

where $\{A_t\}$ denotes the instantaneous activity of the species A at a certain moment of time (t) and so on for the other species. The reaction quotient is taken at a particular instant in time, not necessarily the moment when equilibrium is reached. The reaction quotient is directly related to Le Chatelier's principle. For a reaction at chemical equilibrium, the equilibrium constant, K, may be defined as:

$$K = \frac{\{C_{\varepsilon q}\}^n \{D\varepsilon q\}^p \ldots}{\{A_{\varepsilon q}\}\text{k} \{B\varepsilon q\}^m \ldots}$$

where $\{A_{eq}\}$ is the activity of the species A when the mixture is at equilibrium, etc. By comparing the values of Q_r and K, one can determine whether the reaction will shift to the right, to the left, or if the concentrations will remain the same (equilibrium).

- The relationship of reaction quotient Q_r with the instantaneous derivative of Gibbs energy ($\Delta_r G$) and standard Gibbs energy change ($\Delta_r G^\theta$) is given by

$$\Delta_r G = \Delta_r G^\theta + RT \ln Q_r$$

6.4 HYDROGEN REFERENCE ELECTRODE

6.4.1 The Hydrogen Electrode

The hydrogen electrode is easy to build. All that is required is to bubble hydrogen through an acid solution of known pH so that it is saturated with hydrogen. An (ideal) noble metal electrode is placed into the solution. Platinized platinum is generally used to insure a large electrochemical surface area and rapid equilibrium conditions. The pressure of the hydrogen must also be known. Fairly detailed construction details are given (Fig. 6.3).

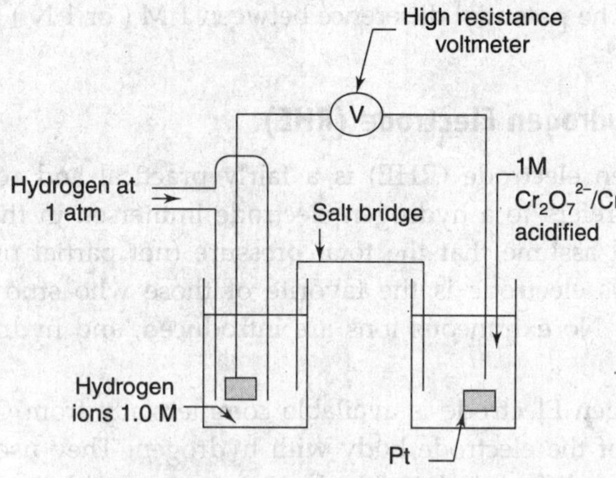

Fig. 6.3 *Demonstrate hydrogen electrode.*

The potential is given by the Nernst equation:

$$2\,H^+ + 2e^- \rightarrow H_2$$

$$E = E^o - RT/2F \ln \mathbf{P}(H_2)/\mathbf{a}(H^+)^2$$

$$= E^o - (RT/F)^*pH - RT/2F \ln \mathbf{P}(H_2)$$

$\mathbf{P}(H_2)$ is the partial pressure of H_2 in atmospheres. $\mathbf{a}(H^+)$ is the activity of hydrogen ion. E^o for this half-reaction is defined to be 0.0V at all temperatures. The potential of this reference is only 0.0V if the partial pressure (actually, fugacity) of H_2 and the activity of H^+ are both unity.

6.4.2 Standard Hydrogen Electrode (SHE)

The standard hydrogen electrode (SHE) is officially the 0.0V thermodynamic reference point for all potential measurements at all temperatures. It is easy to define, but not convenient to implement rigorously.

How much acid do you need to make a solution with a = 1 (1 atmosphere) Compton & Sanders suggest $[H^+]$ = 1.18 M, but that number will be different for sulfuric and hydrochloric acids since the activity coefficients are different.

Things are not much better when you think about the partial pressure of hydrogen. The vapor pressure of water is about 25 mm Hg at room temperature. So, even if the barometric pressure is 760 mm Hg, the partial pressure of water saturated hydrogen is only 735 mm Hg.

Using 1 M H^+ and using a total pressure of one atmosphere each only contributes about 5 mV error. The errors are in opposite directions, so they tend to cancel.

6.4.3 Normal Hydrogen Electrode (NHE)

The SHE is also sometimes called the **NHE**, for Normal Hydrogen Electrode. "Normal" makes the think of the NCE, "Normal Calomel Electrode," which is filled with 1 N KCl. have made the association. The potential difference between 1 M (or 1 N) HCl and HCl @ a = 1 is small.

6.4.4 Reversible Hydrogen Electrode (RHE)

The reversible hydrogen electrode (RHE) is a fairly practical and reproducible electrode "standard." The term refers to a hydrogen electrode immersed in the electrolyte solution actually used. It would assume that the total pressure (not partial pressure of hydrogen) is one atmosphere. This electrode is the favorite of those who study hydrogen UPD on Pt electrodes in H_2SO_4. No extraneous ions are introduced, and hydrogen has to be there anyway.

A Reversible Hydrogen Electrode is available commercially from Gaskatel. This clever design fills the inside of the electrode body with hydrogen. They use a carbon supported platinum catalyst as a gas diffusion electrode that is in contact with the fixed pressure hydrogen atmosphere on the inside and the solution on the outside.

6.5 QUINHYDRONE ELECTRODE

6.5.1 Introduction

In 1920, Billmann showed that it was possible to form with the organic compound, quinhydrone, an electrode which could be used for making hydrogen-ion determinations and with which very constant voltages were possible. The quinhydrone molecule is apparently formed by the union of one molecule of quinine, $C_6H_4O_2$ and one of hydroquinone, $C_4H_4O_2H_2$. In an aqueous solution it is very highly dissociated into its components.

Hydroquinone and quinine, the products of the dissociation of the quinhydrone, are the active bodies of the electrode. According to the equilibrium equitation,

$$C_4H_4O_2H_2 \rightleftharpoons C_4H_4O_2 + H_2$$

The concentration of hydrogen is represented by the formula:

$$(H_2) = \frac{(C_4H_4O_2H_2)}{(C_4H_4O_2)} + K \qquad \text{(A)}$$

6.5.2 Theory

The theory of the quinhydrone electrode has been given by Billmann LaMer and Parsons and by Koithoff and Furman. The solid quinhydrone which is added to the soil-water mixture, dissolves and dissociates into equi-molecular quantities of hydroquinone and quinone. From the equation:

$$\pi(\text{E.M.F}) = \frac{RT}{2F} \text{ in } \frac{(\text{Quinone})}{(\text{hydroquinone})} + \frac{RT}{F} \ln (H^+) + K \qquad \text{(A)}$$

The first term is when the concentration of quinone is the same as the concentration of hydroquinone, and therefore

$$\frac{\pi Rt}{F} \quad \text{Is}(H^+) + K \qquad \text{(B)}$$

If the activity of the ions is disregarded, K is π ($r\ \pi_\alpha$) when the concentration of the hydrogen ion is 1.

Billmann, by checking the quinhydrone electrode against a hydrogen electrode in solutions of the same H^+ concentration, found that at 2°C. π = 0.6990 and at 18°C. 0.7044, the platinum of the quinhydrone electrode being positive to the hydrogen electrode. Thus, at 18°

$$\pi = 0.0577 \log (H^+) + 0.7044 \qquad \text{(C)}$$

and at 25°

$$\pi = 0.0591 \log (H^+) + 0.6990 \qquad \text{(D)}$$

6.5.3 Methods

Electrode vessel is sketched in Fig. 6.4. The capacity is 2-5 c.c. First quinhydrone powder (Kahlbaum's preparation) is put in the vessel and the liquid is poured into it, and then it is stoppered and shaken. The stopper has a sealed platinum wire and mercury.

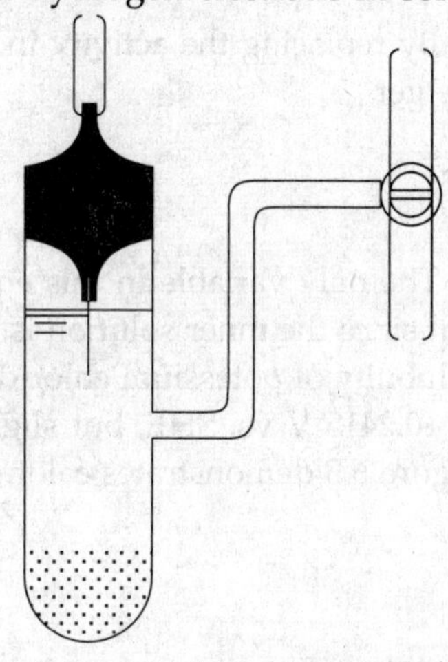

Fig. 6.4 *Quinhydrone electrode vessel*

6.6 SATURATED CALOMEL ELECTRODE

The saturated calomel electrode (SCE) is a reference electrode based on the reaction between elemental mercury and mercury(I) chloride. The aqueous phase in contact with the mercury and the mercury(I) chloride (Hg_2Cl_2, "calomel") is a saturated solution of potassium chloride in water. The electrode is normally linked via a porous frit to the solution in which the other electrode is immersed. This porous frit is a salt bridge.

In cell notation the electrode is written as:

$$Cl^-(4M) \mid Hg_2\,Cl_2(s) \mid Hg(l) \mid Pt$$

6.6.1 Theory of Operation

The electrode is based on the redox reaction

$$Hg_2^{2+} + 2e^- \rightleftharpoons 2Hg(l)$$

The Nernst equation for this reaction is

$$E = E^0_{Hg_2^{2+}/Hg} - \frac{RT}{2F} \ln \frac{1}{a_{Hg_2^{2+}}}$$

where E^0 is the standard electrode potential for the reaction and a_{Hg} is the activity for the mercury cation (the activity for a liquid is 1). This activity can be found from the solubility product of the reaction

$$Hg_2^{2+} + 2Cl^- \rightleftarrows Hg_2\,Cl_2(s),\ K_{sp} = a_{Hg_2^{2+}}\,a_{Cl^-}^2$$

By replacing the activity in the Nernst equation with the value in the solubility equation, we get

$$E = E^{2}_{Hg_2^{2+}/Hg} + \frac{RT}{2F}\ln K_{sp} - \frac{RT}{2F}\ln a_{Cl^-}^2$$

The only variable in this equation is the activity (or concentration) of the chloride anion. But since the inner solution is saturated with potassium chloride, this activity is fixed by the solubility of potassium chloride. When saturated the redox potential of the calomel electrode is +0.2415 V vs. SHE, but slightly higher when the chloride solution is less than saturated. Figure 6.5 demonstrates calomel reference electrode.

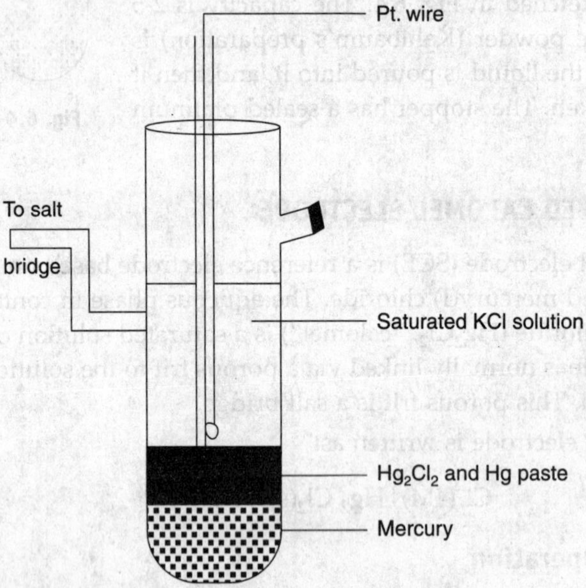

Fig. 6.5 *Calomel electrode.*

6.6.2 Applications

The SCE is used in pH measurement, cyclic voltammetry and general aqueous electrochemistry.

This electrode and the silver/silver chloride reference electrode work in the same way. In both electrodes, the activity of the metal ion is fixed by the solubility of the metal salt.

The calomel electrode contains mercury, which poses much greater health hazards than the silver metal used in the Ag/AgCl electrode.

6.7 GLASS pH ELECTRODE

The glass electrode has been around almost a century. In 1906, Max Cremer during his studies of liquid interfaces blew in a flame a thin bubble at the end of a glass tube and measured the potential difference between one liquid inside and a different one on the outside. Fritz Haber (Nobel Prize winner for the synthesis of ammonia) and Zygmunt Klemensiewicz measured in a series of experiments the potential difference on two sides of a glass membrane; they discovered that the potential is a function of proton activity and coined the name glass electrode.

A glass electrode (Fig. 6.6) is actually a device, not an electrode in an electrochemical sense of the word. It consists of a glass bulb membrane, which gives it its name and an electrically insulating tubular body, which separates an internal solution and a silver/silver chloride electrode from the studied solution. The Ag/AgCl electrode is connected to a lead cable terminated with some connector that can hook up to a special voltmeter, the pH meter. The pH meter measures the potential difference and its changes across the glass membrane. The potential difference must be obtained between two points; one is the electrode contacting the internal solution, a second is obtained by connecting to a reference electrode, immersed in the studied solution. Often, this reference electrode is built in the glass electrode (a combination electrode), in a concentric double barrel body of the device.

In potentiometric measurement one point is connected to internal solution and second is reference point. The completed glass electrode with a reference electrode cell is represented by the electrochemical shorthand

$$Ag/AgCl \mid HCl \mid glass \parallel probed\ solution \mid reference\ electrode \qquad (1)$$

The potential difference relevant to pH measurement builds up across the outside glass/solution interface marked(||). The key functional part, the glass membrane, is manufactured by blowing molten glass into a thin-walled bulb with a wall about 0.1 mm thick. The bulb is then sealed to a thicker glass or plastic tube, and filled, for example, with a solution of HCl (0.1 mol/dm3). In this solution is immersed a silver/silver chloride electrode with a lead to the outside through a permanent hermetic seal. The filling solution has constant Cl-concentration, which keeps the Ag/AgCl inner electrode at fixed potential.

The pH sensing ability of the glass electrode stems from the ion exchange property of its glass membrane. Glass is mostly amorphous silicon dioxide, with embedded oxides of alkali metals. When the surface of glass is exposed to water, some Si–O- groups become protonated

$$\text{Si-O-} + H_3O^+ \equiv \text{Si-O-}H^+ + H_2O \qquad (2)$$

The exchange of hydronium (or written as proton, H^+) between the solid membrane and the surrounding solution, and the equilibrium nature of this exchange, is the key principle

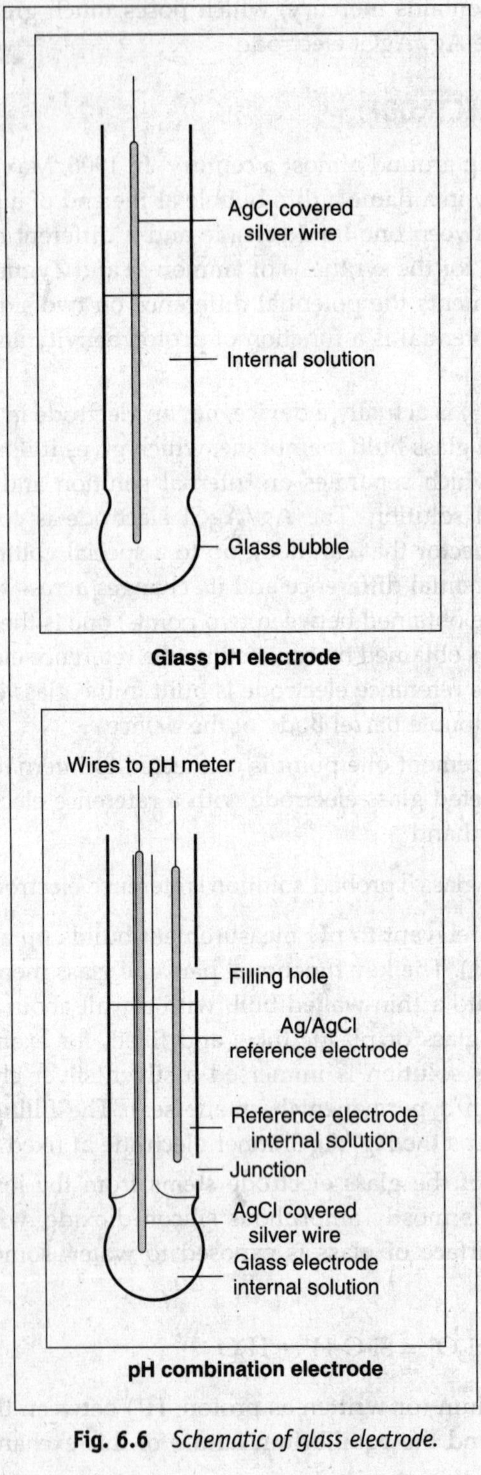

Fig. 6.6 *Schematic of glass electrode.*

of H_3O^+ sensing. As with any interface separating two phases between which ionic exchange equilibrium is established, the glass membrane/solution interface becomes the site of a potential difference

$$^{E}\text{glass wall/solution} \sim |RT/2.303F \log a(H_3O+)| \quad (3)$$

where R is the molar gas constant 8.314 J mol^{-1} K^{-1}, T is the temperature in kelvins, F is the Faraday constant 96,485.3 C, 2.303 is a conversion between natural and common logarithm, and $a(H_3O^+)$ is the activity of hydronium, which can be at lower concentrations equated with its concentration. At 30°C the value of $RT/2.303F$ is approximately 0.060 V.

The glass membrane has two wall/solution interfaces and there is potential buildup on each of them, with opposite polarity. But the pH inside the bulb is constant, because the internal solution is sealed. Therefore, the inner surface potential is constant, adding merely to an offset to the overall potential of the device. Additional contribution to the offset comes from potentials of the inner solution electrode, and the reference electrode, which are also constant. The changes in the device potential are therefore due entirely to the pH changes of the outside solution and the potential of the glass electrode/reference electrode setup is

$$^{E}\text{glass electrode} = E' + RT/2.303F \log a(H_3O^+) \quad (4)$$

where E' represents the sum of the constant offset potentials of the inner glass surface/solution and the two Ag/AgCl electrodes. At 30°C the potential of the glass membrane changes by about 60 mV for each one unit of pH (i.e., a tenfold activity change).

The possible range of hydronium activity encountered in aqueous solutions is large, as much as 10 to 10^{-15} mol/dm^3. To encompass such a wide dynamic range, a logarithmic scale was established, defining the pH of a solution as

$$\text{pH} = -\log a(H_3O^+) \quad (5)$$

The negative sign assures that pH of most solutions, except extremely acidic ones, is always positive. It is fortuitous that the potential of the glass electrode (Eq. 4) is also a logarithmic function of hydronium activity; therefore, the potential of the glass electrode is a linear function of pH

$$^{E}\text{glass electrode} = E' - RT/2.303F \text{ pH} \quad (6)$$

The ^{E}glass electrode is measured with a pH meter, a voltmeter with circuitry allowing direct readout of pH. The meter subtracts the potential offset E'. Although it could be calculated, it is more practical to calibrate the glass electrode using buffers. For example, if two buffers are used in sequence, one at pH 7 and another at pH 4, it is also possible to calibrate the actual slope of Eq. 6. Although it is mathematically given ($RT/2.303F$), the behavior in the presence of interfering ions or with deteriorating glass electrode is not always that ideal. The correction for temperature (T) is also made in a two-buffer calculation. Increasingly, glass electrodes and hand-held pH detectors, have built-in temperature probes, linked to an automated electronic temperature correction. Because the pH of standard buffers is available to two decimal places, measured pH should be reported to no more than two decimal places, despite that most pH meters can display three places or more. Practical reproducibility is about ± 0.05 units.

The glass electrodes of today share with the glass electrodes of yesterday both their working principle and their limitations. A common interference with sensing hydronium comes from ions of similar size, notably the alkali metals. The interference is somewhat alleviated by using specialty lithium glass, with sites too small to fit Na^+ or K^+. Alkali metal interference is not the only complication. In solutions of concentrated alkali hydroxides not only the metal ion concentration is high, the activity of H_3O^+ is small, making equilibration more complicated. And on top of that, strong alkaline solutions etch and possibly permanently damage glass surfaces. Fluoride is another ion that can physically damage glass. Additionally, solutions with proteins, that adsorb on surface, or solutions of ions of more noble metals (Ag, Cu) that could reduce and deposit on the glass surface, should also be avoided.

A glass electrode is perhaps the most successful and ubiquitous electrochemical sensor. It provides information about the activity of hydronium ions, H_3O^+, in water. Because water, which mildly dissociates to H_3O^+ and OH^- ions, is the most common solvent medium, and chemical reactions in water largely depend on H_3O^+ activity, the ability to measure it is essential. And conversely, because H_3O^+ activity, or rather, its negative logarithm, the pH, is so easy to measure. pH is the most commonly monitored and recorded parameter of liquid samples.

6.8 pH SYSTEMS

A successful pH reading dependes upon all components of the system being operational. Problems with any one of the three: meter, electrode, or buffer will yield poor readings.

Meters: A pH meter is a sophisticated voltmeter capable of reading small millivolt changes from the pH electrode system. The meter is seldom the source of problems for pH measurements. Modern pH meters have temperature compensation (either manual or automatic) to correct for variations in slope caused by changes in temperature. Microprocessor technology has created many new convenience features for pH measurement; autobuffer recognition, calculated slope and % efficiency, log tables for concentration of ions, and more.

Electrodes: A pH electrode consists of two half-cells; an indicating electrode and a reference electrode. Most applications today use a combination electrode with both half-cells in one body. Over 90% of pH measurement problems are related to the improper use, storage, or selection of electrodes.

Buffers: These solutions of known pH value allow the use to adjust the system to read accurate measurements. For best accuracy:

- Standardization should be performed with fresh buffer solutions.
- Buffers used should frame the range of pH for the samples being tested.
- Buffers should be at the same temperature as the samples. (For example, if all your samples are at 50°C, warm your buffers to 50°C using a beaker in a water bath.)

Buffer values are dependent upon temperature.

6.9 UNDERSTANDING: pH MEASUREMENT

In the process world, pH is an important parameter to be measured and controlled.

The pH of a solution indicates how acidic or basic (alkaline) it is. The pH term translates the values of the hydrogen ion concentration - which ordinarily ranges between about 1 and 10 × –14 gram-equivalents per liter - into numbers between 0 and 14.

On the pH scale a very acidic solution has a low pH value such as 0, 1, or (which corresponds to a large concentration of hydrogen ions; 10 × 0, 10 × –1, or 10 × –2 gram-equivalents per liter) while a very basic solution has a high pH value, such as 12, 13, or 14 which corresponds to a small number of hydrogen ions (10 × –12, 10 × –13, or 10 × –14 gram-equivalents per liter). Neutral solution such as water has a pH of approximately 7.

A pH measurement loop is made up of three components, the pH sensor, which includes a measuring electrode, a reference electrode, and a temperature sensor; a preamplifier; and an analyzer or transmitter. A pH measurement loop is essentially a battery where the positive terminal is the measuring electrode and the negative terminal is the reference electrode. The measuring electrode, which is sensitive to the hydrogen ion, develops a potential (voltage) directly related to the hydrogen ion concentration of the solution. The reference electrode provides a stable potential against which the measuring electrode can be compared.

When immersed in the solution, the reference electrode potential does not change with the changing hydrogen ion concentration. A solution in the reference electrode also makes contact with the sample solution and the measuring electrode through a junction, completing the circuit. Output of the measuring electrode changes with temperature (even though the process remains at a constant pH), so a temperature sensor is necessary to correct for this change in output. This is done in the analyzer or transmitter software.

The pH sensor components are usually combined into one device called combination pH electrode. The measuring electrode is usually glass and quite fragile. Recent developments have replaced the glass with more durable solid state sensors. The preamplifier is a signal-conditioning device. It takes the high impedance pH electrode signal and changes it into alow impedance signal which the analyzer or transmitter can accept. The preamplifier also strengthens and stabilizes the signal, making it less susceptible to electrical noise.

The sensor's electrical signal is then displayed. This is commonly done in a 120/240 V ac-powered analyzer or in a 24 V dc loop-powered transmitter.

Additionally, the analyzer or transmitter has a man machine interface for calibrating the sensor and configuring outputs and alarms, if pH control is being done.

Keep in mind, application requirements should be carefully considered when choosing a pH electrode. Accurate pH measurement and the resulting precise control that it can allow, can go a long way toward process optimisation and result in increased product quality and consistency. Accurate, stable pH measurement also controls and often lowers chemical usage, minimising system maintenance and expense.

A system's pH electrodes require periodic maintenance to clean and calibrate them. The length of time between cleaning and calibration depends on process conditions and the user's accuracy and stability expectations. Overtime, electrical properties of the measuring and reference electrode change. Calibration in known-value pH solutions called buffers will correct for some of these changes. Cleaning of the measuring sensor and reference junction will also help. However, just as batteries have a limited life, a pH electrode's lifetime is also finite. Even in the "friendliest" environments, pH electrodes have to be replaced eventually (Fig. 6.7).

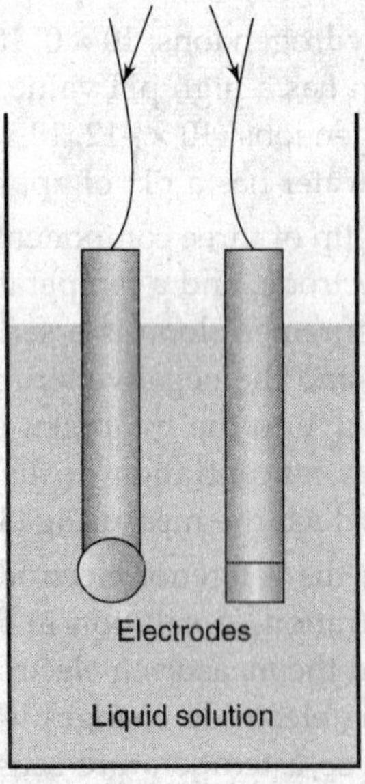

Fig. 6.7 *Voltage produced between electrodes is proportional to the pH.*

The design and operational theory of pH electrodes is a very complex subject, explored only briefly here. What is important to understand is that these two electrodes generate a voltage directly proportional to the pH of the solution. At a pH of 7 (neutral), the electrodes will produce 0 volt between them. At a low pH (acid) a voltage will be developed of one polarity, and at a high pH (caustic) a voltage will be developed of the opposite polarity.

An unfortunate design constraint of pH electrodes is that one of them (called the measurement electrode) must be constructed of special glass to create the ion selective barrier needed to screen out hydrogen ions from all the other ions floating around in the solution. This glass is chemically doped with lithium ions, which is what makes it react electrochemically to hydrogen ions. Of course, glass is not exactly what you would call a "conductor;" rather, it is an extremely good insulator. This presents a major problem if our intent is to measure voltage between the two electrodes. The circuit path from one electrode contact, through the glass barrier, through the solution, to the other electrode, and back through the other electrode's contact, is one of extremely high resistance.

The other electrode (called the reference electrode) is made from a chemical solution of neutral (7) pH buffer solution (usually potassium chloride) allowed to exchange ions with the process solution through a porous separator, forming a relatively low resistance connection to the test liquid. At first, one might be inclined to ask: why not just dip a metal wire into the solution to get an electrical connection to the liquid? The reason that this will not work is because metals tend to be highly reactive in ionic solutions and can produce a significant voltage across the interface of metal-to-liquid contact. The use of a wet chemical interface with the measured solution is necessary to avoid creating such a voltage, which of course would be falsely interpreted by any measuring device as being indicative of pH.

Here is an illustration of the measurement electrode's construction. Note the thin, lithium-doped glass membrane across which the pH voltage is generated (Fig. 6.8):

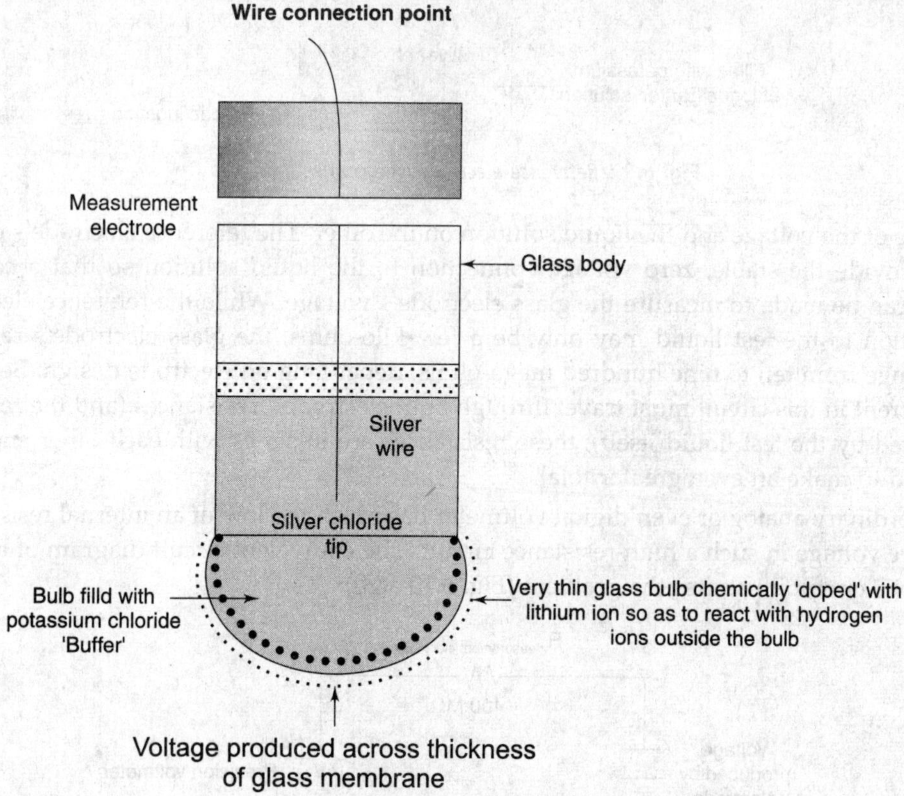

Fig. 6.8 *Measurement electrode wire connection point.*

Here is an illustration of the reference electrode's construction. The porous junction shown at the bottom of the electrode is where the potassium chloride buffer and process liquid interface with each other (Fig. 6.9).

The measurement electrode's purpose is to generate the voltage used to measure the solution's pH. This voltage appears across the thickness of the glass, placing the silver wire on

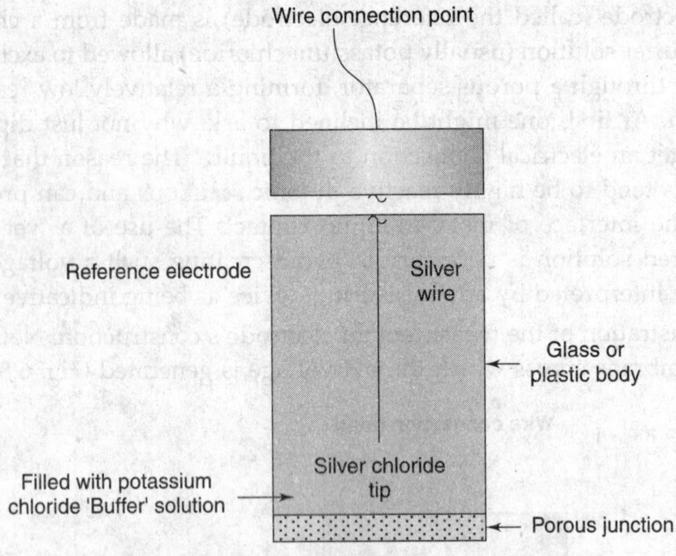

Fig. 6.9 *Reference electrode wire connection point.*

one side of the voltage and the liquid solution on the other. The reference electrode's purpose is to provide the stable, zero voltage connection to the liquid solution so that a complete circuit can be made to measure the glass electrode's voltage. While the reference electrode's connection to the test liquid may only be a few kilo-ohms, the glass electrode's resistance may range from ten to nine hundred mega-ohms, depending on electrode design. Being that any current in this circuit must travel through both electrodes' resistances (and the resistance presented by the test liquid itself), these resistances are in series with each other and therefore add to make an even greater total.

An ordinary analog or even digital voltmeter has much too low of an internal resistance to measure voltage in such a high-resistance circuit. The equivalent circuit diagram of a typical pH probe circuit illustrates the problem (Fig. 6.10 a&b):

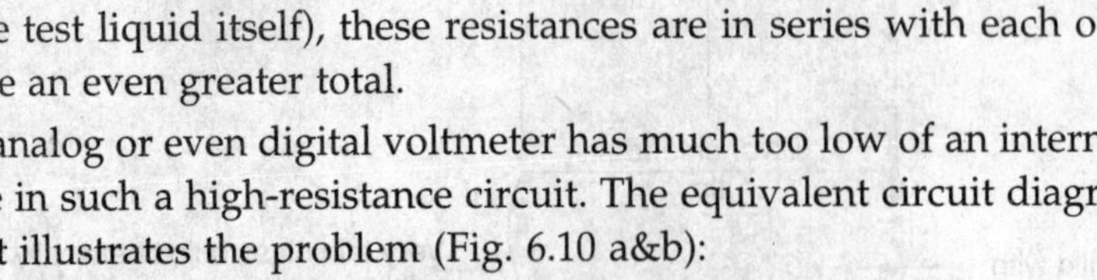

Fig. 6.10(a) *Circuit diagram of a typical pH probe circuit.*

Even a very small circuit current traveling through the high resistances of each component in the circuit (especially the measurement electrode's glass membrane), will produce relatively substantial voltage drops across those resistances, seriously reducing the voltage seen by the meter. Making matters worse is the fact that the voltage differential generated by the measurement electrode is very small, in the millivolt range (ideally 59.16 millivolts per pH unit at room temperature). The meter used for this task must be very sensitive and have an extremely high input resistance.

The most common solution to this measurement problem is to use an amplified meter with an extremely high internal resistance to measure the electrode voltage, so as to draw as little current through the circuit as possible. With modern semiconductor components, a voltmeter with an input resistance of up to 1017 Ω can be built with little difficulty.

Another approach, seldom seen in contemporary use, is to use apotentiometric "null-balance" voltage measurement setup to measure this voltage without drawing any current from the circuit under test. If a technician desired to check the voltage output between a pair of pH electrodes, this would probably be the most practical means of doing so using only standard bench top metering equipment.

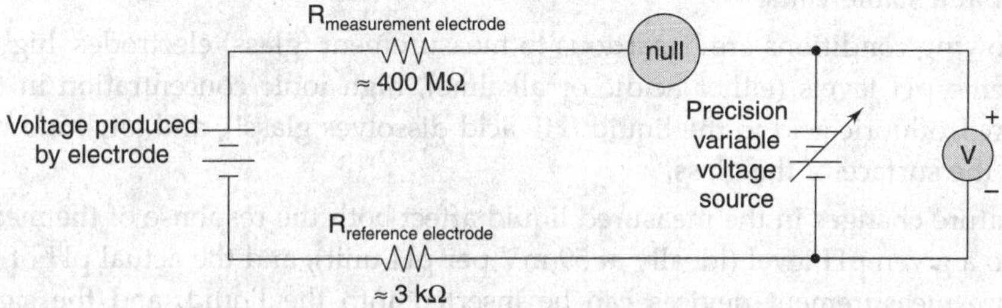

Fig. 6.10(b) *Diagram of a typical pH probe circuit.*

As usual, the precision voltage supply would be adjusted by the technician until the null detector registered zero, then the voltmeter connected in parallel with the supply would be viewed to obtain a voltage reading. With the detector "nulled" (registering exactly zero), there should be zero current in the pH electrode circuit, and therefore no voltage dropped across the resistances of the either electrode, giving the real electrode voltage at the voltmeter terminals.

Wiring requirements for pH electrodes tend to be even more severe than thermocouple wiring, demanding very clean connections and short distances of wire (10 yards or less, even with gold-plated contacts and shielded cable) for accurate and reliable measurement. As with thermocouples, however, the disadvantages of electrode pH measurement are offset by the advantages: good accuracy and relative technical simplicity.

Few instrumentation technologies inspire the awe and mystique commanded by pH measurement, because it is so widely misunderstood and difficult to troubleshoot. Without elaborating on the exact chemistry of pH measurement, a few words of wisdom can be given here about pH measurement systems.

All pH electrodes have a finite life, and that lifespan depends greatly on the type and severity of service. In some applications, a pH electrode life of one month may be considered long, and in other applications the same electrode(s) may be expected to last for over a year.

Because the glass (measurement) electrode is responsible for generating the pH-proportional voltage, it is the one to be considered suspect if the measurement system fails to generate sufficient voltage change for a given change in pH (approximately 59 millivolts per pH unit), or fails to respond quickly enough to a fast change in test liquid pH.

If a pH measurement system "drifts," creating offset errors, the problem likely lies with the reference electrode, which is supposed to provide a zero-voltage connection with the measured solution. Because pH measurement is a logarithmic representation of ion concentration, there is an incredible range of process conditions represented in the seemingly simple 0-14 pH scale. Also, due to the non-linear nature of the logarithmic scale, a change of 1 pH at the top end (say, from 12 to 13 pH) does not represent the same quantity of chemical activity change as a change of 1 pH at the bottom end (say, from 2 to 3 pH). Control system engineers and technicians must be aware of this dynamic if there is to be any hope of controlling process pH at a stable value.

The following conditions are hazardous to measurement (glass) electrodes: high temperatures, extreme pH levels (either acidic or alkaline), high ionic concentration in the liquid, abrasion, hydrofluoric acid in the liquid (HF acid dissolves glass!), and any kind of material coating on the surface of the glass.

Temperature changes in the measured liquid affect both the response of the measurement electrode to a given pH level (ideally at 59 mV per pH unit), and the actual pH of the liquid. Temperature measurement devices can be inserted into the liquid, and the signals from those devices used to compensate for the effect of temperature on pH measurement, but this will only compensate for the measurement electrode's mV/pH response, not the actual pH change of the process liquid.

Advances are still being made in the field of pH measurement, some of which hold great promise for overcoming traditional limitations of pH electrodes. One such technology uses a device called a field-effect transistor to electrostatically measure the voltage produced by a ion-permeable membrane rather than measure the voltage with an actual voltmeter circuit. While this technology harbors limitations of its own, it is at least a pioneering concept, and may prove more practical at a later date.

6.10 pH ELECTRODE OPERATION GUIDE

6.10.1 pH Electrode Preparation

The following is a general procedure for preparing most pH electrodes. Refer to the electrode user guide or instruction manual for an electrode-specific preparation procedure.

1. Remove the protective shipping cap, sleeve or bottle from the electrode pH-sensing bulb and save the cap, sleeve or bottle for storage. If the electrode has a storage bottle covering the pH-sensing bulb, unscrew the storage bottle cap before removing the electrode from the storage bottle.

2. Clean any salt deposits from the electrode exterior by rinsing it with distilled water.
3. If the electrode is refillable, uncover the fill hole and add the appropriate filling solution to the electrode. To maintain an adequate flow rate, the level of filling solution must always be above the reference junction and at least one inch above the sample level. The fill hole should be open whenever the electrode is in use.
4. Gently shake the electrode downward (similar to a clinical thermometer) to remove any air bubbles that may be trapped inside the electrode.
5. Soak the electrode in pH electrode storage solution, or standard pH electrode storage solution, for at least 30 minutes. The storage solution is recommended for enhanced electrode performance.
6. Connect the electrode to the meter.
7. Select at least two pH buffers that bracket the expected sample pH and will be used during the calibration of the pH electrode.

6.10.2 Tips for pH Buffer Use

These solutions of known pH value allow the user to adjust the system to read accurate measurements.

For best accuracy:

- ✦ Standardization should be performed with fresh buffer solutions.
- ✦ Buffers used should frame the range of pH for the samples being tested.
- ✦ Buffers should be at the same temperature as the samples. (For example; if all your samples are at 50°C, warm your buffers to 50°C using a beaker in a water bath.)
- ✦ Buffer values are dependent upon temperature. The following table lists the values of pH buffers at various temperatures.

Table 6.1 *NIST values of ph buffers at various temperatures*

Temperature (°C)	*4.01 Buffer*	*6.86 Buffer*	*9.18 Buffer*
	0.05 M Potassium Acid Phithalate	*0.025M KH_2PO_4 0.025M Na_2HPO_4*	*0.01M Borax*
0°	4.00 pH	6.98 pH	9.46 pH
10°	4.00 pH	6.92 pH	9.33 pH
20°	4.00 pH	6.88 pH	9.22 pH
25°	4.01 pH	6.86 pH	9.18 pH
30°	4.02 pH	6.85 pH	9.14 pH
40°	4.04 pH	6.84 pH	9.07 pH
50°	4.06 pH	6.83 pH	9.01 pH
60°	4.09 pH	6.84 pH	8.96 pH
70°	4.13 pH	6.85 pH	8.92 pH
80°	4.16 pH	6.86 pH	8.88 pH
90°	4.21 pH	6.88 pH	8.85 H

6.10.3 pH Electrode Care

Electrode Calibration

Since glass pH electrodes measure H+ concentration relative to their reference half-cells, they must be calibrated periodically to ensure accurate, repeatable measurements. Our wide selection of commercial pH calibration buffers include solutions standardized against pH references for calibrating meters with resolution up to 0.001 pH.

Although calibration against one pH reference buffer (one-point calibration) typically ensures accurate pH measurement, frequent two-point or even three-point calibrations ensure the most reliable results. Make sure your pH system includes calibration buffers for a range of pH values.

Conditioning

pH electrodes are shipped with the electrodes moist. Prior to using your electrode for the first time, follow these three steps to condition your electrode:

1. Remove the protective cap or rubber boot from the bottom of the sensor and rinse the electrode with distilled or deionized water.
2. Place the electrode in a beaker containing one of the liquids listed below (in order of ionic ability to condition the electrode). Soak for 20 minutes.
   - 3.8 M or 4.0 M KCl
   - 4.0 pH buffer
   - 7.0 pH buffer
3. After conditioning the sensor for 20 minutes, rinse the electrode with distilled or deionized water. The electrode is now ready for calibration and to measure pH.

   Note: Never condition a pH electrode in distilled or deionized water exposure to pure water will damage the special glass membrane.

Handling

Electrodes should be rinsed between samples with distilled or deionized water. Never wipe an electrode—wiping can cause erroneous readings due to static charges. Blot the end of the electrode with lint-free paper to remove excess water.

Refillable Electrodes

The filling solution in refillable electrodes should be filled up to, but not past, the refill hole. Make sure the refill hole is left open when measuring to ensure that the fill solution flows properly through the reference junction.

Storage

Always keep your pH electrode moist. We recommend that you store your electrode in a solution of 4 M KCl. If 4 M KCl is not available, use a pH 4 or 7 buffer solution. Do not store electrode in distilled or deionized water—this will cause ions to leach out of the glass bulb and render the electrode useless.

After storage, you may notice white KCl crystals forming outside your electrode. This will not interfere with measurements. Simply rinse the electrode and blot dry before use.

Protective Rubber Boot

Most electrodes are shipped with a protective rubber boot over the glass bulb to help prevent cracking or scratching. Remove the rubber boot before using your electrode. Keep your electrode in long-term storage with the boot on — just fill the boot with enough 4 M KCl solution to cover the glass bulb and replenish as needed to keep the bulb moist.

6.10.4 pH Electrode Measurements

6.10.4.1 Measurement recommendations

- Check the electrode slope daily by performing at least a two-buffer calibration. The slope should be 92 to 102%.
- Unless otherwise specified, only use the recommended filling solution in refillable pH electrodes. The pH electrode summary shows the electrode filling solutions that are recommended for the scientific pH electrodes.
- If the electrode is refillable, uncover the fill hole during measurements to ensure a uniform flow of filling solution. The filling solution level inside of the electrode must be at least one inch above the sample solution level.
- The sample solution level must be above the pH electrode reference junction when the electrode is immersed in the sample.
- Between measurements, rinse the electrode with distilled water and then with the next solution to be measured. To reduce the chance of error due to polarization, avoid rubbing or wiping the electrode bulb. Use a lint-free tissue and gently blot the bulb.
- Use a magnetic stir plate and stir bar or the scientific stirrer probe, to stir all buffers and samples at a moderate, uniform rate.
- Place a piece of insulating material, such as Styrofoam or cardboard, between the magnetic stir plate and beaker to prevent measurement errors from the transfer of heat to the sample.
- If the electrode is refillable and the electrode is used in dirty or viscous samples or the electrode response becomes sluggish, empty the electrode completely and hold the reference junction under warm running water. Empty any water from the electrode and fill it with fresh filling solution. For a more thorough cleaning.
- Flat surface electrodes may be used on any moist surface or in liquids.
- Handle the micro pH electrodes with care. Do not touch the pH bulb and stem against the bottom or walls of the sample containers.

6.10.4.2 Sample requirements

Electrodes with an epoxy body should only be used in aqueous solutions.

Electrodes with a glass body may be used in non-aqueous solutions and solutions that contain organic solvents. A minimum of 20% water must be present in the sample for the best measurement results.

The standard Ag/AgCl electrodes, micro Ag/AgCl electrodes and economy electrodes contain a single junction silver/silver chloride reference that will become clogged in solutions that contain silver complexing or binding agents such as TRIS buffer, proteins and sulfides. Frequent cleaning may be required when measuring these solutions, which will shorten the electrode life. To better measure pH in these solutions, use the ultra electrodes, electrodes, aquapro electrodes or double junction electrodes. Proteins cause the additional problem of coating the pH-sensing bulb, so extra care should be taken to keep the electrode clean while measuring samples that contain proteins.

6.10.4.3 pH measurement procedure

1. Verify that the pH electrode was prepared and calibrated correctly. If the electrode is refillable, make sure that the fill hole is uncovered and the filling solution level is at least one inch above the sample solution level.
2. Rinse the electrode with distilled water and blot it dry with a lint-free tissue.
3. Place the electrode into the sample. Use a magnetic stir plate and clean the stir bar or the thermo scientific stirrer probe to stir the sample at a moderate, uniform rate.
4. When the measurement is stable, record the pH value and temperature of the sample.

6.11 LABORATORY pH ELECTRODES

Use laboratory electrodes with pH meters to intermittently test the pH of a solution. Electrodes feature glass or epoxy bodies and come in either combination or half-cell styles. Laboratory electrodes are further categorized according to fermentation, general-purpose laboratory, specialty, and ion-selective applications (Fig. 6.11).

Fermentation electrodes resemble industrial electrodes because they can be used continuously for several days. Fermentation electrodes are rugged, can be hooked up to some pH controllers, and are either autoclavable or *in situ* sterilizable.

Ion-selective electrodes determine the concentration of specific ions. Solid state, gas sensing, and liquid membrane electrodes are also vailable.

6.11.1 Application Guide

A variety of laboratory electrodes are available for general-purpose and specialty laboratory applications. The "Application Guide" below provides information to help you find the electrode that best fits your application. Carefully choose the electrode best suited for your application to ensure accurate pH measurements.

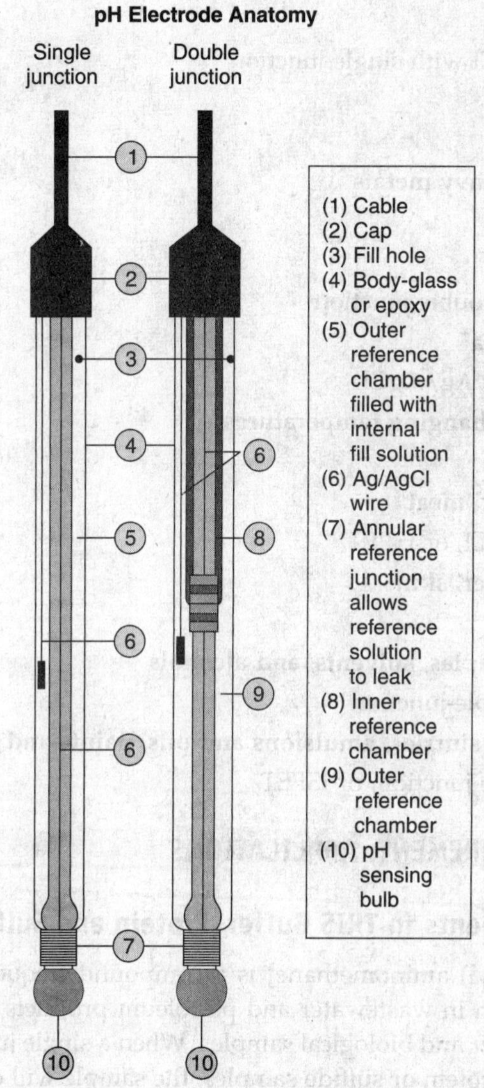

Fig. 6. 11 *pH electrode anatomy*

- **Biological samples, proteins, and tris buffer**
  Calomel (Hg/Hg_2Cl_2) or double-junction
- **Pharmaceuticals**
  Calomel (Hg/Hg_2Cl_2) or double-junction
- **Hydrofluoric acid**
  Antimony or HF electrode
- **Low ionic strength samples, acid rain, boiler feed water and distilled water**
  Sure-Flow™ or ISFET (Ion specific field effect transistor)

- **Drinking water**
  Standard Ag/AgCl with single-junction
- **Wastewater**
  Double-junction
- **Solutions with heavy metals**
  Double-junction
- **Soil samples**
  Soil electrode or double-junction
- **pH>9 and high Na^+**
  Amber glass bulb, Ag/AgCl
- **High or rapidly changing temperatures**
  Ross®
- **Fruits, cheese, and meat**
  Spear-tip, Ag/AgCl, or ISFET
- **Flat surfaces, paper, skin**
  Flat-surface
- **Non-aqueous samples, solvents, and alcohols**
  Sure-Flow or double-junction
- **Viscous samples, slurries, emulsions and oils, paints and inks**
  Sure-Flow, double-junction, or ISFET

6.12 pH MEASUREMENT: APPLICATIONS

6.12.1 pH Measurements in TRIS Buffer, Protein and Sulfide Samples

TRIS [tris (hydroxymethyl) aminomethane] is a compound frequently used as a biological buffer, sulfide is common in wastewater and petroleum products and proteins are often in food products, wastewater and biological samples. When a single junction Ag/AgCl pH electrode is used in buffer, protein or sulfide samples, the sample will complex with the silver in the electrode and clog the electrode junction, preventing stable pH measurements. Proteins also coat the pH-sensing bulb. For the most accurate results and most stable measurements in these samples, use a pH electrode or double-junction Ag/AgCl reference. The AquaPro and Double Junction pH electrodes are recommended for measuring the pH in TRIS buffer, protein or sulfide samples. While measuring protein samples, soak the pH electrode in pH electrode storage solution, between measurements to remove the protein coating.

6.12.2 pH Measurements in Solid Samples and Surface pH Measurements

Solid and semi-solid samples include cheese, meat, powders, paper and agar gels. There are many methods available for measuring the pH of solid and semi-solid samples that include using a flat surface pH electrode, using a spear tip pH electrode, and mixing or blending a fixed amount of sample with distilled water. Once a method is established that is compatible with the sample, the method should be followed exactly with all samples for the most reproducible measurements.

If the sample is soft or semisolid, use a spear tip pH electrode and pierce the sample with the electrode to a fixed depth. If the surface of the sample is moist, use a flat surface pH electrode and measure the surface of the sample. The sample must be moist enough for the pH-sensing bulb and the reference junction to make adequate contact with the sample. If necessary, add a drop of distilled water or potassium chloride to wet the surface before placing the electrode on the sample. For the best reproducibility, all samples should have the same amount of liquid added before measurement.

If the sample is soluble in water or disperses readily in water, mix a fixed percent of sample with a fixed quantity of water and measure the resulting solution. Up to 25% sample by weight or volume is commonly used. Since each sample is different, the details of the method should be determined through trial and error with the sample. For cheese samples – blend 5 grams of cheese with 100 mL of distilled water and stir the solution thoroughly, stop stirring and wait 5 minutes to let the solids settle and then measure the pH of the liquid portion. For paper samples – soak 1 gram of paper pieces in distilled water, wait for 10 minutes and measure the pH of the liquid. For soil samples, mix a fixed ratio of soil and distilled water and stir the solution thoroughly, stop stirring and wait a fixed amount of time to let the solids settle and then immerse the pH electrode at a fixed depth and measure the pH of the liquid portion.

6.12.3 pH Measurements in Pure Water Samples

Pure water samples include distilled water, deionized water, some process waters, well water, some surface waters, treated effluent, boiler feed water and rainwater. In pure water samples the pH electrode response tends to drift and is slow, non-reproducible and inaccurate. Measurements in these samples are more difficult because of the low conductivity of the sample, differences between the low ionic strength solutions and normal ionic strength buffers, changes in the liquid junction potential and the absorption of carbon dioxide into the sample. Since pure water solutions have a low conductivity, the solution will tend to act as an antenna and the electrode response can be noisy.

Calibration of a pH electrode in high ionic strength buffers will increase the time required for stabilization when measuring a low ionic strength sample. In addition, the possibility of sample contamination will be increased. For most precise measurements, buffers and samples should have a similar ionic strength. When two solutions come in contact, diffusion occurs until equilibrium is reached. Since ions have different mobility and diffuse at different rates, a charge imbalance occurs at the point of contact. A liquid junction potential occurs when the electrode filling solution meets the sample. This charge imbalance will be large if there is a large difference in composition between the filling solution and the sample. It is important that the junction potential be constant during measurement. If the two solutions are quite different, normal fluctuations in the boundary layer will produce noise. Constant, reproducible junction potentials are achieved by measuring samples and standards with similar ionic strengths. Since pure waters contain little dissolved material, their buffering capacity is small. Absorption of carbon dioxide from the atmosphere will cause a slow change in pH, observed as a drifting pH reading and a different pH from the original sample. Samples that are not previously saturated with carbon dioxide must be handled with care.

6.12.4 pH Measurements in Sludges, Suspensions, Colloids, Slurries and Viscous Samples

Sludge, suspension, colloid, slurry and viscous samples include wastewater, mud, paper pulp and corn syrup. These samples clog the electrode junction and coat the pH-sensing bulb, resulting in slow electrode response, measurement drift and pH measurement errors. The thickness of the sample can also cause a sample carryover problem – after the pH electrode is removed from the sample, some of the sample may cling to the electrode.

For the most accurate results and most stable measurements in these samples, use a pH electrode or open junction and soak the pH electrode in pH electrode storage solution, between measurements. For the best reproducibility, immerse the pH electrode at a fixed depth in all samples. Sure-Flow junctions resist clogging, provide a constant flow of filling solution and are extremely easy to clean. Open junctions resist clogging, contain a low maintenance polymer gel as the filling solution and are easy to clean. The Ag/AgCl Sure-Flow and AquaPro pH electrodes are recommended for measuring the pH in sludge, suspension, colloid, slurry and viscous samples. The pH electrode storage solution will enhance the pH electrode performance by acting as a very mild cleaning solution. For a more thorough cleaning procedure, refer to the **pH Electrode Cleaning Procedures** section.

6.12.5 pH Measurements in Extreme pH or High Salt Content Samples

Samples with a pH less than 2, pH greater than 12 and salt content greater than 0.1 M include battery acid, copper plating solutions and brines. These samples create measurement issues with the reference portion of the pH electrode, due to the liquid junction potential. The liquid junction potential occurs when there is a large difference in composition between the filling solution and the sample, resulting in measurement drift and slow electrode response.

In samples with a pH of 2 to 12 or salt content less than 0.1 M, the liquid junction potential is rarely a problem since the filling solution is formulated for the best performance under these conditions. In samples with a pH less than 2, pH greater than 12 or a salt content greater than 0.1 M, a pH electrode with double junction Ag/AgCl reference is recommended. The filling solution and refillable double junction pH electrodes can be modified to match the sample composition, which will reduce the liquid junction potential and provide accurate, more stable measurements in these samples.

For samples with a pH greater than 12, add a dilute base to the recommended electrode filling solution to raise the filling solution pH. For samples with a pH less than 2, add a dilute acid to the recommended electrode filling solution to lower the filling solution pH. For samples with a high salt content, use a strong solution of the same salt as the electrode filling solution. For example, when measuring the pH of sodium bromide brines, use sodium bromide in the outer chamber of a double junction electrode.

6.12.6 pH Measurements in Non-aqueous Samples

Non-aqueous samples contain liquids other than water and include oils, alcohols, cosmetics, methanol and acetone. These samples produce unstable readings, slow electrode response and pH measurement errors due to high sample resistance, bulb dehydration, sample carry-over contamination and liquid junction potentials. The high resistance (low conductivity) of many organic solvents causes noise and slow electrode response. To avoid this, an electrode with a low resistance glass bulb should be used. It is often helpful to add a small amount of an inert salt (for example, a quaternary ammonium salt) to increase the conductance of the solution. Addition of a salt will change the activity of the hydrogen ion, causing a small shift in the pH; however, the error is small when compared with measurement drift issues.

Slow electrode response and drift are also caused by the dehydration of the glass pH sensing bulb. For the pH sensing bulb to function, the surface of the glass must be hydrated so hydrogen ions can be absorbed onto the surface of the glass. A non-aqueous solution causes the dehydration of the pH sensing bulb and frequently soaking the pH electrode in water or a pH buffer hydrates the bulb, providing a faster, more stable response. To prevent contamination of samples due to carryover on the electrode bulb, the electrode should be rinsed between measurements with a solvent that will dissolve the sample material from the electrode, and then rinsed with a volatile solvent like acetone to remove the rinse solution, followed by soaking it in a pH buffer.

Large liquid junction potentials that cause unstable readings are a problem when measuring non-aqueous samples. When a reference electrode with an aqueous filling solution comes in contact with a non-aqueous sample, unequal diffusion of ions occur, causing continuous measurement drift. To minimize this problem, the electrode filling solution should be modified to provide compatibility with the sample and minimize junction potential problems. A refillable electrode with a double junction reference, pH electrode, should be used to allow the filling solution to be easily modified. Several alternative filling solutions can be used and include a mixture of methanol and water saturated with KCI, glacial acetic acid saturated with Kl, acetone saturated with KI and methanol saturated with LiCI. When using a solvent in the filling solution, a salt must be added to improve the electrical conductance. KI is often more soluble than KCI and can be substituted as required. When adjusting the filling solution, allow the electrode to reach equilibrium overnight before using it. Solvent-based filling solutions should not be used in epoxy body electrodes. Since epoxy body electrodes should not be used in certain solvents, especially highly polar organic solvents, a glass pH electrode should be used.

The pH value of a non-aqueous solution cannot be accurately compared to an aqueous buffer. The activity of the hydrogen ion varies, depending on the background medium, due to the differences in dielectric constants, solvent acidities and ion mobilities. The pH values obtained when measuring non-aqueous samples can only be used as relative measurements, to compare the acid-base qualities of similar solvents or to indicate when adjustments in pH are needed. The readings cannot be reported as actual pH values, since the pH scale is based on the relative acidity or alkalinity of water.

6.13 REFERENCES

1. Zumdahl, Steven; Zumdahl, Susan (2003). *Chemistry* 6th edn., Houghton Mifflin Company.
2. IUnder certain circumstances (see chemical equilibrium) each activity term such as {*A*} may be replaced by a concentration term, [*A*]. Both the reaction quotient and the equilibrium constant are then concentration quotients.
3. Banus MG (June 1941). "A Design for a Saturated Calomel Electrode". Science (New York, N.Y.) 93 (2425): 601–602. doi:10.1126/science. 93.2425.601-a. PMID 17795970, http://www.sciencemag.org/cgi/ pmidlookup?view=long&pmid=17795970.
4. M. Cremer, *Z. Biol.*, 47, 562 (1906).
5. F. Haber and Z. Klemensiewicz, *Z. Physik. Chem.*, 67, 385(1909).

CHAPTER

7

Digestion Apparatus

7.1 INTRODUCTION

Acid digestion is an important sample preparation technique used for Total Kjeldahl Nitrogen and elemental analysis. Digestion may be defined as the process in which a complex substance is decomposed into volatile gases and simple salts which are soluble in dilute acid solution. This is accomplished by exposing the sample material to a combination of mineral acids or acid and hydrogen peroxide at an elevated temperature. Once the substance under analysis has been dissolved, chemical methods can be applied to measure the concentration of each element.

Research chemists at Hach Company have been studying the chemistry and mechanics of digestion for 14 years. Most of this work has focused on digestion using sulfuric acid and hydrogen peroxide, a modification of the Kjeldahl method, which is suitable for determining a range of metals and non-metals including nitrogen. Digestion was integrated into a total analytical system based on spectrophotometric analysis. Analytes determined include: Ag, AI, Ca, Cd, Co, Cr, Cu, Fe, K, Mg, Mn, N, Ni, Pb, P, S, and Zn.

An early objective was to achieve quantitative recovery of Kjeldahl nitrogen in the digestion. It was discovered that standard natural samples and even certified reference materials could not be used as reliable means of assessing accuracy of the analysis. This stemmed from the fact that the accuracy of assay values are only as good as the laboratory method and technique used to generate the results. Also, there is a certain degree of uncertainty in the composition of a heterogeneous natural substance. Analyses may determine the precision of sampling but not necessarily the precision of the analytical procedure.

Researchers were not aware of any formal standard system based on synthetic compounds. Kjeldahl laboratory standardization is frequently based on an ammonium salt, but this only assures the analyst that ammonia can be retained throughout the procedure. It does not prove the digestion is actually effective in converting organic nitrogen into ammonia. Consider the reaction complexity in the Kjehadl digestion, where nitrogen in various structures and oxidation states must be fully reduced, to (−3) while carbon is oxidized to carbon dioxide. There was clearly a need to establish a means of verification that the digestion was consistently recovering the nitrogen in the sample.

7.2 EXPERIMENTAL DEVELOPMENTS

Digesdahl Digestion Apparatus

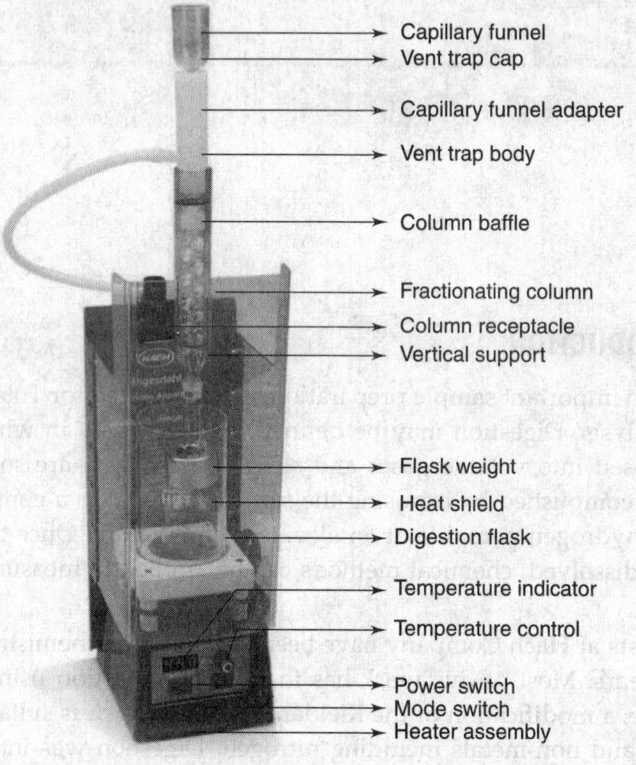

Fig. 7.1 *Digesdahl digestion apparatus.*

7.2.1 Digestion Flask

Initially, a 100 mL standard volumetric flask was used as the digestion flask. Later, a special flask with more consistent dimensions and a controlled flat bottom of uniform thickness was designed. This new design is important for consistent heat transfer so the temperature of the digestion mixture can be better regulated. A weight ring applied to the flask while on the heater provides stability (Fig. 7.1).

7.2.2 Fractionating Column

The column or head serves two main functions. First, it provides a manifold for attachment of the fume evacuation hose and for support of the capillary funnel. Second, it acts as a fractional reflux condenser. Optimum reaction conditions require water to be rapidly boiled while hydrogen peroxide (H_2O_2) and sulfuric acid (H_2SO_4) are retained. With proper condenser design, most of the H_2O_2, at a boiling point of 150°C, is condensed and returned to the flask while steam is swept out of the vent. The fractionating column of about 20 cm length has three rows of dimples, and is equipped with a baffle and a fume trap. The fractionating column is used for reagent addition and fume evacuation. A capillary funnel, inserted into the column, controls the rate of hydrogen peroxide addition during digestion. (See Fig. 7.2)

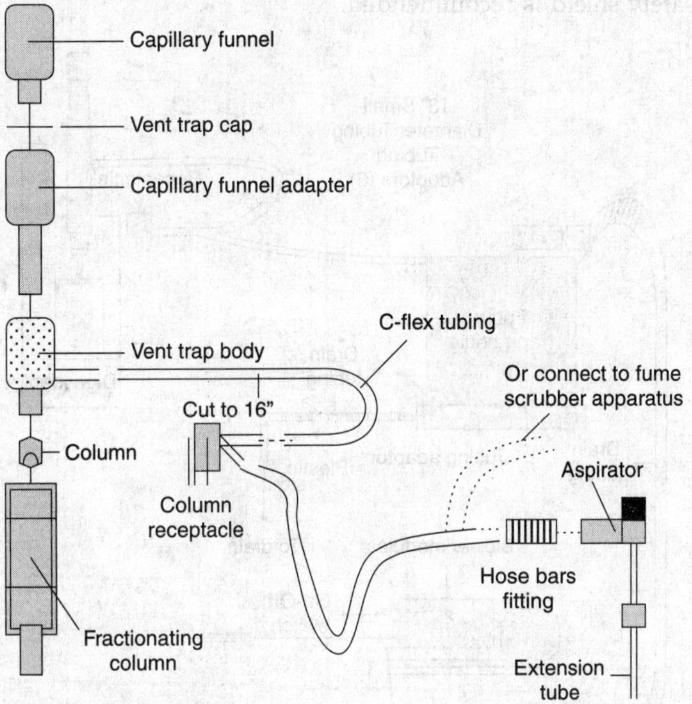

Fig. 7.2 *Vigreux fractionating column with capillary funnel.*

7.2.3 Capillary Funnel

At constant head pressure and temperature, a given liquid flows through a capillary column at a constant flow rate. This principle is applied to delivery of H_2O_2 through the capillary funnel. Although the liquid head changes somewhat, the flow rate averages 3 mL/minute.

7.2.4 Heater Assembly

The Digesdahl's heating apparatus is built around a 250 watt ceramic disc element. Initially, the controller was a proportional time on/off device. Recently, a thermostat was installed which uses a thermocouple to sense the temperature of the metal shell surrounding the ceramic element. The metal-clad element was found to provide better heat transfer to the flask than to the ceramic surface. The heating element and digestion flask are enclosed by a glass heat shield to aid in heat transfer to the sample.

7.2.5 Fume Evacuation

A water aspirator produces the ventilation airstream, and dissolves the exhaust gases from the digestion. Alternatively, a fan-driven fume scrubber has been constructed. It recirculates air back into the room after removing condensable gases in a stainless steel condenser and absorbing the remaining components in a limestone-activated charcoal unit. (Fig. 7.3) Use of a fume hood or safety shield is recommended.

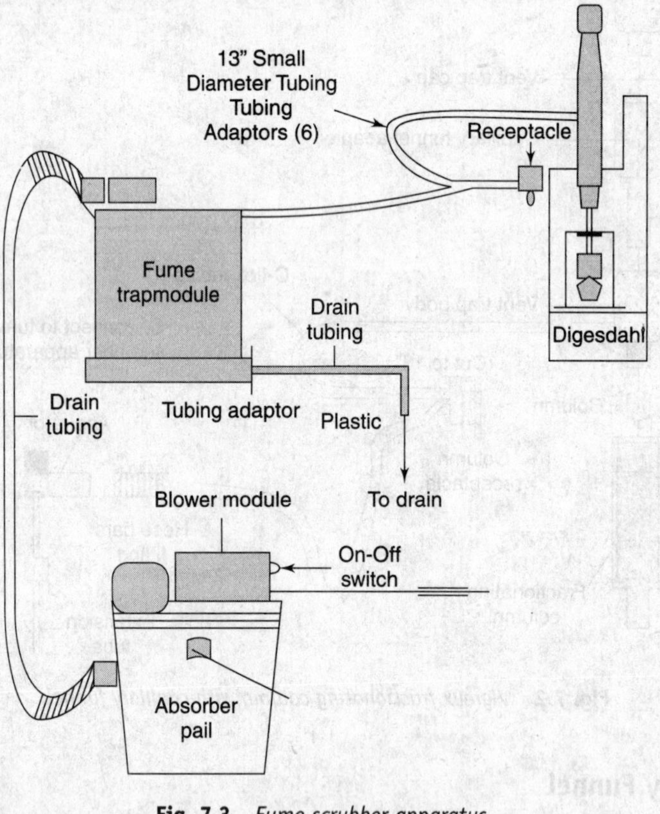

Fig. 7.3 *Fume scrubber apparatus.*

7.2.6 Reagents

Sulfuric acid (H_2SO_4) ACS, 95-98%

50% hydrogen peroxide (H_2O_2), 50%

30% hydrogen peroxide (H_2O_2) ACS, 29-32%

Being less concentrated, 30% H_2O_2 has less oxidizing strength than 50% H_2O_2, and is consequently a less effective reagent for this purpose. When using the 30% reagent, approximately 1/3 more volume is required to get the same results. Most research was done using the 50% reagent, and all the digestion times and volumes reflect its use. When using 30%

peroxide always run a digestion standard, either glycine p-toluenesulfonate or nicotinic acid p-toluenesulfonate, to check completion of the digestion. All recommended safety precautions apply to both strengths of the H_2O_2.

7.3 PROCEDURE

Various procedural modifications were examined in the digestion with H_2SO_4 and H_2O_2. The following is an abbreviated procedure outline for the General Digestion Procedure and is not intended as stand-alone instructions. Refer to Hach's Water Analysis Handbook or the revised third edition of the Digesdahl Digestion Apparatus Instruction Manual for the complete general digestion procedure.

1. Measure or weigh the appropriate size of sample as specified in individual procedures. Transfer the measured sample to a digestion flask.

   Note: Never exceed 0.5 g of sample (dry weight). The maximum volume of a liquid sample is 40 mL.
2. Add concentrated sulfuric acid (specific gravity 1.84) and two or more silicon carbide boiling chips to the digestion flask.

   Note: Refer to specific procedures for recommended sulfuric acid volume. Never use less than 3 mL of concentrated sulfuric acid.

   Note: Always perform digestion behind a safety shield or in a closed fume hood. Safety glasses are mandatory.
3. Place the flask weight followed by the fractionating column with funnel on the flask. Place the flask on the Digesdahl and assemble the heat shield. Turn on the aspirator (mechanical or water).

   Note: Make sure there is suction to the fractionating column.
4. Heat for 3-5 minutes.

   Note: Do not boil sample to near or complete dryness.
5. Add 10 mL of 50% hydrogen peroxide to the funnel with a pour-out dispenser, pipette, graduated cylinder or Hach Hydrogen Peroxide Pump.

   Note:
   - Visually confirm the presence of sulfuric acid in the flask before adding hydrogen peroxide.
   - If the digest does pot turn colorless, add 5 mL increments of hydrogen peroxide until digest becomes clear.
6. Heat the sample for one minute after all hydrogen peroxide has entered the digestion flask.

   Note: Do not heat to dryness.
7. Take the flask off the heater and allow the flask to cool on the cooling pad. Remove the fractionating column from the digestion flask after the flask has cooled.
8. Dilute the digest to the 100 mL mark with demineralized water; mix. Complete the analysis by the appropriate analytical method.

7.4 RESULTS AND DISCUSSION

7.4.1 Conditions for Complete Kjeldahl Digestion Using Peroxide

The scale of apparatus used dictates the maximum sample size. In the 100 mL digestion flask, < 0.5 g of anhydrous material can usually be digested effectively. (As a routine practice 0.25 g of sample were used.) Samples that contain water may be scaled up by a proportional amount. Never exceed 0.5 g of sample (dry weight) or 40 mL of liquid sample.

More than 0.5 g (anhydrous) overloads the capacity of the flask, and is difficult to contain. Larger samples tend to form a charred mass which is hard to oxidize and requires an excessive volume of H_2O_2. There is no restriction on minimum quantity.

7.4.2 Sufficient Sulfuric Acid

The amount of concentrated sulfuric acid (specific gravity 1.84) must be sufficient to prevent the digestion from going to dryness. Too little acid may cause the sample to overheat, which may result in the loss of the analytes due to decomposition. Any part of the flask bottom that becomes dry will overheat and may cause the flask to burst. Use an amount that will leave about 2 mL of residual acid covering the bottom of the flask when the digestion is complete.

Sulfuric acid consumption depends on the anhydrous mass of material and the chemical composition of the substance. Use of 4 mL H_2SO_4 is suitable for many materials, but not all. The analyst must pay close attention to the amount of residual H_2SO_4 for the type of sample digested and adjust the amount of acid or sample accordingly. Never use less than 3 mL of concentrated sulfuric acid. Larger volumes of sulfuric acid may be used, but a large excess should be avoided because sample pH adjustment is required in most subsequent determinations.

7.4.3 Proper Digestion Solution Temperature

The technical literature contains several examples of ineffective use of H_2O_2 due to a combination of timing of the H_2O_2 addition and temperature. If the H_2O_2 is added to a cold digestion mixture followed by heating, the H_2O_2 decomposes before the digest reaches the proper temperature. Addition of H_2O_2 to a digest that is too hot volatilizes most of the H_2O_2 with little benefit. Furthermore, excessive temperature contributes to spray loss of sample as a fine mist.

The proper temperature on the metal-clad heating element as measured by the internal thermocouple is 440-480°C. At this setting, the H_2SO_4 reaches 320°C during preheat carbonization. During H_2O_2 addition, the solution temperature is maintained above 280°C.

Table 7.1 *Sulfuric acid residuals for a variety of sample types**

Sample	*Structure*	*H_2SO_4 Residual, mL*
Starch	carbohydrate	2.32
Cyanoethylcellulose	carbohydrate	1.58
Soybean meal	protein	2.55
Linseed oil	fatty acid	1.21
Benzene	aromatic hydrocarbon	1.40
Paraffin	hydrocarbon	1.71
Motor oil	hydrocarbon	1.08
Polyethylene	hydrocarbon	0.67
Bunker C fuel oil	hydrocarbon	0.00
Sewage (25 mL)		3.44

*Experimental Conditions: 0.25 9 sample, 4.00 mL H_2SO_4' four-minute preheat, 20 mL H_2O_2 at 3 mL/minute, two-minute postheat

7.4.4 Carbonization Period

A carbonization period prior to the addition of H_2O_2 provides a reducing environment which helps convert organic nitrogen to ammonia. In the presence of oxidizable carbon compounds, sulfuric acid reacts to produce sulfur dioxide, which is the active reducing agent. The reaction is:

$$H_2SO_4 \rightarrow H_2O + SO_2 + ½\ O_2$$

In nitroaniline, the nitro-N is fully recovered in as little as two-minute carbonization. Dinitrobenzene yielded about 85% recovery with the 4-minute preheat period. However, when 0.25 g of salicylic acid was added and carbonization was extended to 15 minutes, the recovery was about 98%. Inorganic nitrate was fully recovered using salicylic acid pretreatment in concentrated sulfuric acid according to the reactions:

$$2MNO_3 + H_2SO_4 \rightarrow M_2SO_4 + 2HNO_3$$

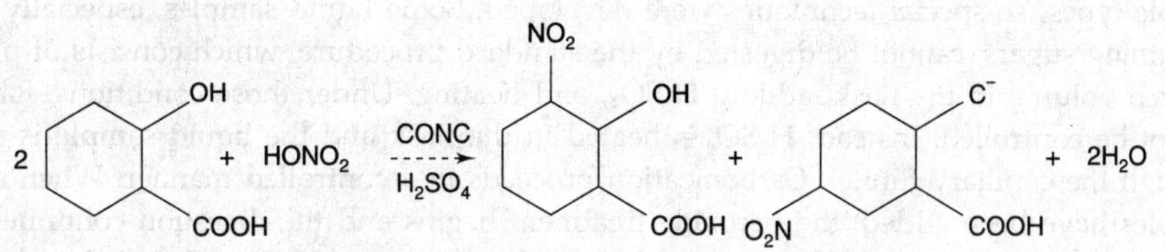

It was interesting to find the amino-N in glycine was fully recovered regardless of whether there was a carbonization period or not. A preheat period of 2 to 5 minutes is recommended for routine digestions.

7.4.5 Adequate Peroxide Concentration for Sufficient Time

Researchers believe that H_2O_2 immediately reacts with H_2SO_4 at digestion temperature to give H_2SO_5 (peroxymonosulfuric acid) by the reaction:

$$H_2SO_4 + H_2O_2 \rightarrow H_2SO_5 + H_2O$$

This is an extremely powerful oxidizing agent toward carbonaceous material. The objective is to maintain an adequate concentration of H_2SO_5 in the hot digestion mixture for a long enough time to complete oxidation of the carbonaceous material. Use the capillary funnel to meter the H_2O_2 into the flask at 3 mL/min.

The amount of peroxide that must be added for complete digestion can be determined by digesting a sample multiple times with incremental increases in the amount of peroxide added (i.e., 5 mL, 10 mL, 15 mL, 20 mL). Results of the analysis for the parameter of interest may then be graphed to determine the minimum amount of peroxide needed for optimum sample digestion.

The preferred and recommended peroxide reagent is 50% hydrogen peroxide. Research studies have shown that the best recovery and reproducibility are achieved using the 50% peroxide reagent. The 50% hydrogen peroxide reagent produces dependable results in research and actual applications. The optimal rate of peroxide addition has been determined as 3 mL/minute using the 50% reagent. This may not hold true if other strengths of hydrogen peroxide are used.

If 50% hydrogen peroxide is not available, 30% peroxide may be used as a last resort with caution. Because of the lesser strength, at least 1.67 times more volume must be used (i.e., 16.7 mL of 30% vs. 10 mL of 50% peroxide). Always run a digestion standard, either glycine p-toluenesliifonate or nicotinic acid p-toluenesulfonate, when using 30% peroxide to check completion of the digestion. All recommended safety precautions apply to both strengths of the hydrogen peroxide.

7.4.6 Containment of Sample

Loss of sample from the digestion flask can occur in two ways: 1) foaming or boiling over, and 2) mist or spray in the ventilation airstream. Foaming is a serious problem with certain sample types, so special techniques were developed. Some liquid samples, especially those containing sugars, cannot be digested by the standard procedure, which consists of placing a given volume in the flask, adding H_2SO_4, and heating. Under those conditions, foaming cannot be controlled. Instead, H_2SO_4 is heated in the flask and the liquid sample is added through the capillary funnel. Carbonization proceeds in a controlled manner. When all the samples have been added, the peroxide treatment begins and the digestion continues normally. Other ways to control foaming include early addition of H_2O_2 and reducing the temperature during carbonization.

Sample loss as spray or mist occurs when small droplets of liquid are swept out of the flask along with gases, and escape in the ventilation airstream. This loss was measured by passing the vent airstream through a glass, water-cooled condenser. The condensate was collected and analyzed. It was discovered that proper fractionating column design reduced spray loss to an insignificant level. Data is presented in Table 7.2.

Table 7.2 *Effect of fractionating column design on the protein loss resulting from mist carryover in the exhaust stream*

Column Design	Protein Loss, %	Relative Loss, %†
Open	0.41	0.89
Vigreux without baffle	0.23	0.49
Helix packed column	0.10	0.22

*Digestion of 0.250 g of Ammonium-PTSA (Primary Standard, 46.26% crude protein) using 18 mL of 50% hydrogen peroxide

† $\frac{\text{\% Protein Loss} \times 100}{46.26\%}$

7.5 SAFETY CONSIDERATIONS

Some samples are more difficult to digest than others. In many cases, complete recovery of the analyte is achieved immediately upon clearing of the digestate (when the digest becomes colorless). However, resistant materials require several minutes of continued peroxide digestion after clearing to obtain 100% recovery. To ensure complete digestion, consider the variables described below.

7.5.1 Sample Size

For solid or organic liquid samples, less than 0.5 gram of dry material can usually be digested. Routinely, use only 0.25 gram. Samples containing water can be increased by a proportional amount. The maximum sample volume for aqueous solutions or suspensions is 40 mL; there is no minimum sample size. When the percent solids exceeds 1% of the sample volume, reduce the maximum sample volume by this formula:

$$\text{mL sample} = \frac{40}{\text{\% solids}}$$

Samples containing suspensions or a high percent of solids are very difficult to reproduce. Mix or homogenize these samples thoroughly before sampling.

7.5.2 Chemicals

Concentrated sulfuric acid and hydrogen peroxide used in the digestion process should be handled correctly and with caution. Sulfuric acid is a strong acid and can cause burns on the skin and permanent damage if eye contact occurs. Hydrogen peroxide (30% or 50%) is a powerful oxidant and should never be stored near flammable materials. Like sulfuric acid, it can cause burns and eye damage. In case of eye or skin contact with either chemical, flush eyes and/or skin with water for 15 minutes. Remove contaminated clothing. Call a physician.

Both sulfuric acid and hydrogen peroxide are highly corrosive and should be cleaned up with water if spilled on instruments or a counter top. Read and observe all warnings on the reagent labels and MSDS.

Proper handling and storage procedures involving hydrogen peroxide should always address two major characteristics of the product: first, it is a strong oxidizing agent (corrosive), and second, it can decompose, releasing heat and oxygen. The chemical nature of hydrogen peroxide makes it an irritant to the skin, to mucus membranes and particularly to the eyes. It will cause chemical burns at industrial concentrations and may cause spontaneous combustion upon immediate or prolonged contact with combustibles.

Hydrogen peroxide decomposes to form water and oxygen. The natural decomposition rate of the normal industrial grade product is very low, but it will accelerate when contaminated by materials such as dust, metallic ions, or alkali. Accelerated decomposition from contamination will result in the significant production of oxygen and liberation of heat. These products will support combustion and will cause pressure bursts in confined spaces. Commercial grades of hydrogen peroxide contain small quantities of additives (termed "stabilizers") to prevent accelerated decomposition from occurring during normal product usage.

Observe the following precautions for handling and storage of hydrogen peroxide.

Store in a cool place away from direct sunlight (preferably in a refrigerator).

Store in the original containers with closures as supplied and keep closed when not in use.

Wear gloves and safety glasses when handling the material.

Use silicon carbide boiling chips when digesting liquid samples.

Wash contaminated skin and body quickly with plenty of water. Remove contaminated clothing and wash well before using again. Wash regularly.

Wash eyes with plenty of water if contaminated and get medical advice quickly.

Get medical advice without delay if the material is ingested.

Flush all spillage with large amounts of water.

Do not store near heat sources or in contact with combustible or organic materials.

Do not allow contact with decomposition catalysts (metals, dust, alkali, etc.) for transfer or storage systems.

Caps on the reagent bottles are made with a special porous liner that allows venting of gas. The venting cap always must be used on the bottle of hydrogen peroxide. As a precaution, the reagent bottles are shipped in a plastic bag. If there is evidence of leakage during shipment, wear gloves when removing the bottle from the bag and rinse the bottle with water when removed from the bag. Rinse the bag before disposal.

7.5.3 Primary Standards

Several compounds were examined as candidates for digestion standards. Researchers decided the compounds chosen should meet the stringent criteria of "primary standards" so they could be synthesized as they contain an absolute nitrogen content. This would provide every laboratory its own internal proof of accuracy.

The selection criteria should be.

1. high molecular weight
2. recrystallization without water of hydration
3. non-hygroscopic
4. thermally stable
5. not air-oxidizable
6. water-soluble
7. high purity
8. nitrogen content in useful range
9. degree of digestion difficulty
10. simple and economical, direct synthesis
11. free-flowing solid

Three primary standard compounds of varying digestibility (Ammonium-p-toluenesulfonic acid, [PTSA], Glycine-PTSA and Nicotinic acid-PTSA) were found to be sufficient to prove the accuracy of the overall nitrogen determination. These p-toluenesulfonate salts were chosen because they are stable, non-hygroscopic materials that can be prepared and purified to primary standard quality, and the protein content is lowered to a practical range compatible with the method of analysis. For example, pure glycine (116.6% protein content) or pure nicotinic acid (71.12% protein content) could be used as a primary standard, but the nitrogen levels are too high to fit the working range of the analysis. The three selected standard compounds have digestibilities ranging from an index of zero for Ammonium-PTSA (no digestion required), to 10 for Nicotinic acid-PTSA. which is the most difficult material commonly encountered. Glycine-PTSA (digestibility index of 3) is intermediate, and is quite easily digested. All three compounds may be oven dried at 100°C without thermal decomposition, although drying is unnecessary, because they do not absorb significant amounts of moisture, even at 90% relative humidity. Also, they are non-oxidizable, free-flowing, crystalline solids that do not stick on weighing paper or spatula, yet they are water-soluble and stable indefinitely.

7.6 APPLICATIONS

7.6.1 General Digestion Procedure

Refer to Hach's Water Analysis Handbook-Second edition or the revised third edition of the Digesdahl Digestion Apparatus Instruction Manual for the complete general digestion procedure.

Cereal Products

Examples: Grains, pasta, flour

Analytes: Protein, Ca, Cu, Fe, Mg, Mn, P, K, Zn

Procedure: General digestion

Cereal Products

Example: Corn starch

Analyte: Protein

Procedure: General Digestion

Fruits/Vegetables

Examples: Fresh, frozen and canned produce, processing samples

Analytes: Protein, Ca, Mg, P, Fe, Cu, Mn, Zn

Procedure: General digestion

Edible Oils

Examples: Corn oil, soybean oil, peanut oil

Analytes: Phosphorus

Procedure: General digestion, except split H_2O_2 addition-5 mL immediately upon heating and 5 mL later.

Meat/Poultry/Fish

Examples: Fresh or processed products, mechanically de-boned poultry

Analytes: Protein, Ca, Mg, P, bone-equivalent, fat

Procedure: Add 50 mL H_2O and 50 mL H_2SO_4. Stir until sample dissolves and fat separates*. Pipette 5 mL sample solution into digestion flask and use general digestion, except no additional H_2SO_4.

Meat/Poultry/Fish

Examples: Fresh meat used for processing

Analyte: Total extractable protein (salt soluble)

Procedure: Treat sample with 150 mL 10% NaCI solution for 2 minutes in a high-speed blender. Filter with Celite. Add 5 mL of the filtrate to the digestion flask. Add 5 mL of H_2SO_4 to the digestion flask. Heat to fuming (480°C, 900°F). Add 5 mL H_2O_2 at 3 mL/minute.

Dairy Products

Examples: Cheese, dry milk, dry whey, fluid milk and whey

Analytes: Protein, Ca, Mg, K, P

Procedure: General digestion

Dairy Products

Example: Cheese

Analyte: Fat

Procedure: Add 50 mL H_2O, 2 mL pentadecane and 50 mL of H_2SO_4 to the sample. Stir until the sample dissolves and the fat separates. The acid solution may be digested for protein and elemental analysis. Attach measuring stem, add 50% H_2SO_4 and read the fat content.

Beverages

Examples: Beer, wine, soft drinks, fruit and vegetable juices

Analytes: Protein, Ca, Mg, P, K, Fe, Mn, Cu Procedure: Preheat 4 mL of H_2SO_4 for I minute (470°C, 875 OF).

Add the sample through the capillary funnel. Add 10 mL of H_2O_2 at 3 mL/ minute.

Animal Feed/Pet Food

Examples: Grains, forage, complete ration, protein supplements

Analytes: Protein, Ca, Cu, Fe, Mg, Mn, P, K, Zn Procedure: General digestion

Fertilizer Solutions

Example: Nitrogen fertilizer solutions

Analyte: Total nitrogen (NH_4 +, NO_3, organic)

Procedure: Add 0.4 g salicylic acid and 4 mL H_2SO_4 to the sample and let stand for 10 minutes. Heat at 80°C in a lab oven for 15 minutes. Use the general digestion, except no additional H_2SO_4.

Animal Waste

Example: Feces, urine, manure

Analytes: N, P, K

Procedure: General digestion

Wastewater

Examples: Sewage, septage

Analytes: Ag, AI, Cd, Co, Cr, Cu, Fe, K, Mn, Ni, Pb, P, TKN, Zn

Procedure: Use 4 mL H_2SO_4 to digest the sample and heat to evaporate the water. Use the general digestion. Neutralize the digested sample with 8 N KOH before analysis.

Sludges

Example: Aerobic, anaerobic municipal sludge, dried sludge

Analytes: Ag, AI, Cd, Co, Cr, Cu, Fe, K, Mn, Ni, Pb, TKN, Zn

Procedure: Use the general digestion, except heat diluted digest for 15 minutes and filter.

Fuels/Lubricants

Examples: Coal, grease, lubricating oil

Analytes: N, Ca, Cu, Cd, Cr, Fe, Ni, Zn Procedure: Use the general digestion, except with 6 mL of H_2SO_4 and 20 mL of H_2O_2.
Heat the diluted digest for 15 minutes and then filter for analysis.

Organic Chemicals

Examples: Aliphatic, cyclic and aromatic amines

Analyte: N

Procedure: General digestion

7.7 REFERENCES

Hach Procedure Manuals	*Catalog Number*	*Literature Code*
Systems for Food, Feed, and Beverage Analysis	24309-00	3120
Plant Tissue Analysis Manual	23260-06	3118
Fluid Fertilizer Analysis Manual.	23677-00	3201
Greenhouse Media Analysis	23260-09	3204
Hach Water Analysis Handbook	23196-00	8353
Digesdahl Digestion Apparatus Instruction Manual, 3rd edition	23130-89	
Agricultural Analysis Handbook	22546-08	

7.7.1 Technical Journals

1. Hach, C.C.; Brayton. S.Y.; Kopelove, A.B. "A Powerful Kjeldahl Nitrogen Method Using Peroxymonosulfuric Acid"; J Agric. Food Chem. 1985, 33 (6).
2. Hach, C.C.; Bowden, B.K.; Kopelove, A.B. "More Powerful Peroxide Kjeldahl Digestion Method"; J Assoc. Off. Anal. Chern. 1987, 70 (5).
3. Norton, G.A.; Adams, N.S.; Markuszewski, R.; Brayton, S.V. "Rapid Dissolution Technique for Colorimetric Determination of Nitrogen in Coals"; Fuel 1987, 66.
4. Treybig, D.S.; Haney, P.L. "Colorimetric Determination of Total Nitrogen in Amines with Selenium Catalyst"; Anal. Chem. 1983, 55 (6).
5. Watkins, K.L.; Veum, T.L.; Krauge, G.F. "Total Nitrogen Determination of Various Sample Types: A Comparison of the Hach, Kjeltec and Kjeldahl Methods"; J. Asso:. Off. Anal. Chem. 1987, 70 (3).
6. Chen, Y.S.; Brayton, S.V.; Hach, C.C. "Accuracy in Kjeldahl Protein Analysis"; American Lab 1988.
7. Chen, Y.S.; Brayton, S.V.; Aspelund, T.G. ''A Fast Method for Determination of Total Extractable Protein in Meat"; Manuscript in preparation.

CHAPTER

8 Flame Photometer

8.1 INTRODUCTION

Flame photometry is also called flame atomic emission spectrometry. It is a branch of atomic spectroscopy in which the species examined in the spectrometer are in the form of atom. Flame photometers read atomic emissions to detect the presence of metal salts, principally sodium (Na), potassium (K), lithium (Li) and calcium (Ca). Flame photometry is suitable for qualitative and quantitative determination of several cations, especially for metals that are easily excited to higher energy levels at a relatively low flame temperature – mainly Na, K, Rb, Cs, Ca, Ba, Cu. They measure the flame emission of solutions containing these metal salts and perform analysis of the species. The process of flame photometry uses a hot flame to evaporate a solution, which atomizes the metal and excites a valence electron to upper state. Light is emitted at characteristic wavelengths for each metal as the electron returns to the ground state. Flame photometers are optical filters to monitor for the select emission wavelength produced by the analyte species. Comparison of emission intensity of unknowns to either those of standard solutions or to those of an internal standard, allows quantitative analysis of the analyte metal in the sample solution.

The flame photometer is an instrument of relative measuring, the concentration value of the sample be tested is a standard sample's concentration relative value of same test condition. So before test one must make a group of relative standard samples, then process the standardized operation. Flame photometer is a new design, small dimension, simple structure, easy operation and reliable instrument.

The flame photometers are relatively simple instruments. There is no need for source of light, since it is the measured constituent of the sample that is emitting the light. The energy that is needed for the excitation is provided by the temperature of the flame (2000-3000°C), produced by the burning of acetylene or natural gas (or propane-butane gas) in the presence of air or oxygen. By the heat of the flame and the effect of the reducing gas (fuel), molecules and ions of the sample species are decomposed and reduced to give atoms, e.g. $Na^+ + e^-$? Na. Atoms in the vapor state give line spectra.

The most sensitive parts of the instrument are the aspirator and the burner. The gases play an important role in the aspiration and while making the aerosol. The air sucks up the sample (according to Bernoulli's principle) and passes it into the aspirator, where the bigger drops condense and could be eliminated.

The monochromator selects the suitable (characteristic) wavelength of the emitted light. The usual optical filters could be used. The emitted light reaches the detector. The photomultiplier produces an electric signal proportional to the intensity of emitted light.

Flame photometry has many advantages. It is a simple, relative inexpensive, high sample throughput method used for clinical, biological and environmental analysis. On the other hand, the low temperature makes this method susceptible to, particularly, interference and the stability (or lack thereof) of the flame and aspiration conditions. Many different experimental variables affect the intensity of light emitted from the flame. Fuel and oxidant flow rates and purity, aspiration rates, solution viscosity, concomitants in the samples, etc. affect these. Therefore, careful and frequent calibration is necessary for good results and it is very important to measure the emission from the standard and unknown solutions under conditions that are as nearly identical as possible.

8.2 PRINCIPLE

The flame photometer is a analysis instrument using the basic principle of emission band spectrum, using the heat energy produced the flame itself to exciting the parts of atom in the alkali condition and to make these atoms jump to upper energy level by the absorbed energy. Flame photometry relies upon the fact that the compounds of the alkali and alkaline earth metals can be thermally dissociated in a flame and that some of the atoms produced will be further excited to a higher energy level. When these atoms return to the ground state they emit radiation which lies mainly in the visible region of the spectrum. Each element will emit radiation at a wavelength specific for that element. Over certain ranges of concentration the intensity of the emission is directly proportional to the number of atoms returning to the ground state. This is in turn proportional to the absolute quantity of the species volatilized in the flame, i.e., light emitted is proportional to sample concentration. If the light emitted by the element at the characteristic wavelength is isolated by an optical filter and the intensity of that light measured by a photo detector, then an electrical signal can be obtained proportional to sample concentration. Such an electrical signal can be processed and the radiant obtained in an analog or digital form. Furthermore, the flame strength is the direct ratio with the concentration with atom in liquid, thus it wakes up the basic of quantitative test. The method is called flame photometry, and this kind of instrument is called flame photometer (Fig. 8.1).

8.3 THEORY

This technique uses a flame that evaporates the solvent and also sublimates and atomizes the metal and then excites a valence electron to an upper energy state. Light is emitted at characteristic wavelengths for each metal as the electron returns to the ground state that makes qualitative determination possible. Flame photometers use optical filters to monitor for the selected emission wavelength produced by the analyte species. Comparison of emission intensities of unknowns to either that of standard solutions (plotting calibration curve), or to those of an internal standard (standard addition method), allows quantitative analysis of the analyte metal in the sample solution.

The intensity of the light emitted could be described by the Scheibe-Lomakin equation:

$$I = K.c^n$$

where

c = the concentration of the element

k = constant of proportionality

$n \sim 1$ (at the linear part of the calibration curve), therefore the intensity of emitted light is directly proportional to the concentration of the sample.

Because of the very narrow and characteristic emission lines from the gas-phase atoms in the flame plasma, the method is relatively free of interferences from other elements. Therefore, the flame photometry (as with other atomic spectroscopy methods) is very sensitive; measuring concentration of ppm magnitude (part per million, e.g., mg kg^{-1}) usually does not cause any problem. The optimal concentration range of the solutions for the measured metal ion is 10-3-10-4 mol dm-3. Typical precision for analysis of dilute aqueous solutions are about ±1-5% relative.

8.4 COMPONENTS

A simple flame photometer consists of the following basic components:

(i) **Burner:** a flame that can be maintained in a constant form and at a constant temperature.

(ii) **Nebulizer and mixing chamber:** a means of transporting a homogenous solution into the flame at a steady rate.

(ii) **Simple color filters (interference type):** a means of isolating light of the wavelength to be measured from that of extraneous emissions.

(iv) **Photo-detector**: a means of measuring the intensity of radiation emitted by the flame.

8.4.1 Specification: Standard Flame Photometer

Ranges: 120-160 mmol/l Na (linearised)

0-10.0 mmol/l K

Limits of Detection:

Na_0.2 ppm	–	
K _ 0.2 ppm	–	
Li _ 0.25 ppm	Li	≤ 0.25 ppm
Ca _ 15 ppm	Ca	≤ 15 ppm
Ba _ 30 ppm	Ba	≤ 30 ppm

Photograph of a flame photometer

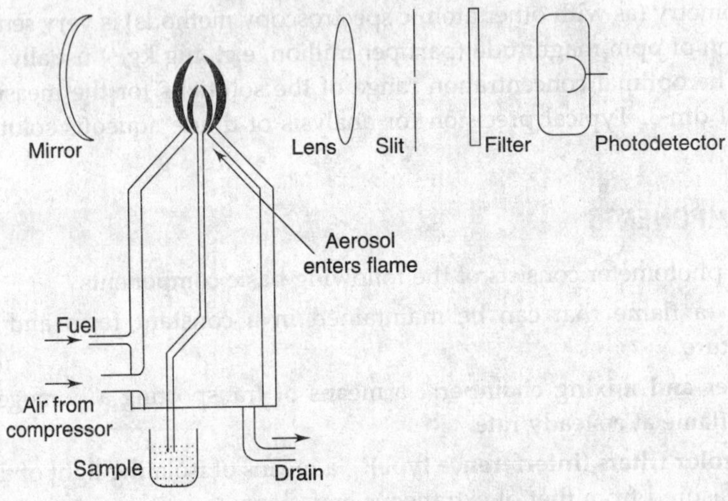

Fig. 8.1 *Flame photometers.*

(Source: http://www.resonancepub.com/atomicspec.htm*)*

Reproducibility: $\leq$ 1% Coefficient of variation (C.V.) for 20 consecutive samples using 10 ppm Na set to read 50.0.
Readings taken at 20 second intervals.

$$\text{CV} = \frac{\text{Sample standard deviation}}{\text{Mean reading}} \times 100$$

And sample standard deviation as:

$$\sqrt{\frac{(\bar{x} - x^2)}{n - 1}}$$

where x is the reading, x is the mean reading of the series and n isthe number of readings.

Linearity: Better than 2% when concentration of 3 ppm Na and K and 5 ppm Li are set to read 100.

Specificity: Interference from Na, K and Li when equal in concentration to the test element will be less than 0.5%.

Stability: Better than 2% over 5 minutes when continuously aspirating 10 ppm, sample set to read 50.0.
Zero drift better than 2% per hour.
Note warm up requirement.

Sample Requirements: Between 2 and 6 mL/minute.

Recorder Output: Nominal 1.00 volt for readout of 100.0.

Warm Up: The flame must be alight for at least 15 minutes to ensure achievement of the above stated specifications.

Services: Electrical: 90-125 V or 190-250 V @ 50/60 Hz.
Air: Moisture and oil-free.
6 liters/minute at 1 kg/cm 2 (14 psi).
Fuel: Propane, butane, natural gas or L.P.G.

Operating Environment: 15°C to 35°C

8.5 SYSTEM DESCRIPTION

8.5.1 Sample System

The sample is taken up by a concentric type nebulizer unit. Air, at a constant pressure, is passed around the end of a small capillary tube forming a partial vacuum inside the capillary tube and drawing sample into it via the nebulizer inlet tubing. The sample leaves the end of the capillary as a fine aerosol. The volume and droplet size of the aerosol is controlled by varying the position of the capillary tube and air jet with respect to each other. This adjustment is the most important single factor affecting the linearity, stability and reproducibility of the instrument.

The aspirated sample is passed into the mixing chamber where the fuel gas is introduced. The sample aerosol then strikes the impact bead which further breaks up any large particles in the spray. The fuel/air/sample mixture is then swirled around two baffles. The majority of sample strikes these baffles and runs out of the mixing chamber, via the drain trap to waste. The remaining sample, consisting of only the smallest droplets mixed uniformly with the fuel, reaches the modified Meker type burner where the mixture burns in a tall, round flame situated in a double skin chimney.

8.5.2 Combustion and Ignition System

This can be used with propane, butane, natural gas or L.P.G. Immediately behind the rear panel fuel bulkhead is the **fuel** solenoid valve. This solenoid valve is normally closed when there is no power to the instrument or where there is power to the instrument but the flame is not alight. The solenoid will be latched in the open position when the flame is alight and has been detected by the flame thermocouple. The fuel solenoid can be held open manually during the flame ignition sequence by holding down the front panel **ignition** switch; fuel then flows through small bore copper tubing to the fuel valve on the front panel. This is a very fine needle valve and to avoid damage, this control should never be screwed hard fully clockwise and should not be used as an on/off valve. From the fuel valve the gas then passes to the mixing chamber and then on to the burner. A restrictor screw is fitted to the mixing chamber fuel inlet for use with propane, butane or L.P.G. but must be removed if natural gas is used.

Air is fed through the rear panel to the air regulator. The air regulator is set to 12 psi. Adjustment should not be necessary but can be achieved by lifting the yellow locking ring and rotating the knob on the side of the regulator until the correct pressure is shown on the pressure gauge. From the air regulator the air passes to the nebulizer where it acts as an atomizing agent.

Ignition is achieved by passing a high voltage spark between the ignition electrode and the center of the burner. The spark is generated by the self-contained ignition unit mounted inside the instrument. This is activated by depressing the front panel **ignition** switch, which simultaneously opens the fuel solenoid, allowing gas to flow through the stream of sparks and hence ignite. The ignition switch must be manually held down until the thermocouple, mounted at the top of the back of the chimney, detects the temperature rise due to the flame, latches the solenoid in the open state and energises the FLM indicator in the display window. If the fuel flow ceases and the flame is extinguished, the thermocouple will detect the fall in temperature and the solenoid will automatically be closed, stopping any further gas escaping to atmosphere.

8.5.3 Optical System Description

Light is emitted from the flame and collected by a plano-convex lens mounted on the inner chimney. The wavelength of light reaching the photo detector is determined by up to 5 different interference filters mounted in a wheel. The filter in the light path is selected by the front panel control and is indicated by this control. Na (589 nm) and K (766 nm) filters are fitted as standard and positions 1, 2 and 3 can be fitted with Ca (620 nm), Ba (520 nm) and Li (670 nm) filters to order.

The filters are held in the wheel by circlips. Should replacements or alternatives need to be fitted, care must be taken to prevent fingerprints getting onto the optical surfaces. If this does happen a small amount of ethanol on a very soft, lint-free cloth may be used to gently wipe away deposits.

Filtered light from the flame falls on the PIN diode mounted on a small PCB on fixing brackets that allow critical adjustment so that the lens on the diode can be set to an optimum position in the light path. The electrical signal from the diode is taken via a 3-pin connector and a screened cable to the main PCB.

8.6 OPERATION

1. Ensure that the drain trap is pushed fully down on its clip. Ensure that the drain trap has solution in it and that no air locks are present. If necessary, purge by adding deionized water and allowing the surplus to run away.
2. Close the **fuel** valve by turning fully clockwise. To avoid damaging the valve, it should not be forced.
3. Turn the **fuel** valve the required number of turns anticlockwise depending on the fuel being used.

Fuel	*Number of Turns*
Propane	3
Butane	4
Natural gas	Fully open
L.P.G.	3.5

4. Turn on the fuel supply at source, i.e., cylinder.
5. Switch on electrical power. If the air compressor is powered separately, switch on the air compressor. Ensure that air is present by listening for the hissing created as it passes through the nebulizer.
6. Depress the ignition switch and hold down. Watch the **FLM** indicator in the display window. When this indicator is illuminated the flame is alight and the ignition switch can be released. If the **FLM** indicator does not light within approximately 20 seconds, release the switch and open the **fuel** valve one turn. Allow the gas to disperse before continuing. Depress the ignition switch for a further 20 seconds. This process may be repeated until successful ignition occurs. If the **fuel** valve has to be opened more than 5 turns more than recommended above.
7. Set the **filter select** control to the desired position.
8. Aspirate deionized water and set the readout to zero by adjusting the **blank** control.
9. Aspirate a standard solution of slightly higher concentration than expected in the samples to be tested. Adjust the **fine and coarse** control until a positive reading is obtained. This is a nominal value to be used in optimizing flame conditions.
10. Adjust the **fuel** valve in a clockwise direction until a peak reading is obtained.

    Note: There is a time delay between adjusting the flow of fuel and seeing the effect of the adjustment. A pause of a few seconds is therefore necessary between making every fine adjustment.

The optimum flame conditions are different for Na and K and retuning is necessary when switching between elements.

11. For optimum performance the instrument should be allowed 15 minutes to warm up. During this warm-up period a blank deionized water sample should be aspirated. It should be noted that results, adequate for most purposes, can be obtained from switch on, although standardization settings will need frequent checking during the warm-up period.

8.7 SAMPLE ANALYSIS

The measurable atomic flame emissions of the alkali and alkaline earth metals in terms of the emission wavelength and the color produced are given below:

Element	*Emission Wavelength (nm)*	*Flame Colour*
Sodium (Na)	589	Yellow
Potassium (K)	766	Violet
Barium (Ba)	554	Lime Green
Calcium (Ca)	622*	Orange
Lithium (Li)	670	Red

*Calcium is measured by using the calcium hydroxide band emission at 622 nm as the calcium main atomic emission occurs at 423 nm.

Since a flame photometer measures the concentration of the element itself in solution, standard solutions prepared from the salts of sodium, potassium, lithium, calcium and barium must be made up to contain the concentrations required in terms of the quality of the elements.

Below are two examples of how to prepare standards of 1 mg Na/100 ml (10 ppm Na) and 1 mg K/100 mL (10 ppm K).

8.7.1 Sodium

Accurately weigh 0.634 g of dry "Analar" quality NaCl, dissolve in pure deionized water and wash into a 500 ml volumetric flask. Fill to the mark with pure deionized water. To prepare the standard solution for use with the flame photometer, this stock solution should be diluted 1 in 50.

Calculation:

Atomic weight Na = 23.0

Molecular weight NaCl = 58.46

Therefore, 0.634 g NaCl contains $\frac{0.634 \times 23}{58.46} = 0.25$ g Na

Thus, in 500 mL of solution there is 250 mg Na or 50 mg Na/100 mL.

Diluting 1 in 50 gives a standard of 1mg Na/100 mL = 10 ppm Na.

8.7.2 Potassium

Accurately weigh 0.477 g of dry "Analar" quality KCl, dissolve in pure deionized water and wash into a 500 mL volumetric flask. Fill to the mark with pure deionized water. To prepare the standard solution for use with the flame photometer, this stock solution should be diluted 1 in 50.

Calculation:

Atomic weight K = 39.1

Molecular weight KCl = 74.56

Therefore, 0.477 g KCl contains $\frac{0.477 \times 39.1}{74.56} = 0.25$ g K

Thus, in 500 mL of solution there is 250 mg K or 50 mg K/100 mL.

Diluting 1 in 50 gives a standard of 1 mg K/100 mL = 10 ppm K.

8.7.3 Storage

Store solutions away from direct sunlight in a cool place, ideally at temperatures below 25°C. Glass containers should not be used for storage as they can affect the sodium concentration levels. Standards should be stored in sealed, plastic vessels and in high concentrations, (e.g., as a stock 1000 ppm solution) and dilutions prepared as required. The long-term storage of low concentration standards is not recommended due to degradation of ionic species.

8.7.4 Sample Preparation

There are several practical points regarding sample preparation, which should be adhered to in order to achieve the required accuracy in the analysis:

1. Avoid handling samples with fingers. This leads to serious contamination, e.g., if a finger is immersed in 20 mL of deionized water the resulting Na concentration will exceed that of a 10 ppm standard.
2. All analyses involve the use of a diluent, which is almost always deionized water. This should be of the highest quality for accurate flame analysis. Sodium, potassium and calcium are present in high concentrations in tap water and thus efficient deionization is essential.
3. Species that cause interference should be removed from samples or the equivalent concentration of the interferant should be present in the standards so as to avoid erroneous results, e.g., if a sample of approximately 10 ppm Na contains approximately 1000 ppm Ca, then Na analysis can only be achieved by removing the Ca with oxalate/oxalic acid or ensuring all standards contain 1000 ppm Ca.
4. Always try to follow a well-documented analytical procedure, which should contain information pertaining to interference removal when applicable.

5. Standards and samples should not be exposed to the atmosphere for long periods due to contamination from airborne particles and the evaporation of the solvent that could lead to elevated concentrations.
6. When in doubt about the equipment or application, the operator should contact Jenway for advice.

8.7.5 Sample Extraction

A number of methods for extracting sodium, potassium, lithium, calcium and barium from a wide variety of raw materials may be obtained.

The sample must be in the form of an aqueous solution, with no solid matter present, to be suitable for direct introduction into the flame photometer. This is achieved by:

- Extracting the salts from solid samples using deionized water or suitable extractants e.g., saturated $CaSO_4$ for sodium in soil. Extraction is more successful using a blender, macerator or shaking machine.
- If the sample is organic then the organic material should be removed by ashing. The remaining oxides are then dissolved using strong acids.
- Filtration/centrifugation is used to remove solid debris.

When aqueous, the sample can then be diluted to a known, accurately measured volume using deionized water. If it is a concentrated sample then the dilution ratio should be increased. If the sample concentration is low then a small volume of diluent and initial extractant should be used. Whichever method of extraction is used, the resultant solution must always be free of any particulate matter that may cause blockages in the nebulizer capillary tube.

8.7.6 Dilution

In order to obtain samples and standards of the right concentration for aspiration into the flame, various levels of dilution will often be necessary. Good quality deionized water should normally be used for carrying out these dilutions and it is recommended that the same batch of water should be used for diluting the samples and standards.

8.8 CALIBRATION

It is important to understand that the principles of flame photometry are such that, over **certain concentration ranges**, light emitted from the flame is **directly** proportional to the concentration of the species being aspirated (Fig. 8.2).

The graph below shows that the direct relationship between the flame emission and concentration is only true at relatively low concentrations. Above these low levels the flame begins to saturate and the flame emission ceases to increase in a linear relationship to concentration.

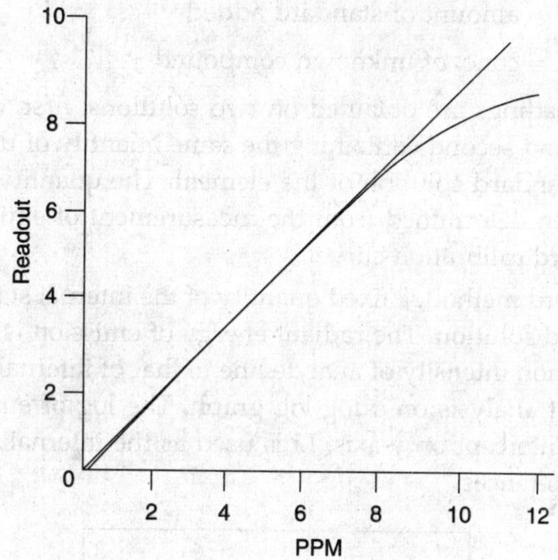

Fig. 8.2 *Relationship between sample concentration and flame emission.*

If the samples being analyzed lie on the linear part of the curve then the user can take direct concentration readings from the digital display. If, however, the concentration of samples are above the levels shown on the graph then the user has the choice of either:

(a) diluting the samples so that they lie on the linear part of the curve,

or

(b) constructing a calibration curve and relating the digital display reading to the concentration by cross-reference to the curve.

A calibration curve is prepared using standard solutions containing known concentrations of the elements to be determined and if necessary, other materials to ensure that the standard and sample backgrounds match. The concentration range covered by the calibration curve will depend upon the expected concentration of the samples so that the sample readings fall somewhere in the middle of the calibration curve.

Once the calibration curve has been plotted, the readings for the sample solutions are compared with the curve to allow the sample concentrations to be established.

8.9 EVALUATION METHODS IN FLAME PHOTOMETRY

There are two methods, one is the standard addition method while the other is called the internal standard method.

$(L_1 - H)\, a = L\, X$ found	(i) If L_1 emission of unknown, a,
$(L_2 - H)\, b = K\,(X + S)$	(ii) If L_2 emission with standard, b
$(L_2 - L_1) = KS$ found	(iii) H background emission

$$\text{X found } \frac{\text{S. added}}{\text{S. found}} = \text{X actual present (iv)}$$

S = amount of standard added

X = conc. of unknown compound

Here net emission readings are obtained on two solutions: first, containing an aliquot of the unknown solution and second containing the same quantity of unknown solution plus a measured amount of standard solution of the element. The quantity of test element in each of these solutions is then determined from the measurement of emission intensities and by constructing the standard calibration curve.

In the internal standard method, a fixed quantity of the internal standard element is added to samples and standard solution. The radiant energy of emission is measured for two solutions. The ratio of emission intensity of analyte line to that of internal standard line is plotted against concentration of analysis on a log-log graph. The log of emission vs. log of C will give straight line with intercept on y-axis. Li is used as the internal standard (Fig. 8.3). This eliminates several interferences.

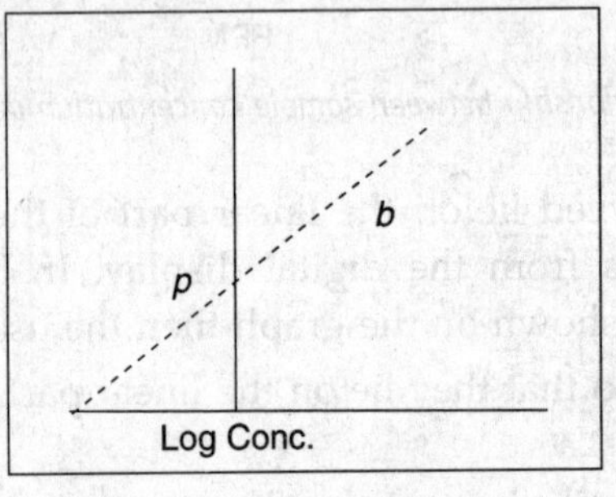

Fig. 8.3 *Standard addition method.*

8.10 FLAME PHOTOMETER – APPLICATION METHODS

The flame photometer using field is very vast, include :

1. Cement, glass, ceramics and fire proofing, etc. construction materials test
2. Fertilizer, earth, drinks test
3. Mine, petrol, metallurgy, chemical products test
4. Drug, food products test
5. Living rubbish of city test
6. Since and study, health, education lab for test.

The following applications methods will enable the user to determine alkali and alkaline earth metal concentrations in a wide variety of sample media.

Determination of alkalis in cement

Determination of calcium in beer

Calcium in biological fluids

Estimation of calcium in milk

Determination of potassium in fertilizers
Determination of potassium in plant material
Determination of available potassium in soils
Sodium and potassium in silicates, minerals and ores
Determination of sodium in raw pine oil
Determination of exchangeable sodium in soils
Flame photometric determination of sulphate
Determination of sodium in fuel oil
Determination of potassium in mixtures containing plant derived resins
Determination of lithium in greases
Simple flame photometric determination of barium
Simple flame photometric determination of calcium
Determination of calcium in fruit juice
Determination of potassium in glass
Determination of sodium in glass
Sodium in straw by flame photometry
Determination of sodium and potassium in fruit juice
Determination of calcium in biscuits

8.11 REFERENCES

1. Sawyer, Heineman, Beebe: Chemistry Experiments for instrumental methods (Wiley, New York, US, 1984
2. D.C. Harris, Quantitative Chemical Analysis (4th edn., W.H. Freeman and Company, New York, US, 1995
3. D.A. Skoog – D.M. West – F.J. Holler: Fundamentals of Analytical Chemistry (Saunders College Publishing, Fort Worth, US 1992)
4. J. Kenkel: Analytical Chemistry for Techniques (Lewis Publishers, Boca Raton, US 1994
5. http://www.chem.usu.edu/~sbialkow/Classes/361/flame.html
6. http://www.designinfo.com/LearnMore/Labware_Scientific_Instruments/analytical_Instruments/Spectrometers/Flame_Photometers
7. http://www.resonancepub.com/atomicspec.htm

CHAPTER 9

Atomic Absorption Spectrophotometer

9.1 INTRODUCTION

Atomic absorption spectroscopy (AAS) is the appropriate term used when electromagnetic radiation (EMR) is absorbed or emitted by atoms and measured. All atoms can absorb EMR, and the wavelengths at which EMR is absorbed or emitted is exclusive for a particular chemical element. AAS is a quantitative method of analysis that is used to analyze many metals and a few non-metals. An instrument used to analyze the sharp resonance line of a sample that, in a flame, emits an atomic vapor. It consists essentially of a light source, a combustion system, a monochromator, a photomultiplier and an electronic readout system.

9.2 THEORETICAL CONCEPTS AND DEFINITIONS

9.2.1 The Atom and Atomic Spectroscopy

The science of atomic spectroscopy has yielded three techniques for analytical use: atomic emission, atomic absorption and atomic fluorescence. In order to understand the relationship of these techniques with each other, it is necessary to have an understanding of the atom itself and of the atomic process involved in each technique.

The atom is, of course, made up of a nucleus surrounded by electrons. Every element has a specific number of electrons which are associated with the atomic nucleus in an orbital structure which is unique to each element. The electrons occupy orbital positions in an orderly and predictable way. The lowest energy, most stable electronic configuration of an atom, known as the "ground state," is the normal orbital configuration for an atom. If energy of the right magnitude is applied to an atom, the energy will be absorbed by the atom, and an outer electron will be promoted to a less stable configuration or "excited state." As this state is unstable, the atom will immediately and spontaneously return to its ground state configuration. The electron will return to its initial, stable orbital position, and radiant energy equivalent to the amount of energy initially absorbed in the excitation process will be emitted. The process is illustrated in Fig. 9.1. Note that step 1 in the process, the excitation, is forced by supplying energy. The decay process in Step 2, involving the emission of light, occurs spontaneously.

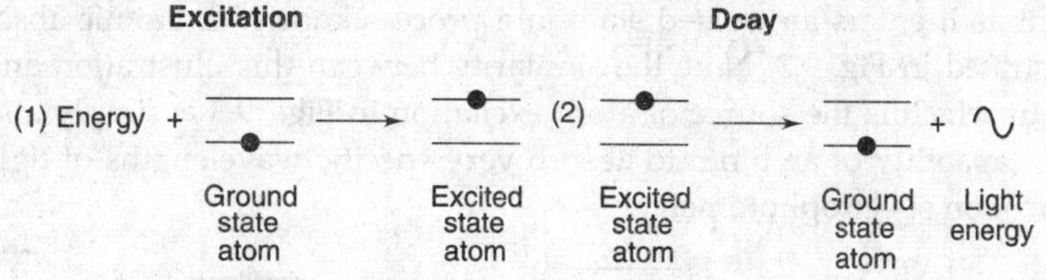

Fig. 9.1 *The process of energy excitation and decay.*

The wavelength of the emitted radiant energy is directly related to the electronic transition which has occurred. Since every element has a unique electronic structure, the wavelength of light emitted is a unique property of each individual element. As the orbital configuration of a large atom may be complex, there are many electronic transitions which can occur, each transition resulting in the emission of a characteristic wavelength of light, as illustrated in Fig. 9.2.

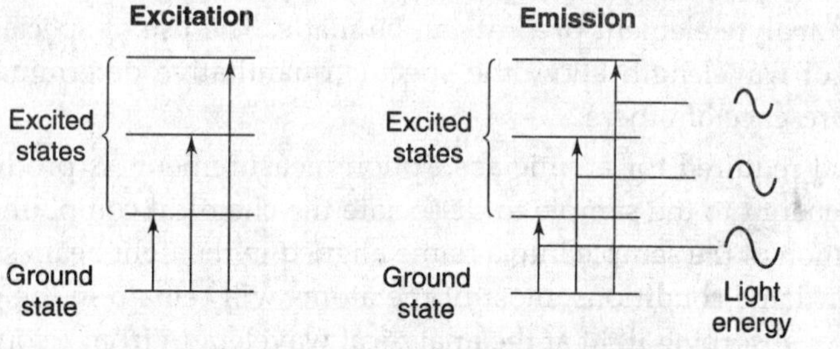

Fig. 9.2 *The process of energy excitation and emission.*

The process of excitation and decay to ground state is involved in all three fields of atomic spectroscopy. The energy absorbed in the excitation process or the energy emitted in the decay process is measured and used for analytical purposes.

In atomic emission, a sample is subjected to a high energy. Thermal environment in order to produce excited state atoms, capable of emitting light. The energy source can be an electrical arc, a flame, or more recently, a plasma. The emission spectrum of an element exposed to such an energy source consists of a collection of the allowable emission wavelengths, commonly called emission lines, because of the discrete nature of the emitted wavelengths. This emission spectrum can be used as a unique characteristic for qualitative identification of the element. Atomic emission using electrical arcs has been widely used in qualitative analysis.

Emission techniques can also be used to determine how much of an element is present in a sample. For a "quantitative" analysis, the intensity of light emitted at the wavelength of the element to be determined is measured. The emission intensity at this wavelength will be greater as the number of atoms of the analyte element increases. The technique of flame photometry is an application of atomic emission for quantitative analysis.

If light of just the right wavelength impinges on a tree, ground state atom, the atom may absorb the light as it enters an excited state in a process known as atomic absorption. This process is illustrated in Fig. 9.3. Note the similarity between this illustration and the one in Step 1. The light which is the source of atom excitation in Fig. 9.3 is simply a specific form of energy. The capability of an atom to absorb very specific wavelengths of light is utilized in atomic absorption spectrophotometry.

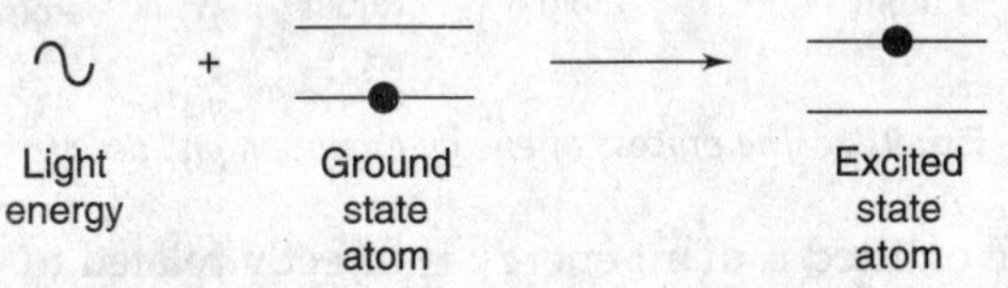

Fig. 9.3 *Atomic absorption process.*

The quantity of interest in atomic absorption measurements is the amount of light at the resonant wavelength which is absorbed as the light passes through a cloud of atoms. As the number of atoms in the light path increases, the amount of light absorbed increases in a predictable way. By measuring the amount of light absorbed, a quantitative determination of the amount of analyte element present can be made. The use of special light sources and careful selection of wavelength allow the specific quantitative determination of individual elements in the presence of others.

The atom cloud required for atomic absorption measurements is produced by supplying enough thermal energy to the sample to dissociate the chemical compounds into free atoms. Aspirating a solution of the sample into a flame aligned in the light beam serves this purpose. Under the proper flame conditions, most of the atoms will remain in the ground state form, which is capable of absorbing light at the analytical wavelength from a source lamp. The ease and speed at which precise and accurate determinations can be made with this technique have made atomic absorption one the most popular methods for the determination of metals.

A third field in atomic spectroscopy is atomic fluorescence. This technique incorporates aspects of both atomic absorption and atomic emission. Like atomic absorption, ground state atoms created in a flame are excited by focusing a beam of light on to the atomic vapor. Instead of looking at the amount of light absorbed in the process, however, the emission resulting from the decay of atoms excited by the source light is measured. The intensity of the "fluorescence" increases with increasing atom concentration, providing the basis for quantitative determination.

The source lamp for atomic fluorescence is out of line with the rest of the optical system, so that the light detector sees only the fluorescence in the flame and not the light from the lamp itself. Lamps are normally much brighter in atomic fluorescence than in atomic absorption, in order to increase the degree of atom excitation and provide high fluorescence sensitivities.

Figure 9.4 illustrates how the three techniques just described are implemented. While atomic absorption technique is the most widely applied of the three techniques and usually offers several advantages over the other two, particular benefits may be gained with either emission or fluorescence in special analytical situations. This is especially true of emission, which will be discussed in more detail in a later chapter.

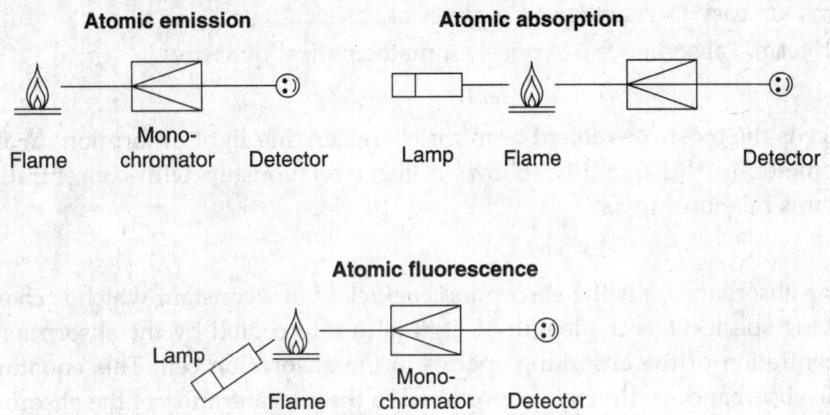

Fig. 9.4 *Atomic emission, Atomic absorption and atomic fluorescence.*

9.2.2 Quantitative Analysis by Atomic Absorption

The atomic absorption process is illustrated in Fig. 9.5. Light at the resonance wavelength of initial intensity I_0 is focused on the flame cell containing ground state atoms. The initial light intensity is decreased by an amount determined by the atom concentration in the flame cell. The light is then directed onto the detector where the reduced intensity I is measured. The amount of light absorbed is determined by comparing I to I_0.

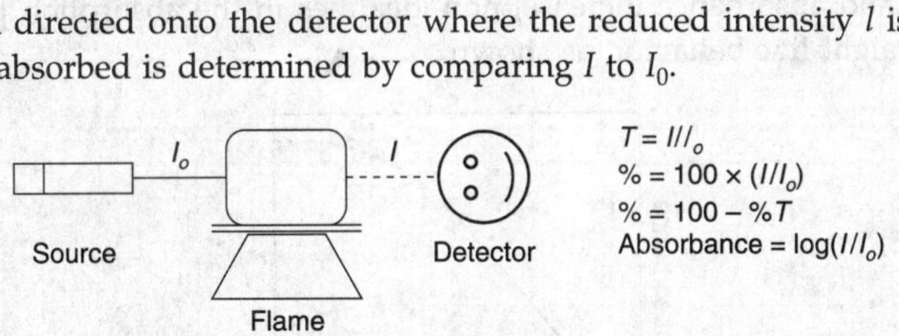

Fig. 9.5 *Atomic absorption process basic concept.*

Several related terms are used to define the amount of light absorption which has taken place. The "transmittance" is defined as the ratio of the final intensity to the initial intensity.

$$T = I/I_0$$

Transmittance is an indication of the fraction of initial light which passes through the flame cell to fall on the detector.

The "percent transmission" is simply the transmittance expressed in percentage terms.

$$\%T = 100 \times I/I_0$$

The "percent absorption" is the complement of percent transmission defining the percentage of the initial light intensity which is absorbed in the flame.

$$\%A = 100 - \%T$$

These terms are easy to visualize on a physical basis.

The fourth term, "absorbance," is purely a mathematical quantity

$$A = \log (I_0/I)$$

Absorbance is the most convenient term for characterizing light absorption. In absorption spectrophotometer, as this quantity follows a linear relationship with concentration Beer's Law defines this relationship as:

$$A = abc$$

where A is the absorbance, a is the absorption coefficient of a constant which is characteristic of the absorbing species; b is the length of light path intercepted by the absorption cell and c is the concentration of the absorbing species in the absorption cell. This equation simply states that the absorbance is directly proportional to the concentration of the absorbing of the absorbing species for a species for a given set of instrumental conditions.

This directly proportional behavior between absorbance and concentration is observed in atomic absorption. When the absorbances of standard solutions containing known concentrations of any analyte are measured and the absorbance data are plotted against concentration, a calibration relationship similar to that in Fig. 9.6 is established. Over the region where the Beer's law relationship is observed, the calibration yields a straight line. As the concentration and absorbance increase, non-idealities in the absorption process cause a deviation from straight-line behavior, as shown.

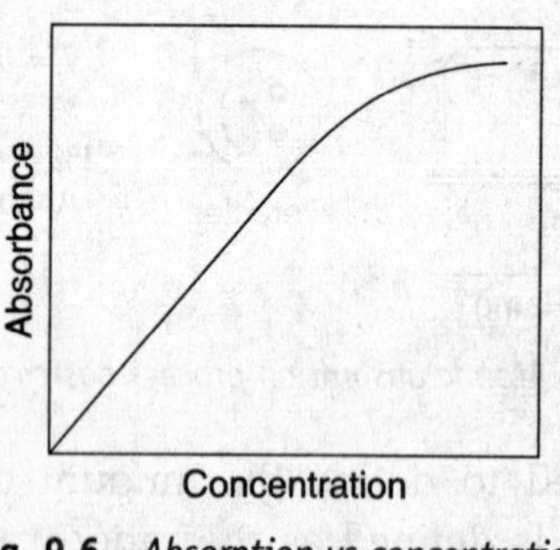

Fig. 9.6 *Absorption vs concentration*

After such a calibration is established, the absorbance of solutions of unknown concentrations may be measured and the concentration determined from the calibration curve. In modern instrumentation, the calibration can be made within the instrument to provide a direct readout of unknown concentrations. Since the advent of built-in microcomputers, accurate calibration even in the non-linear region is possible.

9.2.3 Sensitivity and Detection Limits

"Sensitivity" and "detection limit" are terms which describe two performance characteristics in atomic absorption. The sensitivity is a convention for defining the slope of the absorbance versus concentration calibration for each element. For flame atomic absorption, it is expressed in terms of the concentration of the element in micrograms per milliliter required to produce a 1% absorption; or in terms of absorption units, the sensitivity is the micrograms of element per milliliter which will give an absorbance of 0.0044. As long as measurements are made in the linear working region, the atomic absorption sensitivity of an element can be determined by reading the absorbance produced by a known concentration of the element, and solving a proportional equation for the concentration that would produce an absorbance of 0.0044, as shown below.

$$\frac{\text{Conc of Std}}{\text{measured abs}} = \frac{\text{sensitivity}}{0.0044} \text{ and } \frac{\text{Conc of Std} \times 0.0044}{\text{measure abs}}$$

*UPAC nomenclature uses the term "characteristic concentration" in place of sensitivity. As the latter is currently the more commonly used term, it shall be used here.

The sensitivity values for a standard set of instrumental conditions are usually given for an instrument. Knowing the expected sensitivity allows an operator to determine if all instrumental conditions are optimized and the instrument is performing up to specifications by simply measuring the absorbance of a known concentration and comparing the results to the expected value. A known sensitivity value also allows one to predict the absorbance range which will be observed from a known concentration range or to determine the concentration range which would produce optimum absorbance levels by simply utilizing the relationship shown above.

It should be noted that only the size of an absorption signal can be predicted from the value given for sensitivity. No information is given on how small an absorbance, and hence on how small a concentration can be measured. The limitation on the measurement of small absorbance values is the baseline noise and is not considered in the calculation of sensitivity.

The definition of "detection limit" incorporates a consideration of both signal size and baseline noise to give an indication of the lowest concentration of the element which can be measured. The detection limit is defined as the concentration which will produce a signal-to-noise ratio of two. Figure 9.7 illustrates a recorder tracing an absorption measurement at the detection limit. This is the lowest concentration which can be clearly differentiated from zero.

In summary, the concepts of "sensitivity" and "detection limit" have important distinctions which should be understood. Sensitivity defines only the size of the absorption signal. It serves as a reference for instrument setup. Knowing the sensitivity also makes it possible to determine optimum sample concentrations for analysis.

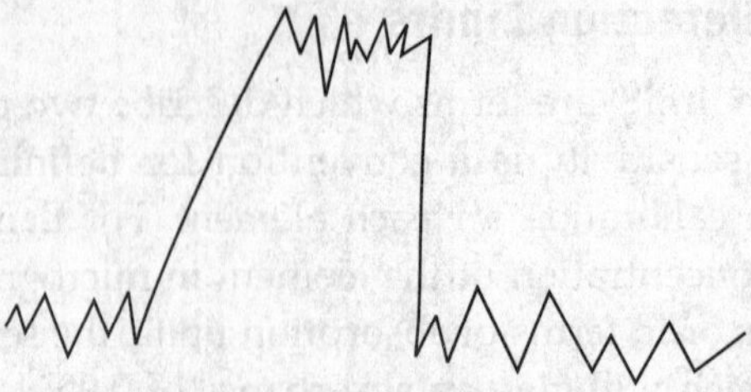

Fig. 9.7 *Detection limit tracing.*

The detection limit describes the signal-to-noise ratio characteristic for the instrument. This term is of greater analytical importance in that it defines the analytical capability of the instrument and provides a means of estimating the lower limit of concentration detection.

9.3 PRINCIPLES

The technique makes use of absorption spectrometry to assess the concentration of an analyte in a sample. It relies therefore heavily on Beer-Lambert law. In short, the electrons of the atoms in the atomizer can be promoted to higher orbitals for a short amount of time by absorbing a set quantity of energy (i.e., light of a given wavelength). This amount of energy (or wavelength) is specific to a particular electron transition in a particular element, and in general, each wavelength corresponds to only one element. This gives the technique its elemental selectivity. As the quantity of energy (the power) put into the flame is known, and the quantity remaining at the other side (at the detector) can be measured, it is possible, from Beer-Lambert law, to calculate how many of these transitions took place, and thus get a signal that is proportional to the concentration of the element being measured.

Atomic absorption units have four basic parts: interchangeable lamps that emit light with element-specific wavelengths, a sample aspirator, a flame or furnace apparatus for volatilizing the sample, and a photon detector. In order to analyze any given element, a lamp is chosen that produces a wavelength of light that is absorbed by that element. Sample solutions are aspirated into the flame. If any ions of the given element are present in the flame, they will absorb light produced by the lamp before it reaches the detector. The amount of light absorbed depends on the amount of the element present in the sample. Absorbance values for unknown samples are compared to calibration curves prepared by running the known samples.

9.4 ATOMIC ABSORPTION INSTRUMENTATION

9.4.1 The Basic Components

To understand the working of the atomic absorption spectrophotometer, let us build one, piece by piece. Every absorption spectrophotometer must have components which fulfill the three basic requirements shown in Fig. 9.8. There must be: (1) a light source; (2) a sample cell; and (3) a means of specific light measurement.

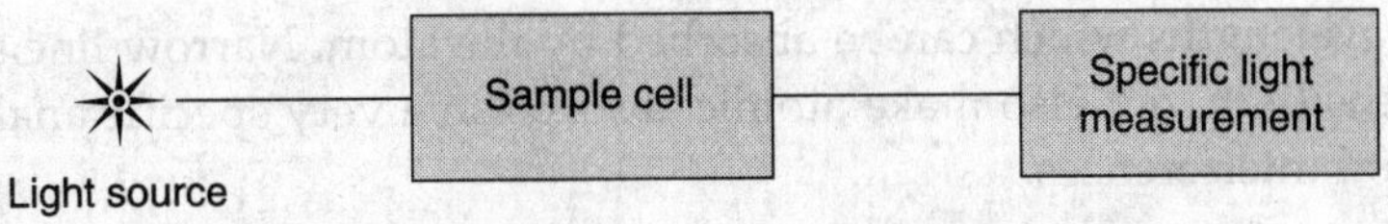

Fig. 9.8 *Requirements for a spectrophotometer.*

In atomic absorption, these functional areas are implemented by the component illustrated in Fig. 9.9. A light source which emits the sharp atomic lines of the element to be determined is required. One of the most widely used sources is the hollow cathode lamp. These lamps are designed to emit the atomic spectrum of a particular element, and the specific lamps are selected for use, depending on the element to be determined.

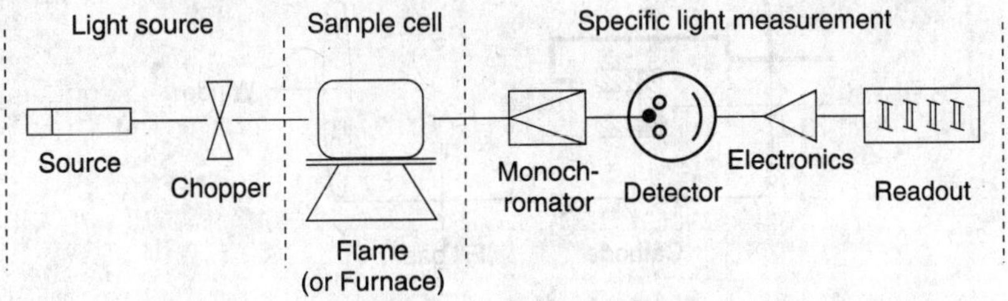

Fig. 9.9 *Single-beam atomic absorption spectrophotometer.*

It is also required that the source radiation be modulated (switched on and off rapidly) to provide a means of selectively amplifying light emitted from the source lamp and ignoring emission from the sample cell. Source modulation can be accomplished with a rotating chopper located between the source and the sample cell or by pulsing the power to the source.

Special considerations are also required for a sample cell for atomic absorption. An atomic vapor must be generated in the light beam from the source. This is generally accomplished by introducing the sample into a burner system or electrically heated furnace aligned in the optical path of the spectrophotometer.

Several components are required for specific light measurement. A monochromator is used to disperse the various wavelengths of light which are emitted from the source and to isolate the particular line of interest. The selection of a specific source and a particular wavelength in that source is what allows the determination of a selected element to be made in the presence of others. The wavelength of light which is isolated by the monochromator is directed. Which serves as the "eye" of the instrument. The electrical current from the photomultiplier is then amplified and processed by the instrument electronics to produce a signal which is a measure of the light attenuation occurring in the sample cell. This signal can be further processed to produce an instrument readout directly in concentration units.

9.4.2 Light Sources

An atom absorbs light at very discrete wavelengths. In order to measure this narrow light absorption with maximum sensitivity, it is necessary to use the line source which emits the very specific wavelengths which can be absorbed by the atom. Narrow line sources not only provide high sensitivity but also make atomic absorption a very specific analytical technique with few spectral interferences.

The hollow cathode lamp is an excellent, bright, line source for most of the elements determinable by atomic absorption. Figure 9.10 shows how a hollow cathode lamp is constructed. The cathode of the lamp is a hollowed-out cylinder of the metal whose spectrum is to be produced. The anode and cathode are sealed in a glass cylinder filled with either neon or argon. A window transparent to the emitted radiation IS fused to the end of the cylinder.

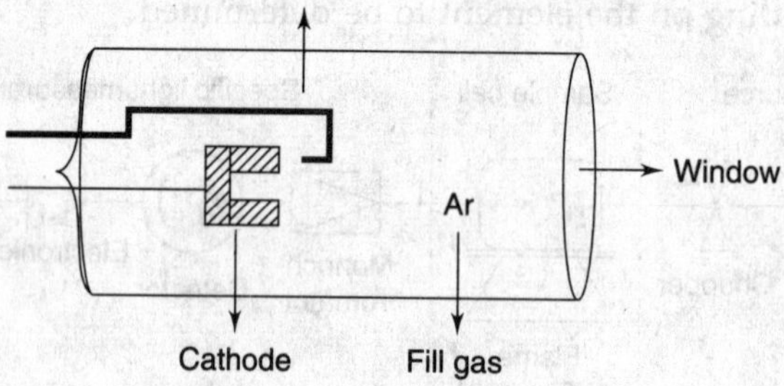

Fig. 9.10 *Hollow cathode lamp.*

The emission process is illustrated in Fig. 9.11. When an electrical potential is applied between anode and cathode, some of the fill gas atoms are ionized. The positively charged ions accelerate through the electrical field to collide with the negatively charged cathode and dislodge individual metal atoms in a process called "sputtering." Sputtered metal atoms are then excited to emission through impact with fill gas ions.

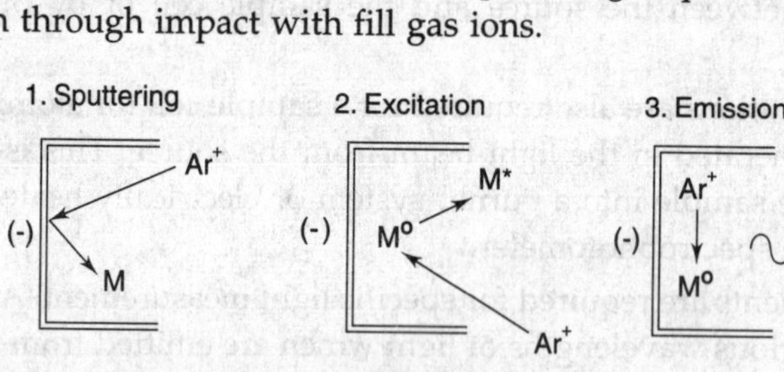

Fig. 9.11 *Hollow cathode lamp process.*

Hollow cathode lamps have a finite lifetime. As the lamp is used, the sputtering process may remove some of the metal atoms from the vicinity of the cathode to be deposited elsewhere. Lamps for volatile metals such as arsenic, selenium, and cadmium age faster due to rapid vaporization of the cathode during use. Hollow cathode lamps may also have a limited shelf life which is unrelated to the amount of use. Adsorption of fill gas atoms onto the inside surface of the lamp housing is a primary cause of this type of lamp failure. Some cathode materials will slowly evolve hydrogen when heated. As the concentration of hydrogen in the fill gas increases, a background continuum emission contaminates the purity of the line spectrum of the element, resulting in a reduction of atomic absorption sensitivity and poor calibration linearity. Lamp failure due to hydrogen contamination can usually be reversed by running the lamp at a few milliamperes of current with reversed polarity for a few minutes. Under these conditions, a tantalum "getter" attached to what is normally the anode will adsorb the hydrogen out of the fill gas and return the lamp to its specified sensitivity and linearity characteristics.

The cathode of the hollow cathode lamp is usually constructed from a highly pure metal resulting in a very pure emission spectrum of the cathode material. It is sometimes possible, however, to construct a cathode from an alloy of several metals. The resulting "multi-element" lamp can be used as a source for all the elements contained in the cathode alloy. Not all metals may be used in combination with others because of metallurgical or spectral limitations. Special consideration should be given before using a multi-element lamp as analytical complications may result. Often the intensity of emission for an element in a multi-element lamp is not as great as that which is observed for the element in a single-element lamp. This loss of intensity could be a serious disadvantage in applications where high precision or low detection limits are required. The increased spectral complexity of multi-element lamps may require that alternate wavelengths or narrower slits be used, which may adversely affect sensitivity and baseline noise.

Each hollow cathode lamp will have a particular current for optimum performance. In general, higher currents will produce a brighter emission and less baseline noise. As the current continues to increase, however, spectral line broadening will occur, resulting in a reduction in sensitivity and linear working range. The current specified for each lamp will usually be the highest possible without producing serious line broadening. This will provide the best signal-to-noise characteristic for the lamp.

Confusion over exactly what current is being used for a hollow lamp may occur as a result of the effect of varying methods in lamp modulation. As explained earlier, the source for atomic absorption must be modulated in order to accomplish selective amplification of the lamp emission signal. This can be accomplished mechanically by using a rotating chopper, illustrated in Figure 9.9, or electronically by pulsing the current supplied to the lamp. Either method produces the same desirable result, however, electronic modulation causes a negative bias in measured lamp current.

Figure 9.12 illustrates the cause for the difference in measured currents in mechanically and electronically modulated systems. For mechanical modulation, the lamp is run at a constant current. Under these conditions an ammeter reading will indicate the actual current flow. For electronic modulation, the current is switched on and off at a rapid rate. An ammeter, which is too slow to follow these rapid fluctuations in current, will indicate a lower current than is actually flowing during the time the lamp is on. The measured current will approximate a time-averaged value. For electronically modulated systems, the actual current can be approximated from the measured current by dividing the measured current by the "duty cycle," the fraction of time the lamp is actually on. For example, for a duty cycle of 50% and a measured lamp current of 10 milliamperes, the actual operating current for an electronically modulated system is:

$$\frac{10 \text{ milliamperes}}{0.5} = 20 \text{ milliamperes}$$

As a result of this phenomenon, specified lamp current settings are lower for atomic absorption instruments which modulate the source electronically. The actual currents used by two different systems cannot be compared without making the adjustment for duty cycle shown above (Fig. 9.12 (a) & (b).

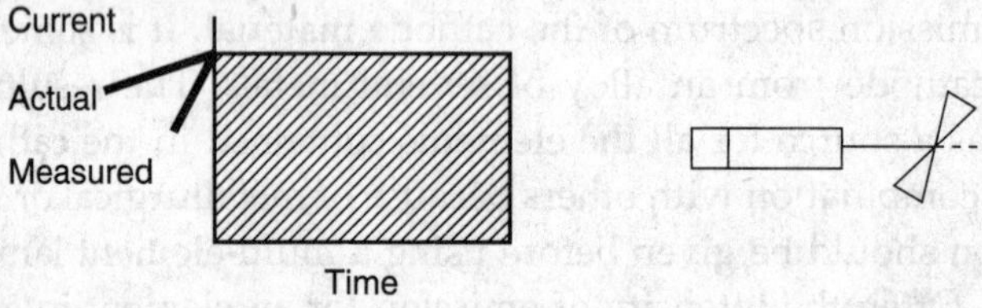

(a)

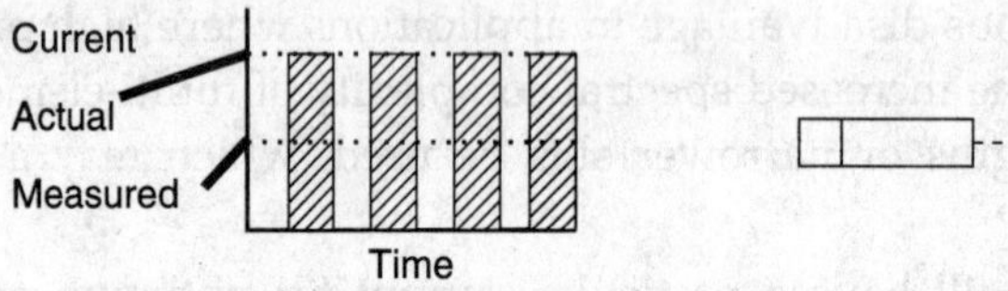

(b)

Fig. 9.12 *(a) Mechanical modulation (b) Electronic modulation.*

For most elements the hollow cathode lamp is a completely satisfactory source for atomic absorption. In a few cases, however, the quality of the analysis is impaired by limitations of the hollow cathode lamp. These are primarily the volatile elements where low intensity and short lamp life are a problem. The atomic absorption determination of these elements can often be dramatically improved with the use of the "electrodeless discharge lamp."

The electrodeless discharge lamp (EDL) design is shown in Fig. 9.13. A small amount of the element or salt of the element for which the source is to be used is sealed inside a quartz bulb. This bulb is placed inside a ceramic cylinder on which an antenna from a radio frequency generator is coiled. When an RF field of sufficient power is applied, the coupled energy will vaporize and excite the atoms inside the bulb into emitting their characteristic spectrum. An accessory power supply is required to operate an EDL.

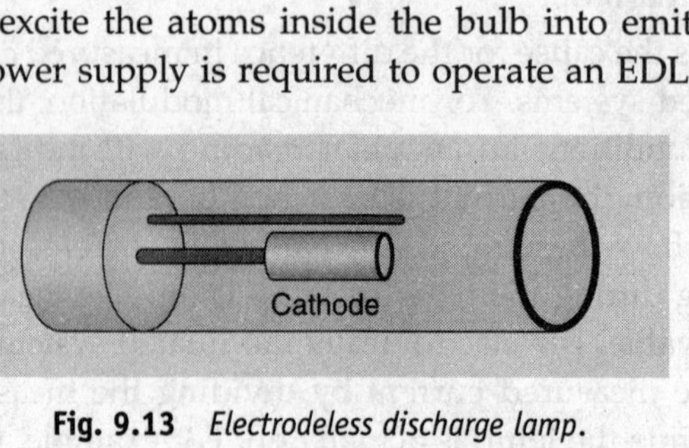

Fig. 9.13 *Electrodeless discharge lamp.*

Electrodeless discharge lamps are typically much more intense and in a few cases more sensitive than comparable hollow cathode lamps. They therefore offer the analytical advantages of better precision and lower detection limits. In addition to providing superior performance, the useful lifetime of an EDL will exceed that of a hollow cathode lamp for the same element. It should be noted, however, that the source of light in the EDL is considerably larger than that in the hollow cathode lamp. As a result, the performance benefits of the EDL can only be observed in instruments with optical systems designed to take advantage of these benefits.

Electrodeless discharge lamps are available for the 17 elements listed in Table 9.1 and others are under development. Hollow cathode lamps also exist for these elements. The EDLs, however, offer definite performance and lifetime benefits for instruments which are optically compatible with the EDL source.

Table 9.1 *Perkin-Elmer electrodeless discharge lamps*

Elements	
Antimony	Potassium
Arsenic	Rubidium
Bismuth	Selenium
Cadmium	Tellurium
Cesium	Thallium
Germanium	Tin
Lead	Titanium
Mercury	Zinc
Phosphorus	

9.4.3 Optical Considerations

The instrument diagrammed in Fig. 9.9 represents a fully functional "single-beam" atomic absorption spectrophotometer. It is called "single-beam" because all measurements are based on the varying intensity of a single beam of light. A significant convenience benefit can be obtained by incorporating additional optics to produce the "double-beam" system, illustrated in Fig. 9.14. In the double-beam design light from the lamp is divided into a sample beam, which is focused through the sample cell, and a reference beam which is directed around the sample cell to serve as a monitor of lamp intensity. In a double-beam system, the readout does not simply represent the intensity of a single-beam of light, but a ratio of the sample and reference beams, both of which originate from the same source. As a result fluctuations in source intensity, which affect the intensity of both sample and reference beams equally, are not translated into fluctuations in instrument readout. The baseline, or zero-point from which absorbance measurements are made, is therefore much more stable in a double-beam instrument. This freedom from the effects of source intensity drift means that lamps need not be warmed up prior to use, resulting in a time saving for the operator and an extension of useful lamp life. The improved baseline stability also makes it easier to use higher scale expansion for the determination of very low concentrations.

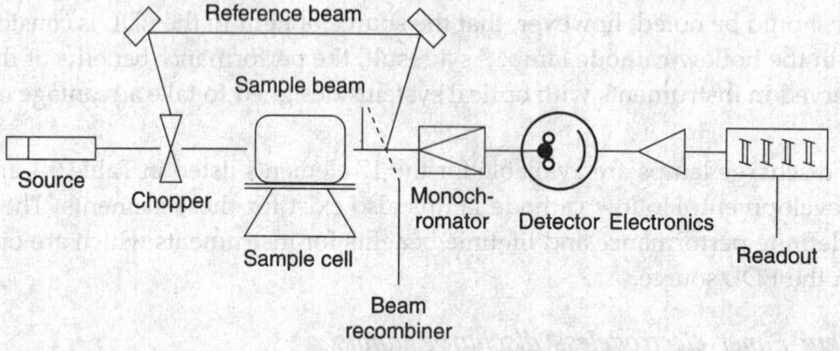

Fig. 9.14 *Double-beam atomic absorption spectrophotometer.*

An important factor in determining the amount of baseline noise that will be present in an atomic absorption analysis is the amount of light energy reaching the photomultiplier tube. As more light energy is focused onto the detector, less electronic gain is required and less noise is present in the signal. As discussed earlier, lamp intensity is optimized to be as bright as possible but line broadening complications place a limit on the intensity which can be obtained from a lamp. Therefore, the photometer system must be designed to make optimum use of the available light from the source in order to produce an instrument with low noise characteristics.

Light from the source must be focused on the sample cell and directed to the monochromator, where the wavelengths of light are dispersed and the analytical line is focused onto the detector. Some energy is lost at each optical surface along the way. Front surfaced, highly reflective, toroidal mirrors can be used to precisely control the focus of the source lamp and the field of view of the light detector with minimal light loss. Alternately, focusing can be accomplished by refraction instead of reflection by using a lens system. However, since the focal length of a lens varies with wavelength, additional adjustable optics, which further reduce energy throughput, must be utilized to obtain proper focus over the full spectral range for atomic absorption.

Particular care must be taken in the monochromator area of the optical system to avoid excessive light loss.Wavelength dispersion is accomplished with a grating, a reflective surface ruled with many fine parallel lines very close together. Reflection from this ruled surface generates an interference phenomenon known as diffraction, in which different wavelengths of light diverge from the grating at different angles.

A typical monochromator is shown in Fig. 9.15. Light from the source enters the monochromator at the entrance slit and is directed to the grating where dispersion takes place. The diverging wavelengths of light are directed towards the exit slit. By adjusting the angle of the grating, a selected emission line from the source can be allowed to pass through the exit-slit and fall onto the detector. All other lines are blocked from exiting.

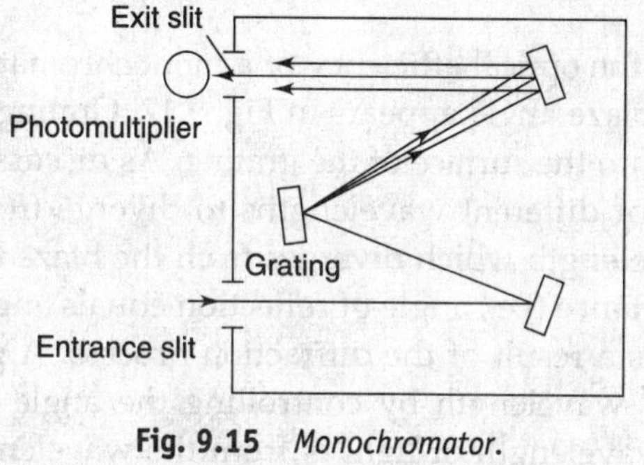

Fig. 9.15 *Monochromator.*

The angle of dispersion at the grating can be controlled by the density of line rulings. Higher dispersion will result from greater line density. High dispersion is important to good energy efficiency of the monochromator. Figure 9.16 illustrates this point. The image of the source focused on the entrance slit and dispersed emission lines at the exit slit are shown for both a low-dispersion and high-dispersion grating. In order to isolate a desired line from nearby lines. It is necessary to use a smaller exit slit in the low-dispersion example than is required in the high-dispersion case. For good resolution (distinct separation of lines) it is necessary that the entrance and exit slits be of the same size. This means that for a low-dispersion grating, the size of the monochromator entrance slit is limited to the narrow size demanded of the exit slit to exclude nearby lines. Thus, much of the available light energy is prevented from ever entering the monochromator. In contrast, for a high-dispersion grating, greater line separation allows the use of wider slits, which make use of more of the available light without a sacrifice in resolution.

A high-dispersion monochromator allows the use of wider slits which make greater use of lamp intensity.

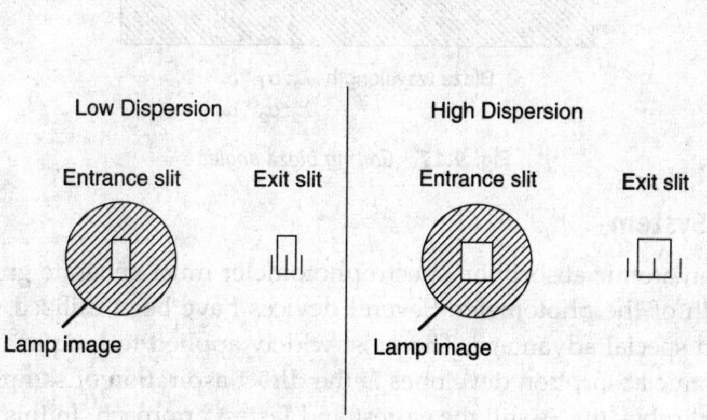

Fig. 9.16

Large, quality gratings of high dispersion are difficult and expensive to make. Therefore, the incentive is great to accept smaller gratings with less line density and poorer dispersion for atomic absorption instrumentation. To obtain the full energy benefit of high dispersion, it is necessary to use a grating of larger surface area to capture all the light from the magnified large slit images.

Another factor affecting the optical efficiency of a monochromator is the blaze angle of the grating. An illustration of blaze angle appears in Fig. 9.17. Grating rulings are in the form of V-shaped grooves carved into the surface of the grating. As discussed, earlier, an interference phenomenon causes light of different wavelengths to diverge from the grating at different angles. The particular wavelength which diverges from the blazed surface at an angle corresponding to specular reflectance (i.e., angle of reflection equals angle of incidence) will suffer the least loss in intensity as a result of the diffraction process. A grating can be constructed for a blaze at any desired wavelength by controlling the angle of cut during ruling. The farther removed a given wavelength of light is from the wavelength for which a grating is blazed, the greater will be the extent of monochromator light loss at that wavelength. The useful atomic absorption wavelength range runs from 180 to 900 nanometers. With one grating blazed somewhere in the middle of this range, significant energy fall-off occurs at the wavelength extremities due to energy inefficiencies in the diffraction process. The technique used to overcome this problem and to provide enhanced energy throughput at the wavelength extremities is to equip the instrument with two gratings, one blazed in the ultraviolet and the other blazed in the visible region of the spectrum. Then by choosing the grating blazed nearest the working wavelength, the optimum energy throughput can be achieved.

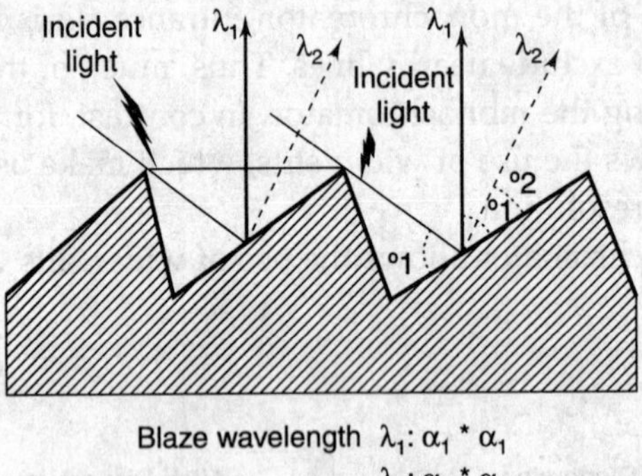

Fig. 9.17 *Grating blaze angle.*

9.4.4 Burner System

The atomizer of an atomic absorption spectrophotometer must generate ground state atoms in the optical path of the photometer. Several devices have been utilized for this purpose, each with its own special advantage. The most widely applied technique around which the routine use of atomic absorption developes is the direct aspiration of sample solution into a flame. Where applicable, this is still the easiest and fastest approach. In this section. We willl discuss the burner system for conventional flame atomization.

Figure 9.18 shows a cutaway view of an atomic absorption burner system. In this "premix" system, sample solution is aspirated through a nebulizer and sprayed as a fine aerosol into the mixing chamber. Here the sample aerosol is mixed with fuel and oxidant gases and carried to the burner head, where combustion and sample atomization occur.

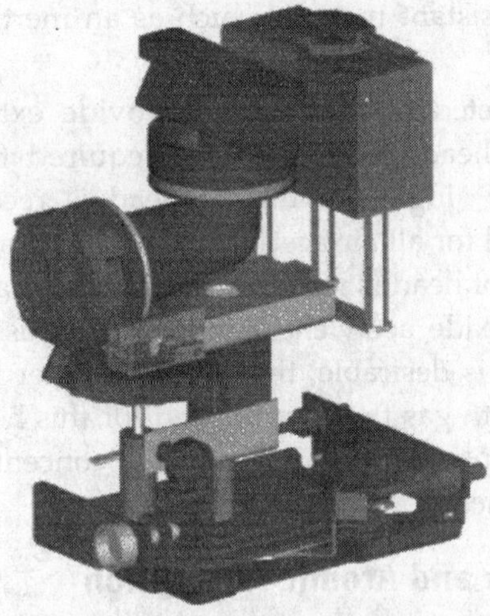

Fig. 9.18 *Premix burner system.*

Fuel gas is introduced into the mixing chamber through the fuel inlet, and the oxidant enters through the nebulizer sidearm. It is advantageous to have in addition an auxiliary oxidant inlet directly into the mixing chamber. This allows the oxidant flow adjustments to be made through the auxiliary line, while the flow through the nebulizer remains constant. Thus, for a burner system with an auxiliary oxidant line, the sample uptake rate is independent of flame condition, and the need to readjust the nebulizer after every flame adjustment is eliminated.

The sample aerosol is of varying droplet size as it is sprayed into the mixing chamber. Upon entering the flame, the water in these droplets is vaporized. The remaining solid material must likewise be vaporized, and chemical bonds must be broken to create free ground state atoms. Where the initial droplet size is large, the sample vaporization and atomization process is more difficult to complete in a short time in which the sample is exposed to the flame environment before passing through the light beam. Incomplete sample atomization will lead to increased susceptibility to analytical interferences. Therefore, a flow spoiler is placed inside the mixing chamber directly in front of the nebulizer. Larger sample droplets, which are not carried around the spoiler by the gas flow, impinge on the spoiler and fall to the bottom of the chamber, where they are removed from the system through the drain. The drain utilizes a liquid trap to prevent combustion gases from escaping through the drain line. The inside of the burner chamber is either constructed of, or coated with, a wettable inert plastic material to provide free drainage of excess sample and prevent burner chamber "memory." When the burner is properly constructed, absorbance response to a change in sample solution should take not more than one or two seconds to reach equilibrium.

Several important factors enter into the nebulizer portion of the burner system. In order to provide for the most efficient nebulization for varying sample solution systems, the nebulizer should be adjustable. Stainless steel has been the most common material used for construction of the nebulizer, but this has the disadvantage of being susceptible to corrosion from samples with a high content of acid or other corrosive agents. For such cases, nebulizers constructed of a corrosion resistant material, such as an inert plastic or platinum-rhodium alloy, should be used.

Burner heads are constructed of titanium to provide extreme resistance to heat and corrosion. Different burner head geometries are required for various flame or sample conditions. A ten-centimeter single-slot burner head is recommended for air-acetylene flames, and this head is useful for all air-acetylene work. For samples containing a high level of dissolved solids, a triple-slot head is recommended. A special five-centimeter burner head is required when a nitrous oxide acetylene flame is to be used. When an especially wide working concentration range is desirable, the five centimeter head offers advantages as an air-acetylene burner head. With gas flows optimized for this head, almost no detection limit loss is observed while the upper limit of determinable concentration is extended compared to the ten-centimeter burner head.

9.4.6 The Microcomputer and Atomic Absorption

A revolution in electronics which has dramatically affected the field of analytical instrumentation is the development of the microprocessor. A microprocessor is a single integrated circuit capable of providing centralized control and data manipulation for a number of attached devices. It is the heart of a tiny computer. Add to it a set of preprogrammed instructions, a memory for data, and some form of input and output for communication with the outside world, and a fully functional microcomputer is created.

A few electronic components which make up the microcomputer may replace hundreds of thousands of discrete components which would be required to do the same control and computational tasks. As a result of the reduced electronic complexity, the cost of instrument electronics has been reduced. This cost benefit along with the innate data handling capabilities of a computer allow more operator convenience and data manipulation features to be incorporated into the instrument without dramatically affecting the price.

One of the most important contributions of the microcomputer to atomic absorption is the ability to conveniently calibrate and accurately compute concentrations from absorbance data, even for non-linear calibration curves. In the linear region, data on one standard and a blank are sufficient for defining the relationship between concentration and absorbance. Where the relationship becomes non-linear, however, more standards are required. The accuracy of a calibration computed for a non-linear relationship depends on the number of standards and the equations used for calibration. For the equation format which optimally fits atomic absorption data, it has been experimentally shown that accurate calibration can be achieved with three standards plus a blank even in cases of severe curvature. Figures 9.19 and 9.20 show microcomputer fitted calibrations and illustrate the accuracy of the calculated concentrations.

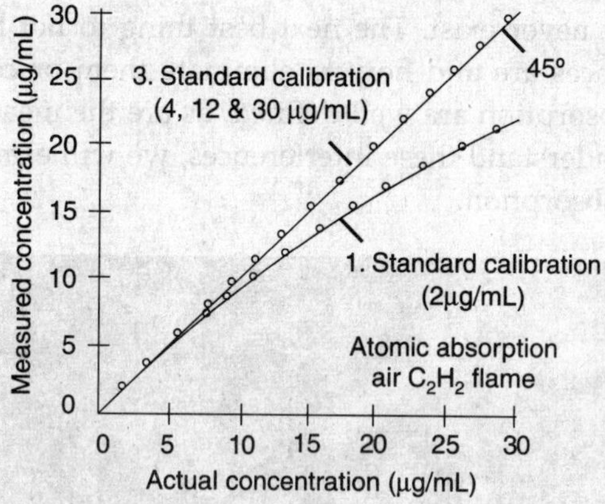

Fig. 9.19 *Cobalt calibration model 560.*

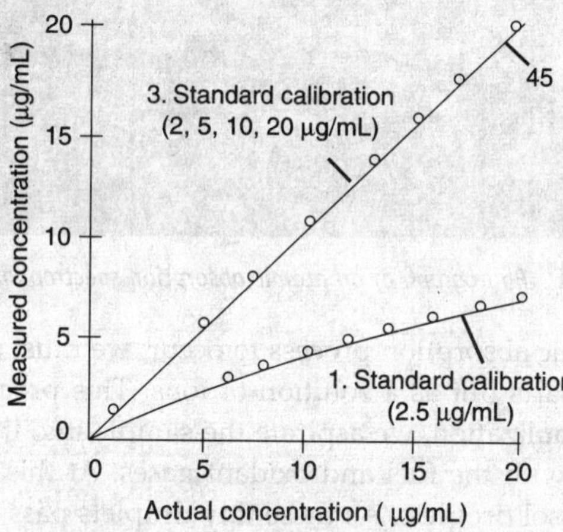

Fig. 9.20 *Sodium by flame emission model 560.*

One of the most dramatic benefits of the microcomputer age in atomic absorption has been the adaptability of a totally microcomputer-controlled instrument to automation. The Perkin Elmer Model 5000 is the first instrument to take full advantage of the technology for this purpose. After recalling prerecorded instrument conditions from a magnetic card into the instrument's internal program memory, the Model 5000 will sequentially analyze up to fifty samples for six different elements under the optimum conditions for each, all without operator's the attention.

9.5 CONTROL OF ANALYTICAL INTERFERENCES

9.5.1 The Flame Process

Atomic absorption is known as a very specific technique with a few interferences. The ultimate analytical method which is absolutely free of any interferences from the nature of the sample will probably never exist. The next best thing to not having interferences is to know what the interferences are and how to eliminate them or compensate for them. The interferences in atomic absorption are well-defined, as are the means for dealing with them. In order to thoroughly understand these interferences, we will examine what goes on in the flame process of atomic absorption.

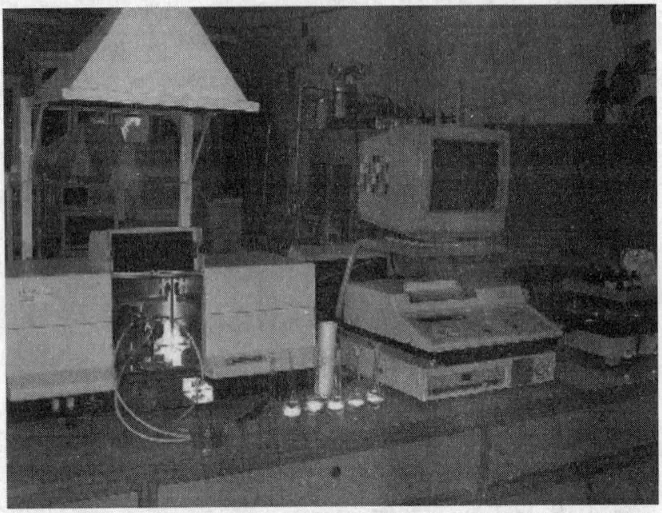

Fig. 9.21 *Photograph of an atomic absorption spectrophotometer.*

In order to get the atomic absorption process to occur, we must produce individual atoms from our sample which starts out as a solution of ions. This process is shown in Fig. 9.22. First, by the process of nebulization, we aspirate the sample into the burner chamber, where it mixes as a fine aerosol with the fuel and oxidant gases. At this point, the metals are still in solution in the fine aerosol droplets. As these tiny droplets pass into the heat of the flame, the process of evaporation or desolvation removes the solvent and leaves tiny solid particles of sample material. As more heat is applied, liquefaction will take place and, of course, additional heat will vaporize the sample. At this point the metal of interest, called the analyte, is bound up with some anion to form a molecule which does not exhibit the atomic absorption phenomenon we wish to measure. By applying still more heat energy, this molecule is dissociated into the individual atoms which make it up.

Since the thermal energy from the flame is responsible for producing the absorbing species, flame temperature is an important parameter governing the flame process. Temperatures for some flames used in atomic absorption are listed in Table 9.2. Cooler flames are subject to more interference problems resulting from insufficient energy for complete atomization.

	$M^+ + A^-$	(Solution)
1. Nebulization ↓		
	$M^+ + A^-$	(Aerosol)
2. Desolvation ↓Δ		
	MA	(Solid)
3. Liquefaction ↓Δ		
	MA	(Liquid)
4. Vaporization ↓Δ		
	MA	(Gas)
5. Atomization ↓Δ		
	$M^* + A^*$	(Gas)
6. Excitation ↓Δ		
	M^*	(Gas)
7. Ionization ↓Δ		
	$M^+ e^-$	(Gas)

Fig. 9.22 *Flame process.*

While the air-acetylene flame is satisfactory for the majority of elements determined by atomic absorption, the hotter nitrous oxide-acetylene flame is required for many refractory-forming elements, and nitrous oxide-acetylene is effective in interference control in other situations.

Table 9.2 *Temperatures of premix flames*

Oxidant-Fuel	°C
Air-Methane	1875
Air-Natural Gas	1700-1900
Air-Hydrogen	2000-2050
Air-Acetylene	2125-2400
N_2O-Acetylene	2600-2800

* J.A. Dean and T.C. Rains, Flame Emission and Atomic Absorption Spectrophotometry. Vol. 1, Marcel Dekker (New York) 1969.

The number of ground state metal atoms formed in step 5 of the flame process (Fig. 9.22) will determine the amount of light absorbed. Concentration is determined by comparing the absorbance of the sample to that of known standard concentrations. The relationship between the number of atoms in the flame and the concentration of analyte in solution is governed by the flame process. If any constituent of the sample alters one or more steps of this process from the performance observed for standards, an interference will exist, and an erroneous concentration measurement will result if the interference is not recognized. There are three areas where this process is vulnerable to interference.

9.5.2 Matrix Interference

The first place in the flame process subject to interference is the very first step, the nebulization. If the sample is more viscous or has considerably different surface tension characteristics than the standard, the sample uptake rate or nebulization efficiency may be different between the sample and the standard. If samples and standards are not introduced into the process at the same rate, it is obvious that the number of atoms in the light beam and, therefore, the absorbance will not correlate between the two. Thus, a matrix interference will exist.

An example of this type of interference is the effect of acid concentration on the absorbance. From Figure 9.23, it can be seen that as phosphoric acid concentration increases, the sample viscosity increases slowing the introduction rate and reducing the sample absorbance. Increased acid or dissolved solid concentration would lead to negative error if not recognized. Matrix interferences can also cause positive error. The presence of an organic solvent in a sample will produce enhanced nebulization efficiency, resulting in increased absorption. One way of compensating for this type of interference is to match as closely as possible the major matrix components of the standard to those of the sample. Any acid or other reagent added to the sample during preparation should also be added to the standards in similar concentrations.

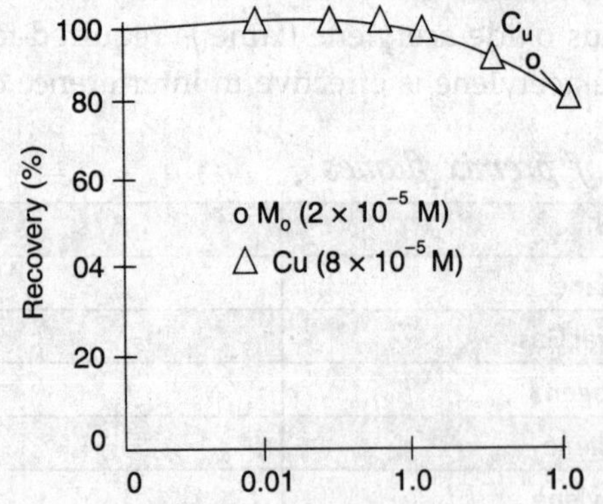

Fig. 9.23 *Matrix interference from viscosity effects.*

9.5.3 Chemical Interference

A second place where interference can enter into the flame process is in step number 5 of the atomization process. In this step, sufficient energy must be available to dissociate the molecular form of the analyte to create free atoms. If the sample contains a component which forms a thermally stable compound with the analyte which is not completely decomposed by the energy available in the flame, a chemical interference will exist.

The effect of phosphate on calcium, illustrated in Fig. 9.24, is an example of chemical interference. Calcium phosphate does not totally dissociate in an air-acetylene flame; and therefore, as phosphate concentration is increased, the absorbance due to calcium atoms decreases. There are two ways of dealing with this problem. One is to eliminate the Interference by adding an excess of another element which will also form a thermally stable compound with the interferent. In the case of calcium, lanthanum is added to tie up the phosphate and free the calcium to be atomized, making the calcium-absorbance independent of the amount of phosphate.

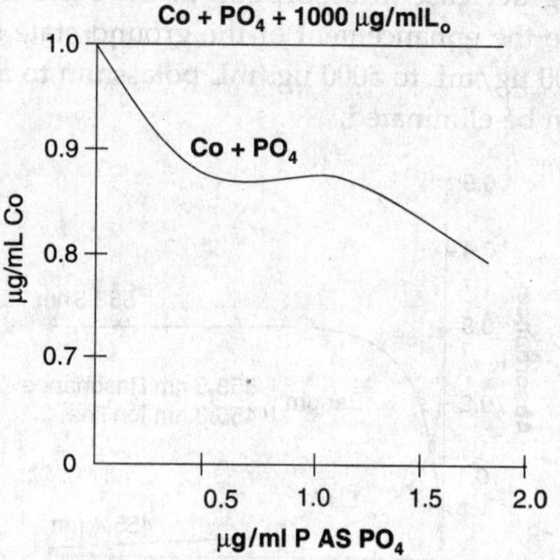

Fig. 9.24 *Interference of phosphate on calcium.*

There is a second approach to solve the chemical interference problem. Since the problem arises because of insufficient energy to decompose a thermally stable analyte compound, the problem can be eliminated by increasing the amount of energy, that is, by using a hotter flame. The nitrous oxide-acetylene flame is considerably hotter than air-acetylene and can often be used to minimize chemical interferences for elements generally determined with air-acetylene. The phosphate interference on calcium, for instance, is not observed in nitrous oxide-acetylene, eliminating the need for the addition of lanthanum.

9.5.4 Ionization Interference

There is a third major interference, however, which is often encountered in hot flames. As illustrated in Fig. 9.3, the dissociation process does not necessarily stop at the ground state atom. If additional energy is applied, the ground state atom can be thermally raised to the excited state; or if thermal energy is sufficient, the electron may be totally removed from the atom, creating an ion. As these electronic rearrangements deplete the number of ground state atoms available for light absorption, atomic absorption at the resonance wavelength is reduced. When an excess of energy destroys the ground state atom, ionization interference exists.

Ionization interference can be eliminated by adding an excess of an element which is also very easily ionized, creating a large number of free electrons in the flame and suppressing the ionization of the analyte; sodium or potassium is commonly used as an ionization suppressant. Figure 9.25 shows ionization suppression for the determination of barium in a nitrous oxide-acetylene flame. The increase in absorption at the barium resonance line and the corresponding decrease in absorption at the barium ion line as a function of added potassium, illustrate the enhancement of the ground state species as the ion form is suppressed. By adding 2000 μg/mL to 5000 μg/mL potassium to all standards and samples, the effects of ionization can be eliminated.

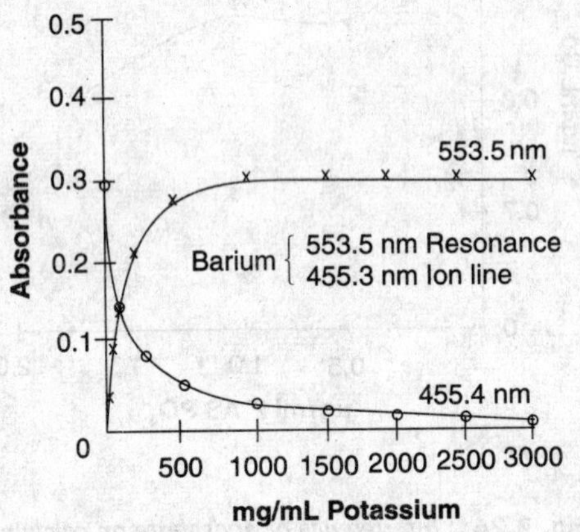

Fig. 9.25 *Ionization interference effect of added potassium.*

9.5.5 Method of Standard Additions

Three major interferences (matrix chemical and ionization and ways of eliminating them have been discussed. There is a very useful technique which often makes it possible to work in the presence of an interference without eliminating the interference itself, and still make an accurate determination of analyte concentration. The technique is called the method of standard additions. Accurate determinations are made without eliminating interferences by making the concentration calibration in the presence of the interference. The standards are added to portions of the sample, thereby letting any interferent present in the sample also affect the standard.

The technique is illustrated in Fig. 9.26. The solid line passing through the origin represents a typical calibration line for a set of aqueous standards. For a water blank, zero absorbance is defined, and as the concentration of the analyte increases, a linear increase in absorbance is observed.

Let us now take a sample and split it into three portions. Nothing is added to the first portion; different amounts of standard are added to portions 2 and 3. Finally, all portions are diluted to the same volume so that the final concentrations of the original sample constituents are the same in each case. Only the amount of added analyte differs by a known

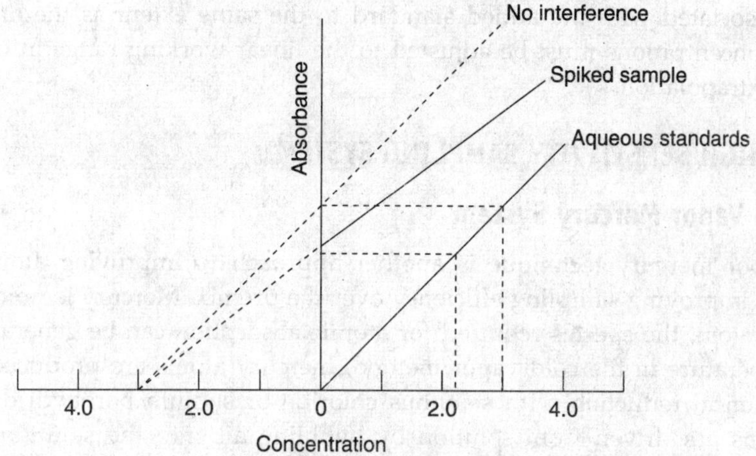

Fig. 9.26 *Method of standard additions.*

amount. If no interference were present in this sample, a plot of measured absorbance versus the concentration of added standard would be parallel to the aqueous standard calibration and offset by an absorbance value resulting from the analyte present in the unspiked sample. If some material is present in the sample causing either a matrix, chemical, or ionization interference, the number of ground state atoms producing atomic absorption will be reduced. The absorbance from the analyte in the unspiked sample will, therefore, be reduced by this interference. However, the absorbance increase from added standard will also be reduced by the same proportional amount, since the concentration of interferent is the same in each solution. Therefore, a straight line will still result, but because of the interference, its slope will be different from that observed for the aqueous standards.

In this situation, if the absorbance of the unspiked sample were to be compared directly to the aqueous calibration, negative error would result. If, however, the slope determined by the standard additions to our sample is used as the calibration slope, an accurate determination of the sample concentration can still be made. By continuing the concentration calibration on the abscissa backward from zero and extrapolating the calibration line backward until it intercepts the concentration axis, the concentration responsible for the absorbance of the unspiked sample is indicated. An accurate determination has been made by calibration in the presence of the interference.

The method of standard additions is an extremely valuable tool in atomic absorption. The presence of an interference can be confirmed by observing the slope of the spiked sample calibration and determining whether or not it is parallel to the aqueous standard line. If it is not, an interference is present. Secondly, if an interference is present, the method of standard additions may allow an accurate determination of the unknown concentration by using the standard additions slope for the calibration. Caution should be used with the technique, however, as it can fail to give correct answers if the interfering species does not become chemically associated with the added standard to the same extent as the original analyte. Further, all concentrations must be adjusted to the linear working range in order to assure an accurate extrapolation.

9.6 HIGH SENSITIVITY SAMPLING SYSTEM

9.6.1 Cold Vapor Mercury System

The cold vapor mercury technique is another approach to improving atomic absorption sensitivity by improving sampling efficiency over the premix. Mercury is unique in that, the ground state atom, the species required for atomic absorption can be generated chemically at room temperature in the cold vapor method, mercury atoms are produced in a reaction bottle by chemical reduction with stannous chloride or sodium borohydride. The volatile mercury atoms are driven from solution by bubbling air into the solution. The air then carries the mercury through a drying tube to take out water vapor and into an absorption cell aligned in the optical path of the atomic absorption spectrophotometer. The absorption cell is sometimes heated in order to eliminate the drying tube, which requires frequent changes of desiccant. As the mercury passes through and out of the cell, a peak-shaped absorption signal is recorded. On some flameless mercury systems, provision is made for recirculating the vapors to allow opportunity for the maximum mercury concentration to be achieved and therefore make the system more sensitive. In this case, the signal would be plateau-shaped. When a constant reading is achieved in this type of system, the vapors are diverted through an activated charcoal scrubber, which takes the mercury out of this system in preparation for running the next sample. This approach to low-level mercury determination is still the best available for making high-sensitivity mercury determinations in samples of a complex nature. Its high sensitivity arises from the fact that all the mercury present in upto 100 milliliters of the sample can be concentrated into the optical path to create the maximum absorbance.

9.6.2 Hydride Generation System

The technique of hydride generation is somewhat similar to the cold vapor mercury sampling system. In hydride generation, gaseous hydrides of certain metals are chemically produced by the addition of sodium borohydride to up to 25 milliliters of sample in a reaction flask. The flask is sealed off, and the gaseous hydrides and hydrogen produced by the reaction are swept by an argon purge into an argon-hydrogen entrained air frame or, alternately, into a heated quartz cell. Both manual and automated equipment for hydride generation are available.

When the sample vapor is atomized in the flame or heated cell, a peak signal is produced, the height of which is proportional to the amount of the analyte in the sample. Figure 9.27 shows absorbance tracings obtained for bismuth by the hydride generation technique. Detection limits for the elements determinable by hydride generation atomic absorption are given in Table 9.3. The extremely low submicrogram-per-liter values are again the result of very efficient atom production for a large volume of sample over a short period of time.

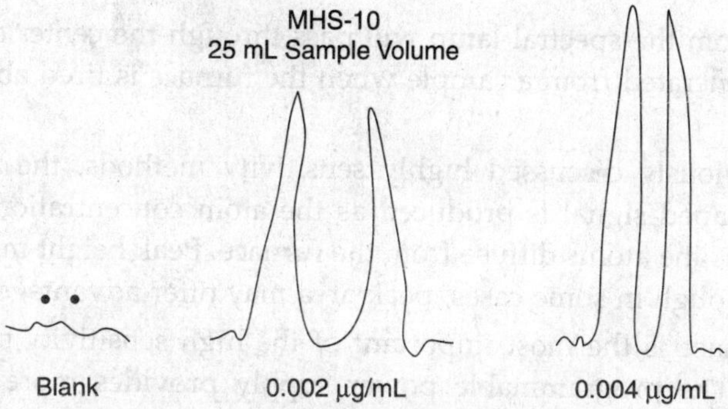

Fig. 9.27 *Bismuth.*

Table 9.3 *Hydride generation technique*

Element	*Absolute Detection Limit (µg)*	*Solution Detection Limit* (µg/liter)*
As	0.003	0.15
Bi	0.005	0.25
Ge	0.200	10.00
Sb	0.005	0.25
Se	0.003	0.15
Sn	0.004	0.20
Te	0.005	–0.25

*Based on 20-mL sample solution

The elements determinable by gaseous hydride production are important environmental and toxicological concerns, and the impressively low detection limits make the method attractive for these applications. While the technique is subject to a variety of interferences, hydride generation is capable of analyzing many samples of a complex nature which are difficult or currently impossible to do by any other method. These facts make hydride generation atomic absorption an important and widely used analytical technique.

9.6.3 Graphite Furnace Atomic Absorption

The search for ways to improve the sensitivity of atomic absorption led ultimately to the graphite furnace. With this device many elements can be determined at concentrations 1000 times lower than what can be detected by flame atomic absorption. The extreme sensitivity of graphite furnace atomic absorption makes it ideal for applications in ultra trace analysis. In addition, since the graphite furnace requires only a few microliters of sample, this technique can be used in situations where only a small amount of sample is available.

The graphite furnace is a flameless sampling device. The energy required for atomization is supplied passing electrical current through a graphite tube into which the sample has been placed. The furnace is aligned in the sampling area of the spectrophotometer in such a way that the light from the spectral lamp will pass through the center of the graphite tube. The atomic vapor generated from a sample when the furnace is fired absorbs light from the lamp.

As with the previously discussed highly sensitivity methods, the absorption signal is transient. A peak-shaped signal is produced as the atom concentration within the furnace rises, and then falls as the atoms diffuse from the furnace. Peak height may generally be used for quantitation, although in some cases, peak area may offer advantages.

The graphite furnace is the most important of the high sensitivity sampling devices for atomic absorption. The programmable power supply provides more versatile and more precise control over heating conditions than any of the earlier devices. The graphite furnace is generally more convenient to use than other high sensitivity sampling accessories. It is applicable to a wide variety of sample types, and in some cases, solid samples can be analyzed directly without prior dissolution. Detection limit capabilities lie in the low picoqram (10^{-12} gram) region, for most elements, and the technique is capable of determining 64 elements.

The ability to use a variety of sampling techniques makes atomic absorption an extremely versatile analytical tool. Table 4.2 lists the elements which are determinable by atomic absorption with the sampling systems which have been discussed.

9.7 APPLICATIONS

Atomic absorption spectroscopy is now a well established technique for the determination of trace elements covering a wide range of analyte types. Spectrometry is the spectroscopic technique used to assess the concentration or amount of a given chemical (atomic, molecular, or ionic) species. In this case, the instrument that performs such measurements is a spectrometer, spectrophotometer, or spectrograph. Spectroscopy/spectrometry is often used in physical and analytical chemistry for the identification of substances through the spectrum emitted from or absorbed by them.

EPA method for metal analysis

Flame atomic absorption methods are referred to as direct aspiration determinations. They are normally completed as single element analyses and are relatively free of interelement spectral interferences. For some elements, the temperature or type of flame used is critical. If flame and analytical conditions are not properly used, chemical and ionization interferences can occur.

Graphite furnace atomic absorption spectrometry replaces the flame with an electrically heated graphite furnace. The major advantage of this technique is that the detection limit can be extremely low. It is applicable for relatively clean samples, however, interferences could be a real problem. It is important for the analyst to establish a set of analytical protocols which is appropriate for the sample to be analyzed and for the information required. Table 9.4 lists the available methods for different metal analyses provided in EPA manual SW-846.

Table 9.4 *EPA methods for determination of metals by direct aspiration*

Analyte	Method Number	Analyte	Method Number	Analyte	Method Number
aluminum	7020	antimony	7040	barium	7080A
beryllium	7090	cadmium	7130	calcium	7140
chromium	7190	cobalt	7200	copper	7210
iron	7380	lead	7420	lithium	7430
magnesium	7450	manganese	7460	molybdenum	7480
nickel	7520	osmium	7550	potassium	7610
silver	7760A	sodium	7770	strontium	7780
thallium	7840	tin	7870	vanadium	7910
zinc	7951				

The concentrations of La, Pr, Nd, Sm, Eu, Gd, Tb, Dy, Ho, Er, Tm, Yb and Y can be determined in rare earth ores and concentrates by the flame atomic absorption method, those of Eu, Gd, Dy, Ho, Er, Tm, Yb and Y also by the flameless method with a graphite cell. Accuracy can be checked by using a standard lanthanides mixture. The results of the determinations in concentrates (by flame and by furnace AAS, both direct measurement and standard addition method) may be compared with those obtained by spectrophotometry. Detection limits were defined in a HGA-70 graphite cell with application of a standard graphite tube. The results of the examinations revealed that both the AAS methods are suitable for lanthanide determinations in the above materials. Heavy lanthanides can be determined in lower concentrations than by spectrophotometry.

9.8 REFERENCES

1. B. Welz, M. Sperling (1999), Atomic Absorption Spectrometry, Wiley-VCH, Weinheim, Germany, ISBN 3-527-28571-7.
2. A. Walsh (1955), The application of atomic absorption spectra to chemical analysis, Spectrochim. Acta 7: 108–117.
3. J.A.C. Broekaert (1998), Analytical Atomic Spectrometry with Flames and Plasmas, 3rd Edition, Wiley-VCH, Weinheim, Germany.
4. B.V. L'vov (1984), Twenty-five years of furnace atomic absorption spectroscopy, Spectrochim. Acta Part B, 39: 149–157.
5. B.V. L'vov (2005), Fifty years of atomic absorption spectrometry; J. Anal. Chem., 60: 382–392.
6. H. Massmann (1968), Vergleich von Atomabsorption und Atomfluoreszenz in der Graphitküvette, Spectrochim. Acta Part B, 23: 215–226.
7. W. Slavin, D.C. Manning, G.R. Carnrick (1981), The stabilized temperature platform furnace, At. Spectrosc. 2: 137–145.

8. B. Welz, H. Becker-Ross, S. Florek, U. Heitmann (2005), High-resolution Continuum Source AAS, Wiley-VCH, Weinheim, Germany, ISBN 3-527-30736-2.
9. H. Becker-Ross, S. Florek, U. Heitmann, R. Weisse (1996), Influence of the spectral bandwidth of the spectrometer on the sensitivity using continuum source AAS, Fresenius J. Anal. Chem. 355: 300–303.
10. J.M. Harnly (1986), Multielement atomic absorption with a continuum source, Anal. Chem. 58: 933A–943A.

CHAPTER

10 Gas Chromatography

10.1 INTRODUCTION

Chromatography is a collective term for separation of mixtures based on the partition of substances between two immiscible heterogeneous phases, one of which is a stationary or fixed phase with a large surface area, and the other a moving or mobile phase which flows over the first phase. Gas chromatography is the most recent branch of chromatography and includes chromatographic processes in which the substance to be analyzed occurs in the gaseous or vapor state or can be converted into such a state.

Gas chromatography is an analytical technique which separates, identifies, qualitates and quantitates the components in a mixture. Considering its tremendous capability, gas chromatography is perhaps the simplest and least expensive among other analytical techniques, due to its simplicity of operation, high separation power and speed. The technique is capable of separating and measuring nanogram amount of substances. In general, gas chromatography is suitable for analysis of substances with vapor pressures of at least 10 millimeters of mercury at the temperature of the column. The wide choice of column packings, detectors and temperature controls allows versatile applications not only to the field of chemistry but also biology and many others.

To assess GC is as proper analytical tool, one must understand, how to achieve the best separation and identification of component mixtures in the sample with reasonable precision, accuracy and speed. Therefore, it is important to examine the type of components that are to be analyzed and certain physical and chemical properties of these compounds. The information regarding the structure and the functional groups, elemental composition, the polarity in the molecule, its molecular weight, boiling point and the thermal stability are very helpful for achieving the best analysis.

10.2 HISTORICAL DEVELOPMENT

Although the first records of gas chromatography go back hundreds of years, its true history began during World War II when a large industrial chemical company instituted a crash program for its development. The first published work appeared in the early 1950s based on the successful experiments by James and Martin, following an earlier suggestion made by Martin and Synage in 1941. In the years between 1952 and 1956, the early apparatus and initial methods of application were developed. In 1956, the first commercial instruments appeared in the market and since then there has been a spectacularly rapid and widespread development in theory, techniques and applications of gas chromatography. Today, it is one of the most widely applied and versatile analytical tools available in basic and applied research and quality control.

10.3 PRINCIPLES OF OPERATION

Gas chromatography is a process by which a mixture is separated into its constituents by a moving gas phase passing over a sorbent. Gas chromatography is divided into two major categories; Gas Solid Chromatography (G.S.C.) where the sorbent is a solid of large surface area and Gas-Liquid-Chromatography (G-L-C) where the sorbent is a non-volatile liquid coated on an inert solid support. The moving gas phase or carrier gas is an inert gas which is made to flow at a constant rate through a packed column in a small diameter tube, containing the sorbent.

When a sample containing a mixture of components is injected into the gas stream, the components move through the column at rates dependent on their respective volatilities and/or interaction with the non-volatile liquid phase. The various molecules tend to dissolve and revaporize as they pass down the column. The molecules of the component having the great solubility or greater interaction with the liquid phase are retarded in their passage. With a column packing of suitable properties and of sufficient length, the partition process between the gas and liquid phases results in a situation in which each component of the sample emerges in time, from the end of the column, as a binary mixture with the carrier gas. In G-S-C, absorptive processes on the solid phase perform the analogous function in retarding the various components. Activated alumina, molecular sieve, silica gel and activated charcoal are examples of materials used in G-S-C.

The experimental procedure for obtaining a gas chromatogram is relatively simple. With a hypodermic syringe, a few microliters of the organic liquid are injected rapidly through a rubber diaphragm in a heated sample port into a carrier gas stream. With a suitable differential detector at the exit of the column, the signal obtained is proportional to the instantaneous concentration of the dilute component in the binary gas mixture. The result on the chart of an electronic recorded is a chromatogram with a series of peaks for various components.

It is very simple to perform the GC analysis of component mixtures. The efficiency of the chromatographic system can be determined from the number of the theoretical plates per meter. Although this term primarily describes the property and resolution efficiency of a column, other extra column variables, such as the detector, inert, injection technique and the carrier gas velocity that can also affect the theoretical plates. It is calculated from the following equation:

$$\text{Theoretical plates/m} = 5.54\,\frac{\left(\dfrac{t}{w}\right)^2}{L}$$

Theoretical plates/m = ____________

where t = retention time of the test compound

W = width of the peak at half height, and

L = length of the column (m).

The number of theoretical plates also depends on the partition ratio of the test compound and its solubility in the liquid phase. Substances that have higher K values have lower plate numbers. Greater plate numbers indicate great resolution or better separation of the component mixture. To achieve better results, use an appropriate column and a proper detector. The components of gas chromatography are described below:

10.4 COMPONENTS OF GAS CHROMATOGRAPHY

Gas chromatographic equipment consists of the following:

1. Carrier gas System
2. Sample Injection System
3. Column Oven
4. Detectors
5. Temperature Control
6. Recording Device
7. Data Processing for Qualitative and Quantitative Analysis.

10.4.1 Carrier Gas System

The carrier gas is the mobile phase used to transport the sample through the column at a selected steady rate. To ensure constant and reproducible conditions, the system is composed of a gas cylinder, pressure and/or flow control, manometer flow meter and pre-heater. The gas which does not interfere either with the statutory phase or the components of the sample is suitable as a carrier gas. The gases generally used are helium, argon, hydrogen and nitrogen. In selecting a carrier gas, detection is the primary factor to be considered since separation can be improved by some means other than changing the carrier gas (Table 10.1).

A constant flow rate is important to eliminate the effect of changes in column resistance. Heating of the gas before it enters the sample injector is necessary in the case of detectors and advisable in other detection methods and is accomplished by means of a pre-heater.

10.4.2 Sample Injection System

The sample injector, located ahead of the block which is heated independently to a temperature approximately 50 to 100 degrees above the column temperature. The solvent and sample are therefore; flash evaporated when injected, without changing the gas flow or the thermal conditions of the column. The quantity of liquid sample ranges from 1 to 10 microliter and gas sample vary from 0.05 to 50 milliliters. Most samples are introduced by means of a small calibrated syringe or a microsyringe, such as the Hamilton Syringe. Samples may also be

Table 10.1 *Carrier gas-selection criteria*

Detector	*Carrier Gases*	*Function*
Thermal Detector	Helium, Hydrogen	The thermal detector measures the difference between the heat conductivity of the pure carrier gas and that containing the components of the sample. The greater the difference between the two heat conductivities, the greater will be the observed signal. The thermal conductivity of gases is immensely proportional to the square root of their molecular weight and consequently hydrogen and helium are the most suitable carrier gases for thermal conductivity detection. Since hydrogen is so reactive and flammable, helium is the ideal carrier for use with thermal conductivity detectors.
Flame Ionization and Thermal Detectors	Air and Oxygen	Their use is limited by the possibility of reactions with the statutory or the components of the sample.
Flame Ionization Detectors (Beta ionization)	Argon	Argon is the most generally used gas with radiation detectors such as Beta ionization, and with flame ionization detectors, with limited use in thermal detectors.
Flame ionization (Gas density balance detectors)	Carbon dioxide	CO_2 gas is used with flame ionization and gas density balance detectors, with limited use in thermal detectors.
Flame Ionization Detector	Carbon monoxide	Carbon monoxide is used with the flame ionization detector.
Gas density and thermal conductivity detectors, flame ionization detector	Hydrogen	Hydrogen is used with the gas density and thermal conductivity detectors. It is used as a fuel with the flame ionization detector.
Radiation Detector	Neon	Neon is used with radiation detector.
Flame Ionization and Gas Density Detectors	Nitrogen	Nitrogen is used with radiation flame ionization and gas density detectors. It has limited use with the thermal conductivity detector. Nitrogen, though readily available, is not too useful for these detectors since its thermal conductivity does not differ too greatly from that of many sample components.

introduced to the column using a gas sampling value. The gas sample may be introduced to the gas sampling even under pressure or by drawing the sample into the loop using a small vacuum pump or a two-way squeeze bulb. A third method of injection is by the use of a sealed ampule which is broken inside a chamber, which replaces the gas sampling value so that the sample is swept into the column.

10.4.3 Column

The heart of chromatography is the column containing the packing on which separations are effected. A wide variety of suitable liquid phases and absorbents provide a large degree of selectivity.

Column packings are either a finely divided adsorbent G-S-C or a finely divided solid support on which a thin layer of liquid adsorbed G-L-C. Adsorbents most frequently used are alumina, silica gel or activated carbon in particle size ranging from 20 to 120 meshes. These are used for gas mixtures and mixtures of low-boiling, relatively non-polar compounds, such as hydrocarbons. Solid-liquid packings are most effective for higher boiling liquids. The most common supports of the liquid phase are "celite" or crushed fire brick. The liquid phase is placed on the solid support by simply mixing the desired proportions or by adding the liquid as a solution in a volatile solvent which is subsequently removed by vacuum filtration.

The basic requirements of the liquid phase are that it has a low vapor pressure at the temperature used and that it provides an adequate separation of the components of the mixture to be analyzed. The first qualification can be met if the liquid has a boiling point approximately 200°C above the temperature at which it is to be used.

At high temperatures the liquid phase expansion reduces the free volume of the column and the viscosity of the carrier gas increases. Thus, the pressure drop increases and the maintenance of a given carrier gas flow requires a higher inlet pressure. At higher temperatures the retention time or volume for a solute tank is reduced.

There are two basic types of columns – the packed column and the capillary or open tubular column. The packed column consists of length of tubing 1-6 millimeters I.D. and usually 0.5 to 6 meters in length, which is normally coiled or looped to be accommodated in the space provided in the instrument. This tubing is packed with a finely divided inert-solid support, which is coated with a thin layer of a non-vaporizable liquid, referred to as the liquid phase. The capillary column is an empty tube, the inner walls of which serve as the support or adsorbent. This type of column has an I.D. 0.1 – 1.0 millimeter and usually is 30 – 100 meters in length.

The basic difference between gas-liquid and gas-solid chromatography, is the column packing. In gas-liquid chromatography the column is packed with a solid material made up of particles on which is deposited a known volume of non-volatile liquid constituting the liquid phase. In gas-solid chromatography the packing is generally an active solid (adsorbent) which separates the sample components by differences in their adsorbent characteristics.

10.4.4 Detectors

Detectors must sense continually, rapidly and with high sensitivity the components which appear in the carrier gas as it emerges from the column, by means of changes in physical or chemical property of the effluent gas stream. The corresponding electrical response is amplified and fed to a recorder. One of the chief factors of the widespread application of gas chromatography is the availability of a great variety of highly efficient detectors. The detectors functions are given below (Table 10.2).

Table 10.2 *Detectors and their functions*

Detectors	*Functions*
Thermal Conductivity	Measures change in heat capacity.
Gas Density	Measures change in density.
Flame Ionization	Measures difference in flame ionization due to combustion of the sample.
Beta Ray Ionization	Measures current flow between two electrodes caused by ionization of the gas by a radioactive source.
Photo Ionization	Measures current flow between two electrodes caused by ionization of the gas by ultraviolet radiation.
Glow Discharge	Measures the voltage change between two electrodes caused by the change in discharge by different gas compositions.
Flame Temperature	Measures change in temperature caused by difference in gas composition in the flame.
Dielectric Constant	Measures the change in the dielectric constant caused by difference in composition of gas between pleates of a capacitor.
Thermal Conductivity Detector	The most widely used detector. Uses either the hot wire or thermistor as the sensing element. The resistance change due to sample components, and the detector measures the thermal conductivity of the sample stream as opposed to the reference stream of the pure carrier gas. The response is approximately proportional to the concentration of the sample component in the detector.
Flame Ionization Detector	Uses hydrogen air or hydrogen/oxygen flame. This flame ionizes the organic sample, material and the ions are collected by an electrode which is positive in relation to the flame. This electrical potential causes a current flow which is an instantaneous measurement of the component concentration. This detector has a high sensitivity about 500 – 2000 times that of the thermal detector. It also has a fast response time, a very small effective cell volume and a high signal-to-noise ratio.
Electron Capture Ionization Detector	This detector utilizes an ion chamber containing a gas with fine electrons at an applied potential just great enough to collect completely the free electrons generated by a radio active source. Molecules from the sample (which have an affinity for the free electrons) will capture the free electrons and become negative ions. The detector current decreases in the presence of the electron capturing molecules.

10.4.5 Temperature Control

Each instrumental component can have a separate control system or the temperature of the assembly may be controlled. For precise and producible gas chromatographic analysis there must be temperature control of (i) the injection system, (ii) the column, (iii) detector and (iv) fraction collector, if used. The injector system must be heated to a point that will volatilize the sample instantaneously and keep it in the vapor state until it reaches the column, which is also heated to assure that the sample components remain in the gaseous state for the passage through the separation column. The detector likewise must be heated to keep the sample components in the vapor phase. The temperature of each component part of the gas chromatograph must be precisely controlled and reproducible so that constant sample retention times may be attained. The programme temperature control must be very reproducible.

10.4.6 Recording Devices

The response of the detector is plotted as chromatogram by the millivolt recorder. Chromatogram is a plot of detector response versus time. With only carrier gas flowing through the detector, the recorder is adjusted to read zero. This zero reading is referred to as the baseline. Each separated component evolves a response by the detector which registers a peak on the chromatogram. The chromatogram will give the different kind of information:

1. identification by retention time, the time it takes for the peak to appear after injection of the sample, and
2. quantitative estimation of a component of the sample, which can be obtained by comparing the area of the peak with that produced by a standard sample of the same component substance at a known concentration.

There are two types of strip chart recorders in use – the galvanometric and potentiometric. Galvanometric recorders are inexpensive but require an amplifier. Potentiometric recorders are more expensive but do not always require an amplifier. Digital readout may be employed by using one of the several types of printing integrators, which merely print out numbers corresponding to the area under the chromatographic curve. There are several types of integrators, mechanical, electromechanical and electronic. The most generally used is the mechanical or disc integrator which measures mechanically the height of the curve. The electromechanical integrator converts the voltage change from the detector signal to a digital signal directly. The electronic integrator is similar to the electromechanical unit but uses a voltage totaling circuit instead of a direct signal.

10.5 GAS CHROMATOGRAPHY – WORKING

Gas Chromatography (GC) is used to separate volatile components of a mixture. A small amount of the sample to be analyzed is drawn up into a syringe. The syringe needle is placed into a hot injector port of the gas chromatograph, and the sample is injected. The injector is set to a temperature higher than the components' boiling points. So, components of the mixture evaporate into the gas phase inside the injector. A carrier gas, such as helium, flows through the injector and pushes the gaseous components of the sample onto the GC column. It is within the column that separation of the components takes place. Molecules partition between the carrier gas (the mobile phase) and the high boiling liquid (the stationary phase) takes place within the GC column.

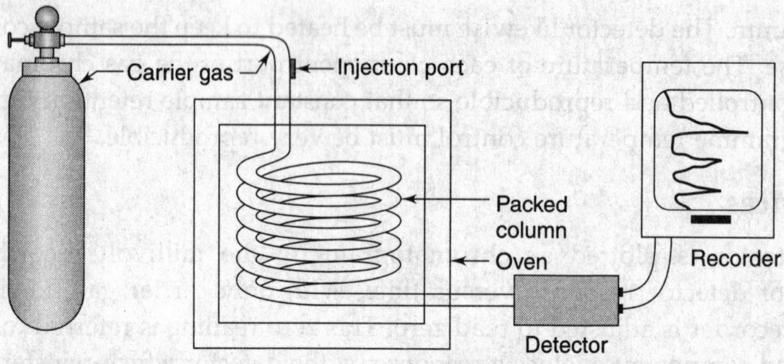

Fig. 10.1 *Single column.*
Source: *Instruments Inc: what is gas chromatography, Tulsa, Oklahoma*

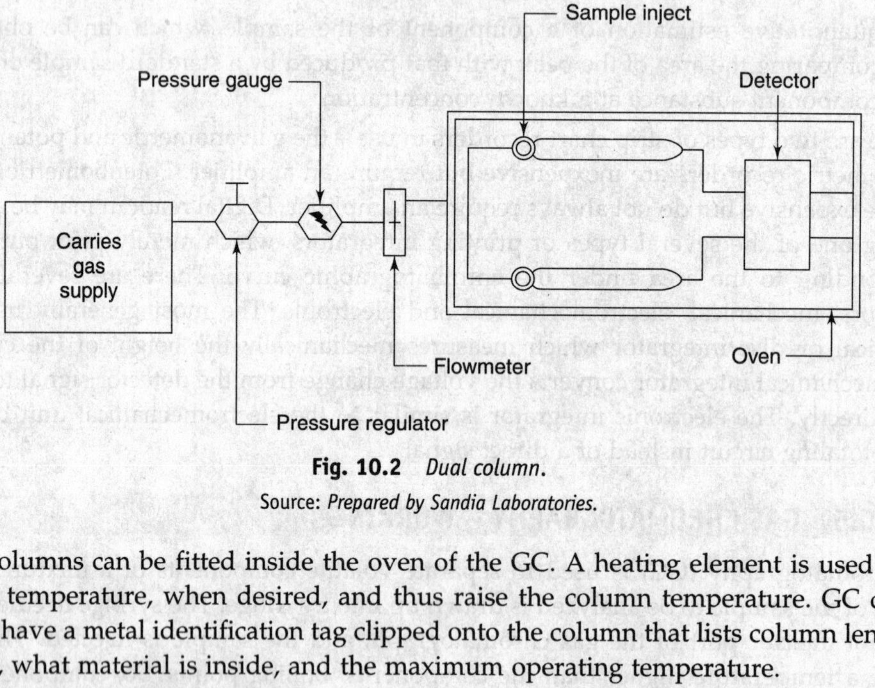

Fig. 10.2 *Dual column.*
Source: *Prepared by Sandia Laboratories.*

Two columns can be fitted inside the oven of the GC. A heating element is used to raise the oven temperature, when desired, and thus raise the column temperature. GC columns typically have a metal identification tag clipped onto the column that lists column length and diameter, what material is inside, and the maximum operating temperature.

After components of the mixture move through the GC column, they reach a detector. Ideally, components of the mixture will reach the detector at varying times due to differences in the partitioning between mobile and stationary phases. The detector sends a signal to the chart recorder which results in a peak on the chart paper. The area of the peak is proportional to the number of molecules generating the signal.

To use the GC, follow these simple steps:

1. Wash a syringe with acetone by filling the syringe completely and ejecting the waste acetone onto a paper towel. Wash 2-3 times.

2. Pull some of your sample into the syringe. That will most likely need to remove air bubbles in the syringe by rapidly moving the plunger up and down while the needle is in the sample. Usually, 1-2 *L of sample is injected into the GC. It is okay to have small air bubbles in the syringe. However, you do not want to inject mostly air or your peaks will be too small on the chart recorder.
3. Make sure the chart recorder is on and set to the appropriate chart speed (Arrow *A*). Set the baseline using the zero on the chart recorder (Arrow *B*). With the pen in place, turn on the chart (Arrow *D*), make sure the pen is down (marking the paper) and the paper is moving.
4. Inject your sample onto either column *A* or column *B* as instructed. Hold the syringe level and push the needle completely into the injector. Once you can no longer see the needle, quickly push the plunger and then pull the syringe out of the injection port.
   (a) The injectors are very hot, so be careful not to touch the silver disk.
   (b) The needle will pass through a rubber septum, so you should feel some resistance. For some of our GCs, the column does not align properly in the injector, so the needle hits the front of the metal column. If you feel that you are pushing against metal, pull the needle out of the injector and try again, perhaps at a slightly different angle. The needle should completely disappear into the injector for proper injection of the sample onto the GC column.
   (c) Inject quickly for best results. Do not hesitate to inject once the needle is properly positioned in the injection port.
   (d) Remove the syringe immediately after injection. (Carrying out notes C and *D* helps to insure that the entire sample enters the GC column at approximately the same time.)
5. Mark your injection time on the chart recorder. This can be done by adjusting the zero just after the sample is injected. It is often convenient for one person to inject the sample while a lab partner marks the injection time at the chart recorder.
6. Clean your syringe immediately after injection. Syringe needles often clog quickly and must be replaced if they are not cleaned after each use.
7. Record the settings of your chart recorder during a run. You need to know the chart speed and the full-scale setting.
8. Record the settings of your GC during a run. A knob on the bottom center of the GC can be turned to read column (or oven) temperature, detector temperature, and injector port temperature in °C. The bridge current is displayed in mA. Note that there are two scales on the display. Be careful to read the appropriate scale.

10.6 ANALYSIS OF THE GAS CHROMATOGRAPH

Report the retention time of each peak (in minutes), and identify each component in the mixture, and the percent composition of the mixture. To determine the percent composition, you will first need to find out the area under each curve.

Area = (height) × (width at 1/2 height)

Mark retention time, height, half-height, and width at 1/2 height on your GC trace. Show your calculations either in your final report or directly on the chromatograph.

You may assume that each component of the mixture causes the same response in the detector. Therefore, the areas under the curves can be used to calculate percent composition of the mixture of alkenes. (This is a reasonable assumption when the components of the mixture are very similar in structure, as are 2-methyl-1-butene and 2-methyl-2-butene.)

% Component 1 = [(area under peak 1)/(total area)] × 100%.

10.7 CALIBRATION

A calibration standard curve is required to be prepared by running standards, prior to the analysis of the unknown samples. There are two ways in which calibration is performed: external standard method and internal standard method. External standard method involves preparation of a calibration curve by plotting area or height response against concentrations of the analyte(s) in the standards. The calibration factor is then calculated as the ratio of concentrations to area/height response and should be constant over a wide range of concentrations. To determine the concentration of the analyte in the unknown sample (extract), the response for the unknown should be compared with that of the standards within the linear range of the curve. Alternatively, an average of response ratio may be calculated which is compared with the response of the analyte. A single point calibration may be used if the area/height response of the analyte is within 20% of the response of the standard.

The internal standard method is more reliable than the external standard method. Equal amounts of one or more internal standards are added onto equal volumes of sample extracts and the calibration standards.

In case of external standard calibration, the area/height response for the analyte peak is compared with that of the standard from the calibration curve or from the calibration factor.

For aqueous samples, the concentration of the analyte is usually expressed in ug/L. All concentration terms including those of the calibration standards and internal standards must be in the same unit. For non-aqueous samples, the concentration of the analyte is calculated in the same way except that weight of the sample W is substituted for the volume of the sample.

10.8 QUALITATIVE AND QUANTITATIVE ANALYSIS: OPERATION CONDITIONS

A number of important points in the operation of a gas chromatograph are:

1. **Selection of Column:** Selection of the optimal column material for the analysis of gas or liquid samples.
2. **Operating temperature:** It must be low enough to separate the low boiling components but high enough to cause the high boiling fractions to come off the

column in a reasonable time. Temperature programme is the only answer to some of these problems. With a mixture of high and low boiling components it is imperative that the temperature is programmed.

3. **Detector:** Proper detector is required for the analysis.

Both qualitative and quantitative analysis can be accomplished on this instrument. In order to be able to duplicate a run at some later date it has been suggested that a definite amount of some known foreign material be added to the sample. The elution time of this material is established and all the components in the mixture are recorded relative to the known material. This allows one at some later date to reproduce the elution time for this added known material which will then allow the analytical parameters to be duplicated. Qualitative analysis can be accomplished by a number of methods. Using peak area, peak height, peak height ×½ peak width, integration by electronic means or any other method that can provide a linear relationship with the concentration.

10.8.1 Qualitative Identification by GC

Based on the discussions given in the preceding sections, one can arrive at an idealized scheme for identification by GC. This is outlined below:

Step I: Retention behavior on columns of different polarity

Step II: Sample modification (a) abstract functional groups, (b) form derivatives to test for the functional group and to obtain more retention information, (c) obtain chromatograms of degradation products.

Step III: Approximate molecular weight using gas density detector

Step IV: Elemental composition from selective/specific detection systems

Step V: Off-line analysis of trapped peaks

Step VI: Use of hyphenated techniques for further characterization

10.8.1.1 Multidimensional chromatography (MDC)

In MDC, different separation mechanisms or systems are coupled together. MD methods are enormously powerful. Giddings has defined MD separation by two conditions.

(i) It is one in which the components of a mixture are subjected to two or more separation steps (mechanisms), in which their displacements depend on different factors.

(ii) When two components are substantially separated in any single step, they always remain separated until the completion of the separative operation. This latter condition rules out simple tandem column arrangements in which components separated in the first column may re-merge in the second. As a consequence of the above-mentioned definition, two or more parameters (e.g., successive t_R) are required to describe the final location of a component after a MD separation. The fact that a difference in only one parameter is needed for success is largely responsible for the unique strength of MD systems.

In spite of all these developments in qualitative identification, it must be conceded that some misidentifications of peaks have occurred (and may continue to occur as well), while analyzing complicated mixtures and environmental samples.

10.8.2 Quantitative Analysis of GC

Quantitative GC is a method comprising the separation, on a GC column, of a n-component mixture to produce "*n*" binary solute mobile phase mixture, and the on-line determination of the solutes in these mixtures by the detector.

Quantitative analysis by GC can be viewed as a self-contained analytical discipline, having its own consistent theory. One may visualize two regions of quantitative GC.

(a) The GC system proper
(b) The nonchromatographic part of the analytical system. This comprises the facilities associated with integration and other data handling operations. Their theoretical background lies in the domain of computerization and other data handling approaches.

Quantitative analysis using GC is based on calibration curves obtained for each of the substances analyzed. Calibrations can usually be made using peak height (for narrow and symmetrical peaks), and peak area in other cases. A large number of papers have shown that the basic principles of quantitation are well established and that the limitations in quantitating chromatographic results are in sampling, sample preparation, lack of specificity at high sensitivities, and the accuracy and validity of calibration standards.

Calibration is obviously needed in all those cases in which a signal (which does not come in the category of fundamental measurements such as those of mass and volume), related to the mass or concentration of a component in a mixture, is obtained. The need to prepare standard mixtures by accurate, reproducible and simple procedures cannot, therefore, be overemphasized. As instrumentation with lower detection limits continues to be developed to monitor lower concentrations of components (as in environmental samples), so calibration technology has no choice other than following these trends. Many authoritative reviews have appeared in the literature dealing with the preparation and use of calibration standards.

The important factors which are to be taken into account to obtain reliable quantitative analyses are:

1. Careful sampling and sample preparation
2. Precise sample injection
3. Appropriate choice of the column and operating conditions, so that adequate resolution of the mixture is obtained
4. Use of a detector having a known and reliable linear response
5. Reliable performance of recording and data handling systems
6. Suitable integration/peak height measurement technique and column. The resolution obtained is fixed; the retention time in CGC will be much smaller.

10.9 APPLICATIONS OF GC

Innumerable applications have been reported. They include analysis (qualitative and quantitative) of a variety of materials (amines, alcohols, acids, esters, petroleum products, glycols, pesticides, dioxins, polychlorinated biphenyls, polynuclear aromatic hydrocarbons, environmental samples, permanent gases, natural gas refinery gases, halogenated compounds, metal chelates, pyrolysis products, head space gases, dissolved gases, pesticide residues, mixtures of isotopes/isomers, clinical/pharmaceutical products, drinking water, wastewater, soil, fragrances, etc.) GC also can be used to monitor reactions, kinetics of reactions and in arriving at thermodynamic parameters.

Elemental analysis (C, H, N, etc., in organic compounds), industrial analysis, preparative applications, clinical applications, analysis of foods/beverages, etc., polymer analysis, evolved gas analysis, surface area determination, detection of bacterial metabolites in spent culture media and body fluids, analysis of soil atmosphere, geological research, gases in solids (like metals, alloys, etc.), liquid samples, gases in electrical insulating oils, etc. are other areas in which GC has found good acceptability.

GC has been applied to solve a wide variety of analytical problems. The applications include qualitative/quantitative analysis, elemental analysis (C, H, N, etc. in organic compounds), forensic science, toxicology, clinical applications, foods and beverages, analysis of polymers, surface area determination, evolved gas analysis, detection of bacterial metabolites in spent culture media and body fluids, head space GC, analysis of metals, inorganic GC, analysis of soil atmosphere, various kinds of analyses needed in the petroleum industry, gases in solid and liquid samples, gases in metals and alloys, gases in electrical insulating oils, pharmaceutical industry, ultra trace analysis of hazardous substances like dioxins, polynuclear aromatic compounds, polychlorinated biphenyls, dibenzofurans, etc., environmental pollution studies (air, water and soil), analysis of fragrances and essential oils, analysis of technical pesticides, pesticide formulations and pesticide residues, speedy drug detection in hospitals, etc.

Compounds analyzed include (besides those specified above) inorganic/organic substances of almost all classes, petroleum products, environmental samples, permanent gases, natural gas, refinery gases, metal chelates, head space gases, dissolved gases, isotopes, isomers, clinical/pharmaceutical products, drinking water, wastewater, soil, fragrances, evolved gases, etc. GC has also been used to study kinetics, reactions, thermodynamic parameters, etc. With the advent of hyphenated techniques like GC-MS, GC-FTIR, GC-AAS, GC-NMR, etc., and multidimensional GC, and the use of multiple detectors, the utility and applicability of GC has been extended to several new areas.

Clinical applications include fast screening of samples for acidic/neutral and basic drugs at therapeutic levels, simultaneous monitoring of drugs at a wide range of concentrations, blood alcohol measurement, barbiturates at low levels, alkaloid analysis, drugs (e.g., anti-epileptic drugs) and their major metabolites without derivatisation, antidepressants at low levels, etc.

In the case of foods and beverages, the applications include resolution of complex mixtures in alcoholic beverages and fermentation products, and analysis of distilled spirits for acetaldehyde, methanol, ethanol, ethyl acetate, n-propanol, isobutyl alcohol, isoamyl alcohol, etc.

Several flavors and fragrances have been analyzed by GC.

A large number of drugs have been analyzed. Different pharmacopoeias recommend chromatographic methods to different extents. Typical applications include, identification, estimation of related substances, isomer content, volatile matter, assay and many other investigations involving volatile components.

GC has been widely used for the analysis of traces of solvents in pharmaceutical products and impurities in solvents.

Pyrolysis GC is used for the identification of non-volatile drugs, polymers and microbiological materials. It is used in forensic investigations also.

Some drugs can be analyzed directly after extraction/dissolution in a solvent. Many require derivatisation—usually silylation, acetylation, or methylation. The retention times of more than 600 drugs, poisons and metabolites are available.

Quality and production control of beer constituents and other brewing products can be effectively carried out by high-speed GC.

Aromatic flavors and trace volatiles in beer, foods and soft drinks are also readily analyzed.

Plastics, packaging and consumer products have been analyzed for residual monomers and other process residues such as O_2, CO_2, H_2O, acetates, aliphatic alcohols, printing inks and solvents.

Analysis of volatile free fatty acids produced by bacteria, particularly anaerobic bacteria, enable a fingerprint of that particular microorganism(s) to be obtained, which assists in the identification of the bacteria.

GC quality control analysis of food products can confirm the presence and quantities of additives, etc. to ensure that the product complies with company and legislative requirements.

Dairy products are analyzed for volatile components (aldehydes, ketones) to determine age and rancidities, fatty acids (as methyl ester derivatives) and milk sugar (as TMS derivatives).

The composition of volatile components on a wide range of natural foods, fruits, vegetables, flavorings and beverages (tea, coffee, cocoa), which gives them their characteristic flavors and odors, has received considerable interest, particularly for the preparation of artificial flavorings and perfumes.

There is a vast amount of literature on the use of GC for studying a wide variety of environmental problems, including estimation of pesticide residues.

Organotin biocides, Tetraalkyl lead compounds, Organomercury compounds, etc., have been analyzed using GC.

CHAPTER 11

High Performance Liquid Chromatography

11.1 INTRODUCTION

High pressure liquid chromatography (HPLC) is a technique used to separate the components of a chemical mixture. These components (or solutes) are first dissolved in a liquid solvent and then forced to flow through a chromatographic column under high pressure. In this column, the mixture is resolved into its components. The amount of resolution is important, and is dependent upon the extent of interaction between the solute components and the stationary phase. The stationary phase is defined as the immobile packing material in the column. The moving part of the system is the mobile phase, which is a liquid. The interaction of the solute with mobile and stationary phases can be manipulated through different choices of solvents and columns. As a result, HPLC acquires a high degree of versatility not found in other chromatographic systems. Thus, HPLC has the ability to easily separate a wide variety of chemical mixtures.

High performance liquid chromatography is basically a highly improved form of column chromatography. Instead of a solvent being allowed to drip through a column under gravity, it is forced through under high pressures of up to 400 atmospheres. That makes it much faster. It also allows you to use a very much smaller particle size for the column packing material which gives a much greater surface area for interactions between the stationary phase and the molecules flowing past it. This allows a much better separation of the components of the mixture. The other major improvement over column chromatography concerns the detection methods which can be used. These methods are highly automated and extremely sensitive.

11.2 BRIEF HISTORY AND DEFINITION

Liquid chromatography was defined in the early 1900s by the work of the Russian botanist, Mikhail S. Tswett. His pioneering studies focused on separating compounds (leaf pigments), extracted from plants using a solvent, in a column packed with particles. Tswett filled an open glass column with particles. Two specific materials that he found useful were powdered chalk (calcium carbonate) and alumina. He poured his sample (solvent extract of homogenized plant leaves) into the column and allowed it to pass into the particle bed.

This was followed by pure solvent. As the sample passed down through the column by gravity, different colored bands could be seen separating because some components were moving faster than others. He related these separated, different-colored bands to the different compounds that were originally contained in the sample. He had created an analytical separation of these compounds based on the differing strength of each compound's chemical attraction to the particles. The compounds that were more strongly attracted to the particles *slowed down*, while other compounds more strongly attracted to the solvent *moved faster*. This process can be described as follows: the compounds contained in the sample distribute, or partition differently between the moving solvent, called the mobile phase, and the particles, called the stationary phase. This causes each compound to move at a different speed, thus creating a separation of the compounds.

Tswett coined the name *chromatography* (from the Greek words *chroma*, meaning color, and *graph*, meaning writing—literally, *color writing*) to describe his colorful experiment. (Curiously, the Russian name Tswett means *color*.) Today, liquid chromatography, in its various forms, has become one of the most powerful tools in analytical chemistry.

Historically, chromatography dates back to 1903 when the Russian chemist, M. Tswett presented results on the separation of chlorophyll pigments using a chalk column. Lacking speed and resolution, the technique of chromatography languished for many years until the field was revolutionized by the Nobel Prize winning work of Martin and Synge in 1941. This work was important because it established a firm theoretical basis for the separation mechanism. No longer was the technique strictly empirical, i.e., utilizing a hit and miss approach for separations. Instead, the theory paved the way for a variety of practical improvements. The first field to capitalize on these developments was gas chromatography (GC). This in part was due to the simple nature of the equipment required, as well as the inherent sensitivity of the technique.

As usage increased, improvements in both the apparatus and theory developed rapidly. From this, it became apparent that many of the drawbacks and limitations of GC could be avoided by using a liquid rather than a gas mobile phase. This marked the birth of HPLC. Initial studies with unrefined equipment led to many technical improvements. For example, high pressure pumps became increasingly accurate. Column packings became more efficient and, at the same time, detection techniques and limits improved.

Today, a very specialized HPLC field has developed that uses a well characterized approach and highly sophisticated equipment. According to most chromatographic theories, the optimum limits of separation potential are now being approached. These limits are quite impressive and, because of this, HPLC enjoys an unprecedented position of deserved popularity.

11.3 LIQUID CHROMATOGRAPHY

The importance of mobile and stationary phase selections in HPLC cannot be overemphasized, since variations in these parameters are used to exploit different types of solute interactions. These interactions underlie all separations and give rise to the following separation modes:

1. The most simple interaction involves separation of solute molecules based on size or shape differences. This is called size exclusion chromatography (SEC), and is widely used in the comparison of complex formulations.
2. Another interaction, based on the dissolving power of one phase over another, causes solutes to partition or equilibrate themselves between the mobile and stationary phases. This is defined as partition chromatography, for which separation is dependent on molecular polarity. These interactions can be so finely tuned, that slight molecular polarity differences are sufficient for most separations.
3. Another means of separating molecules is based on their charge. This approach is referred to as ion-exchange. Very complex interactions can be accessed by this technique, and it is quite sensitive to even minute ionic variations.
4. Finally, adsorption is a liquid-solid interaction of the components with a polar surface alumina or silica gel using nonpolar solvents.

11.3.1 Evaluating an HPLC Separation

As all molecules differ from one another in either shape, polarity or charge, size exclusion, partition and ion exchange chromatography are sufficient for most solute separations. Given a problem, the chromatographer first chooses the interaction type which will retain the species of interest. The primary decision for separation is made with the selection of stationary phase. The separation procedure is finely tuned through mobile phase selection and manipulation. Frequently, this requires changing the mobile phase composition during chromatography. When this process involves more than one solvent it is called gradient programming, and requires some specialized equipment. No matter what the mode is, a careful selection of operating parameters based upon a knowledge of the solute chemistry will usually allow separation of very similar molecular species.

11.3.2 Components of a Liquid Chromatograph

The essential components found in HPLC instrumentation include a pump, injector, column, detector and recorder, connected as shown in Fig. 11.1. The heart of the system is considered to be the column. Since the stationary phase is composed of micron size (10^{-6} M) particles, a high pressure pump is required to move the mobile phase through the column. The chromatographic process begins by injecting the solute onto the top of the column. Separation of components occurs as the solute and mobile phase are pumped through the column. Eventually, each component of the solute elutes from the column. Detection of the eluting components is important, and can be either selective or universal, depending upon the property of the components being measured. The response of the detector to the presence of each component is displayed on a chart recorder and is known as a chromatogram. To collect, store and analyze the chromatographic data, computers, integrators and other data processing equipment are being used more frequently in conjunction with the more commonplace strip chart recorder (Fig. 11.1).

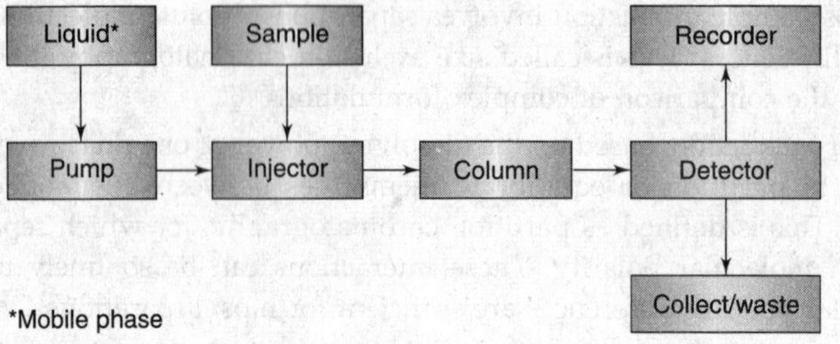

Fig. 11.1 *Major components and interconnections of an HPLC system.*

In HPLC, pump and plumbing leaks cause most of the problems, although trapped or dissolved air may also produce difficulties. Fortunately, these problems are easily detected. Detecting more subtle problems, such as loss of peak resolution on the chromatogram, requires evaluating the chromatographic data. This evaluation may be based on mathematical analysis of the recorded chromatogram.

The chromatography detector output signal should normally produce bell-shaped (Gaussian) peaks representing concentration profiles for the eluting components. The quality of the chromatographic separation can be determined mathematically. One method of mathematical analysis that is quite straightforward requires only a ruler and a calculator. Using these simple tools one can obtain values for the efficiency, selectivity and resolution of the resulting chromatography. Other methods of mathematical analysis will be discussed later.

Efficiency (N) is a very critical measurement, because it describes the theoretical separation potential of the chromatographic system employed. Both reducing the mobile phase flow rate and lengthening the column increase efficiency, but usually at the expense of the analysis time.

Selectivity (α) indicates how well the differences between similar components are utilized. The degree of selectivity depends mainly on the stationary and mobile phase selections. Since mobile phases are easily changed in HPLC, this process provides a means for quickly improving selectivity.

Resolution (R) is the overall estimate of the degree of chromatographic separation. This can be calculated for overlapping bands and depends on both the efficiency and the selectivity. In addition, resolution is also a function of a third term called the capacity factor, k'. 'This term indicates the interaction of the stationary and mobile phase by comparing the elution time of a given component with that of the solvent front. The time required for solvent front elution is also called the dead volume.

These three factors completely describe the separation characteristics and they can be used under defined conditions for testing not only solutes, but columns and stationary phases as well.

11.3.3 Advantages of Modern Liquid Chromatography

The major advantages of HPLC over other chromatographic techniques are speed, resolution and high sensitivity. Rapid resolution of even difficult problems is often routine. In part, this is due to stationary phase improvements which have increased column efficiency so much that shorter columns are becoming increasingly practical. Much work is being done in this area, and the final goal is to increase efficiency so that even shorter columns can be used in HPLC analysis.

Resolution is maintained with these shorter columns by better exploiting the inherent selectivity of different interaction modes. In addition to the increased speed and resolution, standard detectors are now capable of detecting very minute component concentrations. In fact, recent work demonstrated that the analysis of LSD (lysergic acid diethylamide) with HPLC could detect concentrations down to 10^{-15} moles/liter. This makes HPLC one of the best techniques for trace chemical analysis.

In addition to the above advantages, HPLC offers further attractions. From an economic viewpoint, the stability of the equipment and the columns is important. These are now so well designed, they maintain a trouble-free existence for extended periods of time. This is in contrast to other chromatographic techniques where column degradation often occurs. Also important is the growing automation trend in HPLC, which frees the user from some of the more mundane aspects of chromatography. In addition, automation streamlines HPLC methods development, thereby yielding better results in less time.

11.4 TECHNIQUES OF HPLC

11.4.1 Stationary Phases-Mobile Phases

Retention of components on the stationary phase is necessary to effect separation in the chromatographic process. Selective retention arises from differing interactions between the stationary phase and the various components. If this is to be a chemical interaction (as opposed to a physical interaction), different types of molecules will need chemically different types of stationary phases. The choice of stationary phase is governed by the rule, "like attracts like". For example, nonpolar molecules separate best on nonpolar stationary phases, ionic compounds require ion-exchanging phases, and polar compounds are separated on polar stationary phases.

The primary function of the stationary phase in a separation depends largely on its surface characteristics which in part are derived from the underlying substrate (e.g., silica).

In the early days of liquid chromatography, silica gel was used in a chemically unmodified form for polar HPLC separations. However, its unreproducible nature, coupled with other drawbacks, complicated its use. Early attempts at altering this substrate relied on liquid coatings, and these coatings did improve the performance. Since the coatings could wash off slowly, results varied from measurement to measurement. This changed with the advent of the bonded phase, which combined surface coating permanence with high reproducibility.

While both the bonded and coated phases each have their place in the laboratory, the bonded phases are more versatile, and are very widely used.

Recent work has shown it is possible to alter the stationary phase functionality by adding specialized reagents to the mobile phase. Since these reagents are ionic, this technique is called ion pair chromatography. Because of its versatility, and enhancement of chromatographic resolution, it is gaining rapid acceptance among chromatographers.

The stationary phase actions depend on the specific interactions between the solutes and the surface chemistry of the stationary phase. These interactions fall into different classifications or modes. The four HPLC modes are: partition, size exclusion, ion exchange, and adsorption.

11.4.2 Partition HPLC

The partition mode is the most widely used mode for routine HPLC. In this mode, solute components distribute themselves between the mobile and stationary phases. This distribution depends on the relative phase solubility of the components. Because components differ in their relative solubilities, they spend varying amounts of time in the mobile phase and thus elute from the column separately. The extent of interaction with the stationary phase governs the degree of retention (less interactive components elute from the column earlier than more interactive components).

The partition mode is divided into two types: "normal" phase and "reverse" phase. The terms "normal" and "reverse" are derived from the historical origins of chromatography where "normal" partitioning was the first separation technique studied. The conditions for "normal" separations are a polar stationary phase for separation of polar compounds, where the mobile phase is nonpolar. The bonded, "reverse" phase partitioning conditions mirror the normal phase separation conditions. Thus, the stationary phase is nonpolar and the mobile phase is polar. Examples of "normal" and "reverse" phase separations are given in Figs. 11.2 and 11.3.

When the mobile phase polarity is made to vary during a separation, the retention times of the solute components can be manipulated. The process of changing the mobile phase composition, discussed previously, is called gradient programming. It is one of the most significant features of HPLC.

Peak Identity

1. Solvent
2. Impurity
3. a, a-Dlmenthylbenzyl Alcohol
4. a-Methylbenzyl Alcohol
5. 2-Phenyl Ethanol

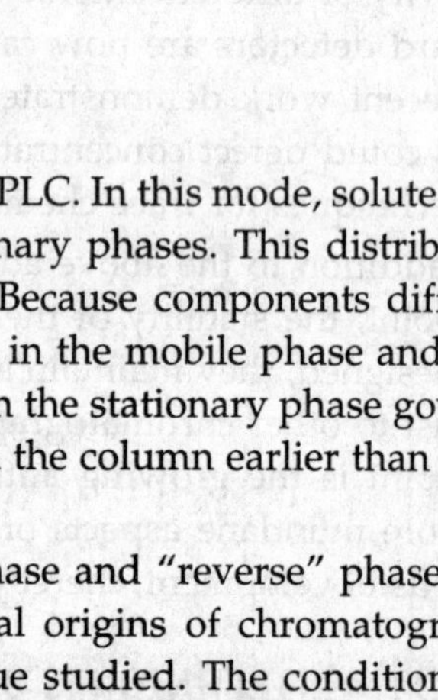

Fig. 11.2 *Normal phase chromatogram.*

6. Benzyl Alcohol
7. Cinnamyl Alcohol

Conditions Column	: ODS 4.6 mm × 250 mm
Column Temperature	: Ambient
Mobile phase	: MeOH/H_2O pH 3.50 with H_3PO_4 (20:80)
Flow Rate	: .73 mL/min
Detection	: UV @ 254 nm
Peaks	: Per chromatograms

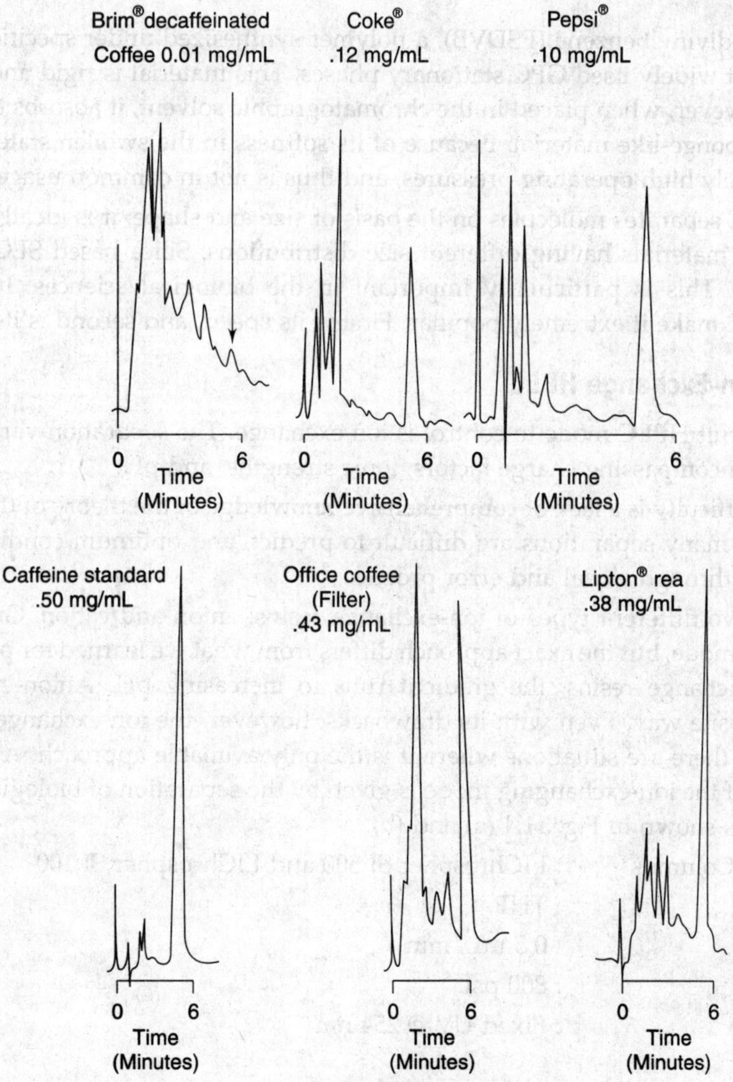

Fig. 11.3 *Reverse phase chromatogram.*

11.4.3 Size Exclusion HPLC

The one mode of HPLC that operates without any chemical interactions is size exclusion chromatography (SEC), also called gel permeation chromatography (GPC). The latter is limited to gels and resinous stationary phases. All interactions are based on the physical characteristics of the component molecules and stationary phases. The stationary phase for SEC is always fully porous with a large internal volume. The interaction depends on a component's ability to penetrate the pore structure of the stationary phase. Smaller molecules exhibit greater penetration, and hence they encounter increased internal volumes when passing through the column. Larger molecules are excluded from extensive retention in the pore structure, and they elute from the column in less time by migrating around the stationary phase.

Polystyrene-divinylbenzene (PSDVB), a polymer synthesized under specific conditions, is one of the most widely used GPC stationary phases. This material is rigid and impermeable when dry. However, when placed in the chromatographic solvent, it absorbs the solvent and swells into a sponge-like material. Because of its softness in the swollen state, it is not used under excessively high operating pressures, and thus is not in common usage now.

Because SEC separates molecules on the basis of size and shape, it is ideally suited for the comparison of materials having different size distributions. Silica based SEC uses aqueous mobile phases. This is particularly important in the biological sciences. In general, two features of SEC make it extremely popular. First is its speed, and second is its ease of use.

11.4.4 MIon-Exchange HPLC

The most difficult HPLC mode to control is ion exchange. The separation variables are complex, usually encompassing charge factors, ionic strengths, and pH.

The main difficulty is a lack of comprehensive knowledge of the theory of the fundamental process. Thus, many separations are difficult to predict, and optimum conditions can often be found only through a trial and error process.

There are two different types of ion-exchange resins: anion and cation. Gradients can be used with this mode, but the exact approach differs from what we learned for partition modes. With cation-exchange resins, the gradient runs to increasing pH. Anion-exchange resins work the opposite way. Even with its drawbacks, however, the ion-exchange mode is quite powerful and there are situations where it is the only available approach. A demonstration of the power of the ion-exchanging mode is given by the separation of biologically important amino acids, as shown in Fig. 11.4 (a) and (b).

Conditions Columns	: LIChrospher SI 500 and LIChrospher SI 100
Solvents	: THF
Flow Rate	: 0.5 mL/min
Pressure	: 800 psl
Detector	: Fixed UV @ 254 nm

Peaks: a. 670.000 MW
b. 110.000 MW
c. 37.000 MW
d. 4.000 MW

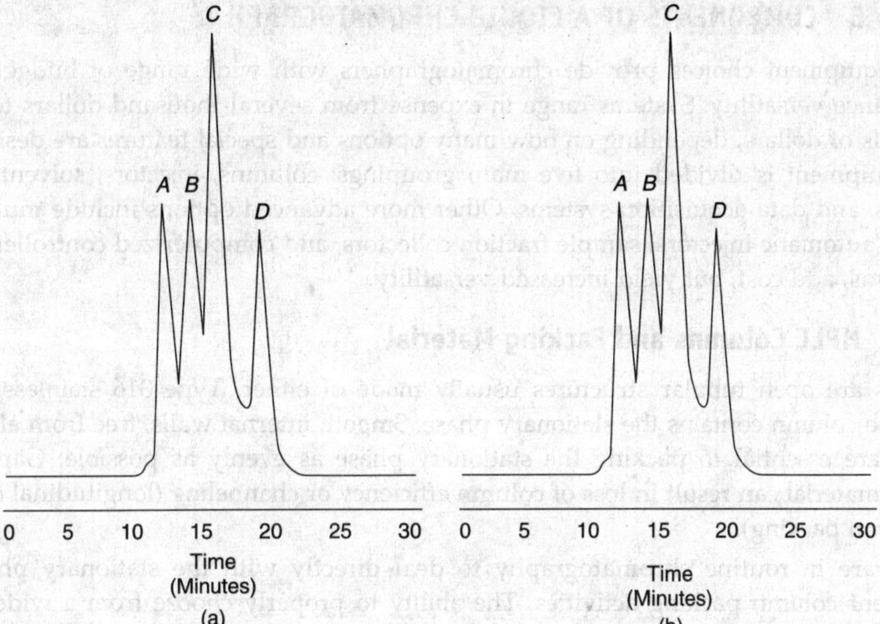

Fig. 11.4 *(a) and (b) Exclusion chromatogram.*

11.4.5 Adsorption HPLC

The adsorption mode of HPLC was at one time the most widely used separation approach. The early adsorbants included silica gel, alumina, and charcoal. The main problem with these stationary phases is their irreproducible nature. In addition, the mechanism of adsorption, whereby the components actually bond to the surface, causing further problems. Certain components bind so strongly to the stationary phase that they are never eluted. The accumulation of these components on the stationary phase also causes many problems. However, there are situations where adsorption can still be used advantageously. The most practical situation is in the separation of components which differ in functionality.

In summary, the four main modes of HPLC each have particular uses and limitations. The partition mode is ideal for separating similar compounds based on chemical polarity differences. The size exclusion mode works best when one is interested in size distributions. Ion exchange, can be used to separate species on the basis of charge differences. Its complex mechanism may be best suited for the separation of biological samples. The adsorption mode, while once widely used, is of decreasing importance today. It lacks both the speed and reproducibility of the other techniques, although it is still useful for class separations.

For the majority of separations, the partition mode, particularly reverse phase, is the most popular. This is followed by SEC and ion exchange. The choice of mode depends on several factors, including the complexity of the separation, as well as the differences between the components to be separated.

11.5 COMPONENTS OF A LIQUID CHROMATOGRAPH

Varied equipment choices provide chromatographers with wide range of budgetary and performance versatility. Systems range in expense from several thousand dollars to tens of thousands of dollars, depending on how many options and special features are desired. The basic equipment is divided into five main groupings: columns, injectors, solvent pumps, detectors, and data acquisition systems. Other more advanced options include multi-pump systems, automatic injectors, sample fraction collectors, and computerized controllers. These latter items, add cost, but yield increased versatility.

11.5.1 HPLC Columns and Packing Material

Columns are open tubular structures usually made of either. Type 316 stainless steel or glass. The column contains the stationary phase. Smooth internal walls, free from all surface defects, are essential to packing the stationary phase as evenly as possible. Gaps in the packing material can result in loss of column efficiency or channeling (longitudinal cracks in the column packing).

It is rare in routine chromatography to deal directly with the stationary phase and subsequent column packing activities. The ability to properly choose from a wide variety of pre-packed columns is of significant importance due to many differences in the particle types, interactive surfaces, and physical characteristics available in stationary phases today.

The three most commonly encountered stationary phases are shown in Figure 11.5. The first type is a rigid material with an impermeable outer structure. This solid surface lends itself to ultra-high speed separations. Its small surface area limits the sample size, and resolution is correspondingly reduced as sample size is increased. Larger surface areas, such as those found in the second particle type, increase both the resolution and the sample load capacity. Because of this, less material is needed for a given separation, and the analysis time does not usually suffer. The fully porous material, the third particle type, offers the utmost in separation capabilities but may increase separation time. The material most commonly used as a stationary phase is silica gel. It is rigid and can be manufactured to give any number of desired surface characteristics. One word of caution about silica gel stationary phases should be mentioned here. While generally stable, they do dissolve under basic conditions. It would not be wise to use these materials when working in high pH (>7.5).

Other particulate materials are also used as stationary phases in HPLC. Porous glasses, which are irregularly shaped and also rigid, are often used in a manner similar to the silica gels, however, they offer slightly different surface characteristics. Softer materials are also used, the most commonly encountered soft gel being the fully porous polystyrene-divinylbenzene (PSDVB) copolymer. It is extensively used in GPC and often serves as a starting point to obtain resins for ion-exchange systems.

Another important feature affecting the separation is the particle size. Small particle sizes are important because they improve the column efficiency. Therefore, the smallest particle sizes should be used whenever possible. Currently, the state-of-the-art particle size is about 5 microns. Theory predicts an optimum size of 3 microns. Larger particles, up to 10 microns, are still widely used, and they give excellent results. But, as the particle size gets larger, efficiency drops. The practical upper limit for a reasonably efficient separation at moderate cost is found with 10 micron particles. The effect of changing particle size on separation is quite dramatic, as shown in Fig. 11.5

Column : 3.1 × 250 mm
Solvent : 80: 15: 5; Heptane: $CHCl_3$: MaoH
Flow Rate : 3.8 mL/min
Temperature : Ambient
Sample : 10 µL
o-, m-, p-nitroaniline
o-, m-, p-phenylenediamine

Column Pressure

5 µm spherisor 10 µm spherisorb 20 um spherisorb
3020 psl 920 psl 470 psl

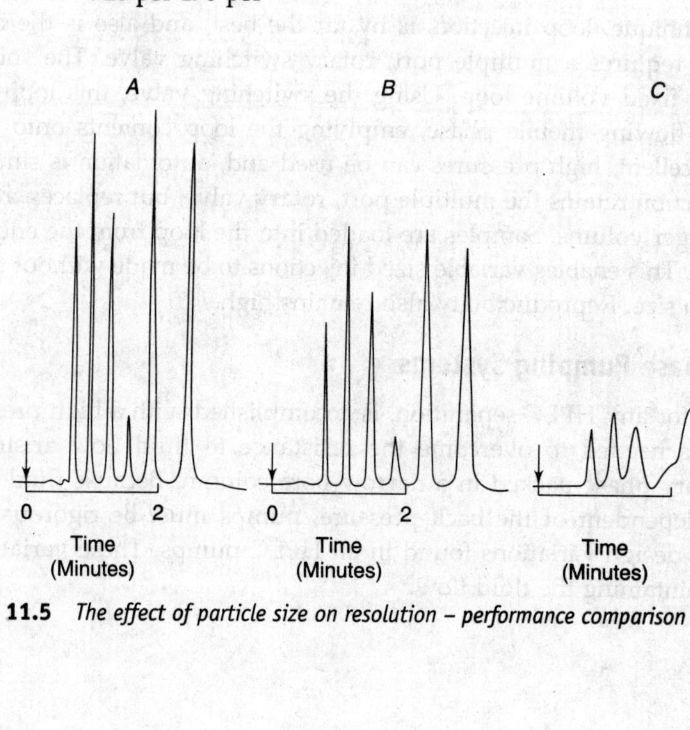

Fig. 11.5 *The effect of particle size on resolution – performance comparison of LC packing.*

An important feature of any mass of particles is size homogeneity. Wide variations in size are to be avoided because, when packed, the different sized particles create variable flow paths for the components. This leads to mixing in and around these particles, creating a decrease in resolution.

11.5.2 Sample Injectors

All separations depend on getting the solute onto the column. This requires an injector, which is located between the column and the high pressure solvent pump. The most simple approach to injection, both conceptually and operationally, is called stopflow injection. As implied, the first step in the process is stopping the solvent flow. With the solvent stopped, the top fitting on the column is removed and the sample placed directly onto the stationary phase. The column is then recapped, and chromatography continues when the mobile phase flow is restarted. This approach, while cumbersome to apply, is quite inexpensive and absolutely reliable. However, the placement of materials directly onto the delicate and expensive stationary phase is not recommended.

A more practical approach to introducing the solute, septum injection, avoids the removal of the top fitting and, in addition, is done while the mobile phase remains flowing. Compared to the previous technique, septum injection is simpler. The septum is a resilient seal at the column inlet where injection is accomplished by inserting a syringe needle through this septum and injecting its contents directly onto the column. The major drawbacks to this approach include: short septum life, necessitating frequent replacement; inability to use high pressures; and septum susceptibility to solvent attack.

The remaining technique, loop injection, is by far the best, and also is the most widely used. Loop injection requires a multiple port, rotary switching valve. The solute mixture is first loaded into a fixed volume loop. Using the switching valve, this loop is switched into the path of the flowing mobile phase, emptying the loop contents onto the column. Reproducibility is excellent, high pressures can be used and, automation is simplified. One variation of loop injection retains the multiple port, rotary valve, but replaces the fixed loop with one of much larger volume. Samples are loaded into the loop from the end of the loop closest to the column. This enables variable sized injections to be made without the necessity for changing the loop size. Reproducibility also remains high.

11.5.3 Mobile Phase Pumping Systems

Fluid flow, essential for any HPLC separation, is accomplished with a high pressure pump. The high pressure is needed to overcome the resistance to fluid flow arising from the microporous stationary phase packed in a narrow bore column. Because fluid flow should remain constant, independent of the back pressure, pumps must be rigorously designed. There are two major design variations found in all HPLC pumps. These variations depend on the method of maintaining the fluid flow.

The continuous displacement (syringe) pump forces solvent from a single large reservoir into the column. This approach offers a smooth and pulseless solvent flow, and is useful in operations requiring high detector sensitivity. Since the total amount of fluid flow during any given pump stroke depends entirely on the size of the reservoir, periodic shut-down of the system is required whenever the reservoir is emptied.

The intermittent displacement pump avoids this refill problem by repeatedly refilling and pumping from a small chamber. A large external reservoir feeds this chamber, so shut-down for refilling is not necessary.

During the refill stroke of the intermittent displacement pump, no solvent flows to the column. As these refill strokes occur in a regular manner, they impart a pulselike characteristic to the solvent flow. This pulsation appears at the detector output as tiny fluctuations on the chromatogram baseline. This is undesirable, and pump manufacturers have taken three different approaches to eliminate or reduce its effect.

One approach uses a baffle between the pump and the injector. A type of baffle now in use contains a column of air that compresses and decompresses with each full stroke cycle. While this does reduce the pulsations, it adds additional problems. First, the air is slowly absorbed into the flowing mobile phase. This air may later reappear as a bubble in the detector cell. Secondly, if the column effluent is recycled through the pump and into the column for a second pass (called recycle chromatography), the extra volume from the baffle adds dead volume into the system. This is never desirable. One alternative to the air baffle is to use long coils of narrow bore steel or Teflon® tubing. This tubing dampens solvent pulses like the air baffle, but usually causes fewer problems. Both techniques are practical and inexpensive, and they are used with many intermittent displacement pump systems.

Another approach to flow noise reduction involves using reciprocal pistons on the pump. These pistons are mechanically connected so that one piston is always pumping the solvent into the column while the other piston is filling for the next stroke. This approach is widely used, even though the pumps become much more complex and increasingly expensive to buy and maintain.

A third approach to flow noise reduction is done with a single piston pump. At the end of a pump stroke, the pump speed rapidly increases to refill the pump chamber. This rapid speed is maintained through the refill stroke and into the pumping stroke until the previous cycle pressure has been attained. Carefully monitor the flow rate, keeping it within set limits. This approach offers reasonable accuracy but is also quite expensive.

The composition of the mobile phase during a run is important for controlling the separation selectivity and retention time. A HPLC system is termed isocratic whenever the solvent does not change composition during a given run. Most routine separations are generated isocratically. In research or methods development work, it may be desirable to change the mobile phase composition during a run (the approach we previously defined as gradient programming).

Solvent gradients can be formed in two ways. If the solvent mixing is performed before the pumps, then the system is a low pressure gradient mixer. These systems tend to be quite simple and inexpensive. When mixing is performed following the pumps, then it is called high pressure gradient mixing. These latter systems tend to be sophisticated and easily controlled. Both approaches are widely used.

Low pressure gradients can be formed in two ways. With step gradients, a feed valve selects from any number of different solvent reservoirs, with the composition of each reservoir being determined prior to the start of a run. When smoother gradients are desired, a second pump is used. This pump feeds from a second reservoir into the primary reservoir feeding the chromatographic pump.

In high pressure gradient systems, two precision pumps operate in conjunction with a gradient controller. This gradient controller, regulates the percent flow of each pump. The controller offers the user a variety of preset options, including the rate of gradient change, total flow rate, starting and final conditions and choice of gradient profile. These controllers are quite easy to use and offer very precise control over the gradient system

11.5.4 HPLC Detectors

Because the solvent constantly changes in a gradient system, it is important to know how different detectors will respond to a variable mobile phase. Certain detector types are more suitable than others.

The HPLC detector provides electrical output signals proportional to the column effluent concentration. Not only can this detector monitor component elution, the electrical signals also provide quantitation. Because of its electrical nature, the detector is often one of the most sophisticated components of a liquid chromatograph.

There are many different approaches to HPLC detection. The most widely used techniques measure either refractive index changes or spectrophotometric absorption. Less common, but still commercially available, are fluorescence, electrical conductivity and electrochemical detectors.

Basic HPLC detector performance features include sensitivity, detection method, signal range and linearity. The sensitivity indicates the minimum detectable concentration of a given component to which the detector will respond. High sensitivity is essential for trace work. The main determinant of sensitivity is the detection principle. The detection principle relates to which type of chemical and/or physical characteristics are detected. For example, a universal detector would measure a characteristic displayed by all molecules. More selective detectors measure a specific characteristic possessed by fewer molecules. Fluorescence is an example of a highly selective detection method.

In general, as the detector type becomes increasingly specific, the sensitivity also increases. Range refers to the upper and lower limits of detection through which the detector response remains linear to the effluent concentration. Universal detectors often have the highest linearity ranges. This favorable characteristic tends to offset their lack of sensitivity.

The detector sample cell design is most important in determining detector performance. Minimal cell volumes avoid losses in resolution or chromatographic "bandspreading". Cells should also be designed to reduce or eliminate air bubble formation as well as provide a clear flow-path. State-of-the-art detectors are usually designed to avoid these problems, often having cell volumes under 10 microliters.

11.5.5 Refractive Index

The Refractive Index (RI) detector is the most widely used universal type detector. It works by measuring the difference between the refractive index of the column effluent and a reference solution. The mobile phase is usually the reference since it contributes the main background signal other than that due to the components. The typical operation of the RI detector requires filling the reference chamber with solvent. This reference is not changed during the run. Thus, the use of gradient programming is not advised because the solvent gradient will cause a drifting baseline, making peak analysis more difficult. The most common application for the RI detector is in isocratic liquid chromatography. Also, it is not usually suitable for high sensitivity work. Besides the reference requirement, the user should be aware that the refractive index of any component changes with temperature. For best performance, the RI detector should be used in a draft-free environment, protected from even minor temperature fluctuations.

11.5.6 Spectrophotometric Absorption

The spectrophotometric detector is much more selective than the RI detector. It works by measuring the absorption of either visible (VIS) or ultra-violet (UV) light through the column effluent. UV absorption is widely used since UV absorbing species are more common in organic chemistry. Absorption of UV energy by a molecule depends on very specific chemical features, as well as the location of these features with respect to the rest of the molecule. Different components absorb different quantities of light. The strongly absorbing chemical chromophores, their absorption wavelengths (λ_{max}.) and their emission strengths (for use in fluorescence) are listed in Table 11.1. Because not all molecules possess these chromophores detection is specific.

Table 11.1 *Absorbing UV chromophores*

Chromophore	λ_{Max}	*Absorbing Strength*
acetylide	177	medium
aldehyde	210	strong
azido	190	medium
amine	195	weak
ether	185	weak
ethylene	190	medium
ketone	190	weak

Contd...

Contd...

nitrite	225	weak
nitro	210	strong
xime	190	medium
thiol	195	weak
thioketone	205	strong
thioether	194	medium
conj. ring	varies	strong

Strength is related to the absorption coefficient

weak = 0-4000 medium = 4000-8000 strong = 8000

Operation of the UV detector usually employs a reference measurement similar to RI. However, the reference is most frequently air. Because certain solvents are UV absorbers, they cannot be used with a UV detector since their use would cut off any detector response. A list of common HPLC solvents and their transmittance cutoffs is given in Table 11.2. Useful HPLC solvents have cutoffs below the detector absorption wavelength. The typical transmittance wavelength of HPLC UV detectors is 254 nm. Additional wavelengths of 214, 229, 280, 308, 360, 410, 440, and 550 nm are commonly available for specific applications.

Table 11.2 *UV transmittance cutoffs (nm) for typical HPLC solvents*

Solvent	*Cutoff*
acetone	330
acetronitrile	190
benzene	280
butyl acetate	255
butyl ether	235
carbon disulfide	380
carbon tetrachloride	265
chloroform	245
cyclohexane	210
1,2 dichloroethane	230
dimethylformarnide	270
p-dioxane	220
ethyl ether	220
ethyl acetate	260
heptane, hexane, pentane	210
methanol, ethanol	210
methylcyclohexane	210

Contd...

Contd...

methylene chloride	241
nitromethane	380
tetrahydrofuran	220
toluene	285
zylene	290

Source: "Instrumental Methods of Analysis" Willard, Merritt and Dean, 4th Ed., D. Van Nostrand Co. Inc., 1965.

The detector output signal is usually the difference between the two signals. This approach can be used to subtract out a broad range background from a more specific signal. Dual wavelength detection is found on some of the more expensive detectors.

11.5.7 Electronic Data Processing in HPLC

Since the detector signal is electronic, use of modern data acquisition techniques can aid in the signal analysis. In addition, some systems can store data in a retrievable form for highly sophisticated computer analysis at a later time.

The main goal in using electronic data systems is to increase analysis accuracy and precision, while reducing operator attention. There are several types of data systems, each differing in terms of available features. For example, some systems only offer stand-alone capability. These systems become the final link between the operator and the data. They are often called turnkey systems because nothing additional is required to make them work. One major distinction applicable to all systems is whether data is to be taken "on-the-fly" or recorded for later analysis. Both have uses in the chromatographic laboratory. Although the recording system may be more beneficial in terms of both record keeping and repeat-analysis capabilities.

In routine analysis, where no automation (in terms of data management or process control) is needed, a pre-programmed computing integrator may be sufficient. If higher control levels are desired, a more intelligent device is necessary, such as a microprocessor or minicomputer.

The advantages of intelligent processors in chromatographs are found in several areas. First, additional automation options become easier to implement. Secondly, complex data analysis becomes more feasible. These analysis options include such features as run parameter optimization and deconvolution (i.e., resolution) of overlapping peaks. Finally, software safeguards can be designed to reduce accidental misuse of the system. For example, the controller can be set to limit the rate of solvent switching. This acts to extend column life by reducing thermal and chemical shocks. In general, these stand-alone, user programmable systems are becoming less expensive and increasingly practical.

Other more advanced features can also be applied to a chromatographic system. These features include computer controlled automatic injectors, multi-pump gradient controllers and sample fraction collectors. These added features are not found on many systems, but they do exist, and can save much time and effort for the chromatographer

11.6 PRINCIPLES OF CHROMATOGRAPHIC SEPARATION

Separation of a mixture depends on the selective retention of the migrating components in the column by the stationary phase. Retention, arising from selective interactions of each component with the stationary phase, is effected by simultaneous migration with the flowing mobile phase. This migration in the mobile phase is adjusted by changing the strength of the mobile phase, in turn controlling the solute-column affinity. Assuming similar component-stationary phase polarity, when the solvent polarity greatly differs from that of the stationary phase, almost all of the components will be retained. This will produce the effect of similar components having nearly identical stationary phase interactions. A closer solvent-stationary phase polarity match will decrease overall component retention. However, similar components will exhibit more selective migration. While these interactions occur on the molecular level, the entire process is analogous to classic zone migration.

At the start of a chromatographic process, a mixture of components is applied via injection to the column head as a single band or zone. As the solvent begins driving this band through the column, individual components begin to migrate at different fractional rates of the solvent flow. The exact fractional rate depends on the extent of component-stationary phase affinity. These interactions are described statistically by a Gaussian function, thus the concentration profile of a given component takes on a typical bell-shaped profile.

Two features of these bell-shaped curves are important. The first is the location of the maximum concentration. This is often taken to be the highest point in the curve. In addition to the band location given by this maximum point, the second feature is the width of the peak at its base, shown in Fig. 11.6. In general, bands become wider as they migrate, due to mobile phase dilution of the components. This band-broadening phenomenon is fundamental to chromatographic science.

In liquid chromatography (as contrasted with other chromatographic techniques), in-column band-broadening is attributed to two factors. The first factor is the contribution from non-instantaneous equilibrium. As the mixture of components migrate down the column, the interactions cause continuous partitioning between mobile and stationary phases. Each partition requires finite time and results in a shuffling process that continuously dilutes the band edges; making it wider. There are two simple approaches to reducing this effect, i.e., allowing a better equilibrium to be established. First is the use of smaller particles, second, is the use of slower flow rates, shown in Fig. 11.6. This latter approach is widely used when only a small efficiency increase is needed to achieve an acceptable separation.

The second factor contributing to band-broadening is called eddy diffusion. Eddy diffusion can be thought of as unevenness of solvent flow within the column. It occurs primarily in irregularly packed columns having either non-homogeneous particle sizes or grossly uneven packing densities. Typically, user-packed columns suffer most from this effect since commercial columns pass through very strict quality control procedures before they are released for sale. It should be noted that when examining a chromatogram it is very difficult to differentiate between these two band spreading effects.

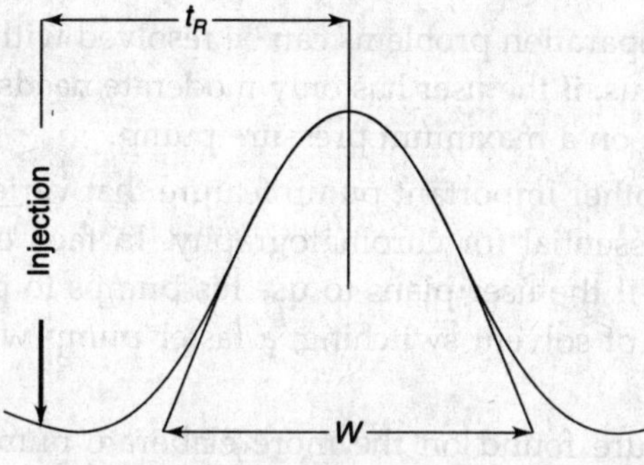

Fig. 11.6

Features of a Bell-shaped curve – Width (*W*) and Retention time (t_R)

If this added chemical is maintained in the mobile phase, then the "reverse" phase column can act like an ion exchanger. This feature is quite attractive, and many applications based on this principle are being developed. The overall separation mechanics of ion pair "reverse" phase chromatography are similar to ion-exchange techniques.

11.7 HPLC HARDWARE COMPONENT SELECTION

A chromatograph is usually purchased to separate and quantify a specific mixture or family of mixtures. Since the user may not initially be aware of all of its potential uses, it becomes important to "build" the system from the standpoint of upward compatibility. While complete stand-alone systems can in most instances be readily obtained, greater versatility at lower cost can be found through the purchase of individual components. The reason for this is somewhat analogous as to why a component stereo system offers improved performance capabilities. The wide range of component choices allows optimum usage of any desired feature.

A chromatograph has five major components. In order to support the operation of these components, additional parts are necessary. A comprehensive starter list of these parts is given in Table 11.3. We will treat each section of equipment, tools, and miscellaneous items separately.

11.7.1 Pumps

The pump is the first item listed in Table 11.3. The most distinguishing factor among pump types is pressure limit. The higher this pressure limit, the more expensive and complex a pump will be. For this reason, the user must carefully consider his overall separation needs.

High pressure pumps are needed to force solvents through packed stationary phase beds. Smaller bed particles require higher pressures to force solvent through the column. There are many advantages using smaller particles, but they may not be essential for all separations. The most important advantages are: higher resolution, faster speed, and increased load capacity. However, only the most demanding separations require these advances in significant amounts. Many separation problems can be resolved with larger particle packings that require less pressure. Thus, if the user has only moderate needs and a restricted budget, his money need not be spent on a maximum pressure pump.

Maximum flow rate is another important pump feature that varies between pumps. High flow rates are usually not essential for chromatography. In fact, efficiency decreases with higher flow rates. However, if the user plans to use his pumps to pack his own column, or does a considerable amount of solvent switching, a faster pump will certainly facilitate the process.

An additional pump feature found on the more elaborate pumps is external electronic control. Although it adds to the expense of the pump, external electronic control is a very desirable feature when automation or electronically controlled gradients are to be run. Alternatively, this becomes an undesirable feature, since it is an unnecessary expense when using isochratic methods.

The degree of flow control also varies with pump expense. More expensive pumps include such state-of-the-art technology as complex electronic feedback and multi-headed configurations.

These features provide greatly enhanced flow rate stability. For the user who does not operate at the highest sensitivity, it may be wise to purchase a more basic pump and add a pulse dampener. These pump systems work very well, are relatively inexpensive, and provide good performance in spite of their simplicity.

In general, the pump choice depends on the needed versatility and performance criteria. The user may find his money better spent on two or three less expensive lower pressure pumps, rather than on one top-line model. Despite the old adage "you get what you pay for", the user may not wish to buy an initial pump that goes beyond his most immediate needs. When these needs expand, the user can upgrade the system by buying a more suitable pump, keeping the original pump available for a variety of other tasks.

11.7.2 Injectors

The injector is next on the list. Here, too, there are several choices to be made. As with pumps, one major selection decision is high or low pressure capability. There are advantages to both. A low pressure injector is usually inexpensive and simple. An expanded diagram of a typical low pressure injector is shown in Fig. 11.7. Low pressure injection fittings and components are under less strain, and thus need not be so rigorously machined. This simplicity allows the user to repair and replace internal components without returning them to the factory. This advance is not often found with the higher pressure steel injectors. The higher pressure stainless steel injectors are more rugged, and machined to high tolerances. In selecting an injector, the user should consider that the high pressure steel injectors are substantially more expensive than their low pressure counterparts. The low pressure plastic units are definitely recommended where Teflon® tubing and glass columns are employed. Because of these choices, the user must define whether his needs require high pressure (for difficult separation problems), or low pressure (for routine separations).

11.7.3 Columns

The column is considered the heart of the system. No separations can occur without it. Stainless steel high pressure columns can be used up to 10,000 psi. Glass columns are more fragile and should never be used above 1,500 psi. In general, it is advisable to maintain minimal pressure in all systems since the higher pressures cause increased component wear and fatigue.

Column selection may be made from either prepacked or hand-packed columns. The prepacked columns are always more expensive; however, they provide a level of quality that the user would be hard pressed to duplicate. Prepacked columns are available with a variety of stationary phase packings. These provide for the users' most demanding needs. For the low and budget user, having no stringent separation requirements, user packed columns can be a very acceptable way of getting into the separation game at minimal cost. The loose stationary phases are less expensive and can be purchased in bulk. Of course, packing the column is not a trivial procedure. The user should be prepared for a variety of problems that only experience can solve.

11.7.4 Detectors/Recorders

The detector is the one item of equipment where quality rather than price should be the determining factor, since the detector qualities directly influence the separation results. Most people start with a simple fixed wavelength UV detector. Originally, this type of detector was available with wavelengths of 254 and 280 nm. Almost all UV absorbing compounds can be detected with one or the other of these wavelengths. Now, more sophisticated fixed wavelength detectors are available with a multitude of wavelength selection.

A chart recorder is the last essential item of equipment. It is used to record the separation results. The most basic recorder has a constant chart speed and one input voltage. Ten mV is the most common input voltage. Gain and offset damping controls are usually included for fine-tuning the recorder. Variable voltage ranges and chart speeds are also available, but at higher costs. In general, chart recorders are the least expensive of all the chromatographic components.

11.7.5 Miscellaneous Items

Some miscellaneous items may also be essential in assembling a user-built chromatograph. In-line solvent filters are highly recommended to provide additional filtering. Two types of filters are available. One type attaches to the end of the solvent intake line. In addition to filtering, this unit acts as a weight and holds the intake line below the surface of the solvent. The second filter type is an in-line filter located near the pump inlet. The use of at least one in-line filter (5-10 μm) is highly recommended.

A pressure relief valve is another highly recommended item for a low cost user-system. This item fits between the pump and the injector and operates by venting pressure in excess of preset value. These relief valves are not necessary with the more expensive pumping systems, because they generally include some sort of pressure relief system internal to the pump, either electrical or mechanical.

Positioned between the pump and injector might be other miscellaneous items: a pulse dampener and pressure sensor. In simpler models, the pressure sensor is merely a gauge. In more complex liquid chromatographs, the pressure sensor is an electronic transducer that feeds a digital readout module. Pulse dampeners vary in size and design, depending on the pump type, operating pressure and flow rate.

Guard columns are primarily used in chromatographic setups to protect the chromatographic column. The guard column can be placed in two different positions. When placed between the pump and the injector, it helps clean the solvent of potential chemical impurities. When placed between the injector and the column, it removes strongly retained, and potentially damaging solute components. The use of a guard column may be desirable in applications involving dirty samples, etc., such as encountered in clinical or environmental situations.

Some other miscellaneous items are also essential for a separation. Column end frits, tubing and unions, already detailed elsewhere, are necessary in all setups. In addition, a small amount of glassware to purify and hold the solvent is needed. Besides glassware and connectors, an array of small tools will be needed to connect and troubleshoot the equipment. A suggested list of tools is found in Table 11.3. This is only a partial list - other items may become necessary as the user progresses in the development of chromatographic expertise.

Table 11.3 *Equipment, tools and miscellaneous supplies*

Equipment	*Miscellaneous*	*Tools*
Pump	In-line solvent filter	4-Assorted screw drivers
Injector	Pressure relief valve	2-6" Adjustable wrench
Column	Pulse dampener	2-1/4" × 5/16" Open end wrench
Detector	Pressure gauge	2-3/8" × 7/16" Open end wrench
Chart Recorder	Guard column	1-1/2" × 9/16" Open end wrench
	Column end frits	Hex key set
	Steel tubing	Triangular file
	Teflon tubing	Emery Cloth
	Unions	
	Glassware for solvents	
	Heating mantle	
	Distillation/reflux	

11.8 GETTING STARTED

In the previous chapter, we investigated selection and testing of the equipment. Assuming that all of the equipment is operational, we would then connect it together. The solvents and samples would be prepared, followed by selection of separation conditions based on solute composition. Finally, the system is turned on, allowed to stabilize, and the solutes injected. This chapter details the particular steps in the process of starting up an HPLC system.

11.8.1 Tubing and Union Connections

The connection between components for high pressure fluid flow usually requires the use of small bore stainless steel tubing. Teflon® tubing can also be used, but it is not good for trace work, nor can it be used if pumping pressures exceed 1,000 psi. Despite these limitations, Teflon® does have positive features, such as ease of handling and flexibility. We will examine the use of both types of tubing.

Separate components are connected together with pieces of steel tubing, as pictured. These connections commonly utilize either Swagelok® or similar fittings. An increasingly popular alternative approach to column connection, having no internal dead volume, is shown in Fig. 11.7. A ferrule, used with both fitting types, is placed on the end of the steel tubing with its tapered end pointing toward the union.

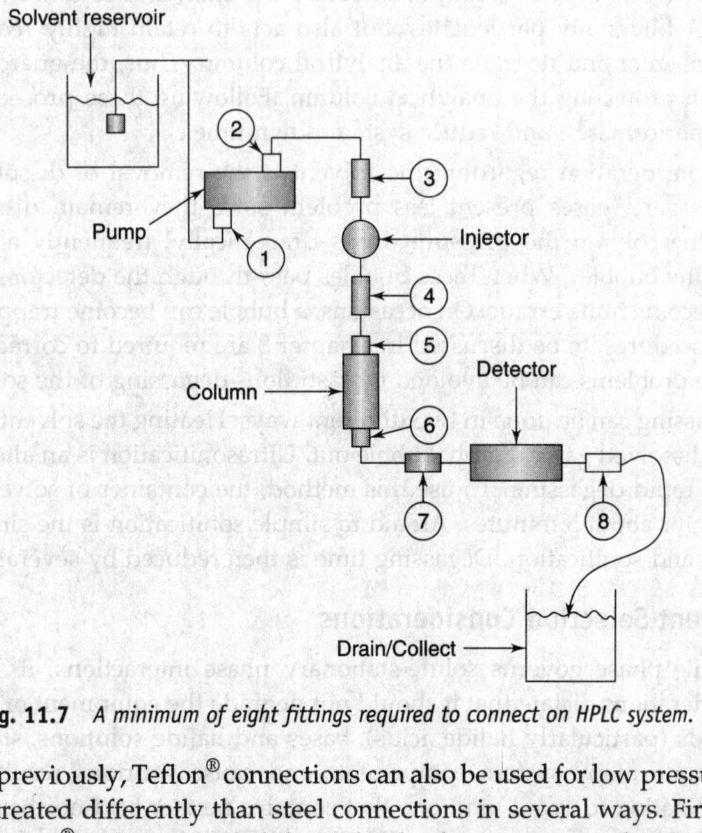

Fig. 11.7 *A minimum of eight fittings required to connect on HPLC system.*

As mentioned previously, Teflon® connections can also be used for low pressure chromatography. They are treated differently than steel connections in several ways. First, the union is female, and the Teflon® ends are guided through a bushing as pictured in Figure 11.7. A flat flare on each tubing end is made with a special apparatus that has a small heated nipple fitting into the tip of the tubing end. The flat flares are butted together in the female union, and the bushings tightened. Practice is usually needed to make perfect Teflon® flares. When carried out correctly, however, they are usually leak-proof and add no dead volume. Once all the equipment is connected together for fluid flow, the user can begin the next phase. This involves preparing the solvent and the solutes.

11.8.2 Solvent Preparation

The solvent plays an essential role in the separation. Therefore, the solvent must be handled carefully to maintain optimum chromatographic performance. The most important preparation consideration is for all samples and solvents to be free of particulates before they enter the chromatographic systems. Clogging of filters and column frits could disable the system. The pumps and columns are also potentially sensitive to particulates.

Two approaches are recommended for particulate removal. The first is filtering an inert porous material such as a 0.2 micron microporous filter is usually employed. Both the solvents and solutes should be filtered. In addition to filtering, some LC column manufacturers recommend placing a "guard" column just before the analytical column. The guard column not only further filters any particulates, but also acts to retain highly reactive species that might otherwise enter and degrade the analytical column. Thus, the guard column serves a dual purpose in protecting the analytical column. Following these procedures will help to maintain high performance and reduce system down-time.

One other consideration regarding the solvent is the removal of dissolved gases. While under high pressure, gases present less problem since they remain dissolved. However, when exiting the column the pressure drops dramatically, frequently allowing dissolved gases to form tiny bubbles. When these bubbles pass through the detector cell, signals to the recorder will become quite erratic. On occasions, a bubble can become trapped in the detector cell. Special procedures, to be discussed in Chapter 8 are required to correct this problem. In general, bubble problems can be avoided by fastidious degassing of the solvent.

Solvent degassing can be done in two different ways. Heating the solvent reduces gas solubility, and the dissolved gases simply bubble out. Ultrasonification is an alternative approach to obtain more rapid degassing. To use this method, the container of solvent is placed in an ultrasonic bath for about 5 minutes. An aid to simple sonification is the simultaneous use of gentle vacuum and sonification. Degassing time is then reduced by several minutes.

11.8.3 Solvent Selection Considerations

Since the mobile phase governs solute-stationary phase interactions, its choice is critical. Practical considerations dictate that it should not degrade the equipment or the column packing. Strong acids (particularly halide acids), bases and halide solutions, should be avoided. These can corrode many system components and cause spurious results. A particularly corrosive combination to avoid (and one that contains neither acid nor base) is a mixture of a peroxide with a halogen solvent (e.g., CH_2Cl_2 and $CHCI_3$).

There are other specific situations to avoid. For example, basic mobile phases should not be used with silica gels. This is also true for most bonded phases, because the bonding can be broken under alkaline conditions. Chemical purity of solvents is another important factor. Since large volumes of solvents are inevitably pumped through the column, trace solvent impurities can easily concentrate in the column and eventually produce detrimental results. Spectrograde or HPLC grade solvents are highly recommended. Solvent purity cannot be over emphasized.

Solvent compatibility with the detector is also important. This is particularly true for ultraviolet and fluorescence detectors. All aromatic solvents are incompatible with these detectors, and cannot be used. For the RI detector, the highest sensitivity will be found with the solvent that has a widely different refractive index from that of the solute components.

Solvent volatility should also be considered if the sample is to be recovered. This is of particular importance in preparative HPLC separations. Highly volatile solvents are recommended in these applications.

Solvent viscosity is another important consideration. High viscosity is a drawback in many separations, since it reduces solute diffusion coefficients, thus reducing mass transfer between solvent and stationary phase. Also, higher pump pressures are required. In general, the mobile phase viscosity should be less than 0.5 centipoise.

11.8.4 Sample Considerations

While solute components should be collectively soluble in the mobile phase, the solubility of each component must differ slightly for selective stationary phase interactions to occur. Sometimes it is difficult to find a solvent that fulfills the earlier requirements, yet remains sample-compatible. Occasionally, the sample must be dissolved in a solvent different from the mobile phase and then injected. But when this is done, care must be taken that the sample does not precipitate in the column. This problem is most commonly encountered in preparative separations. As with solvents, solutes should always be filtered.

Proper sample concentration is also very important. The proper concentration value varies for different detectors. With low sensitivity detectors, such as the RI detector, higher sample concentrations are necessary. Very sensitive detectors, such as fluorescence detectors, require correspondingly lower sample concentrations.

11.8.5 Choosing Separation Conditions

In general, the slower the solvent flow rate, the better the separation. A typical flow rate for analytical separations is 0.5 to 2.0 mL/minute. Faster rates can be used, but above 2 mL/minute the chances for establishing equilibrium conditions in the column are unfavorable. Pressures higher than 4000 psi are not normally encountered at analytical flow rates. Higher pressures may indicate a blockage or plug somewhere in the system.

Figure 11.19 gives an outline approach toward choosing separation conditions based on the nature of the solute components. The process relies on a series of logical decisions regarding the solute component characteristics. The first division is between high and low molecule weight (MW). High MW compounds (>2000) are best separated by the SEC mode, under either aqueous or non-aqueous conditions. While this can also be used for some low MW compounds, these usually require a partitioning type of chromatography.

If the low MW compounds are water soluble, then one has two further choices. The first, ion-exchange method, is usable when water soluble components are electrolytes. Either an anion or a cation exchanger may be applicable, depending on the ionic charge of the components. The non-electrolytic water soluble compounds can be separated by either "normal" or "reverse" phase partitioning, based on differences in polarity.

Compounds soluble in nonpolar solvents form another large class of solutes. These are generally separated by partitioning or adsorption chromatography. As before, there are two partitioning modes: "reverse" phase for nonpolar organics, and "normal" phase for polar organics. In adsorption, one can also choose between silica and alumina stationary phases. Alumina is used preferentially for slightly basic compounds, aromatics, or those containing either halogens or sites of unsaturation. Silica is preferred for all other situations.

11.8.6 Gradient Programming

Solvent selection guidelines are given for each of the above cases in Table 11.2. Variations in the solvent selection are based on changing the solvent strength. In general, changes in the solvent strength affect the resolution by varying k'. The stronger the eluting power of the solvent, the lower the k'. It is often found that isocratic solvent conditions cannot effectively separate all of the components in a mixture. When this is the case, either a gradient separation or change in chromatographic method is indicated.

An example of how a gradient can be selected is given in Fig. 11.3. A single split peak is observed when 0% water (100% methanol) is used in this reverse phase separation. Increasing the water content to 30% (70% methanol) gives increased separation, but bunching near the first peak is still observed. With 60% water (40% methanol), this bunching disappears, but the end peaks become lost. Based on these isochratic HPLC results, the applicable gradient starts with 40% methanol and increases to 100% at a fixed rate. This will produce a complete separation.

With ion-exchange systems, solvent selection is very critical. Generally, the solvent contains a buffer of counter ions, whicn compete with the component ions for ion-exchange sites on the stationary phase. Increasing the buffer concentrations lowers the k' values, and thus solutes elute quicker. A typical gradient used for separating the constituents in urine starts with 0.015 M sodium acetate increasing to 6.0 M. For compounds that are either acidic or basic (amphoteric), pH gradients become useful. The pH controls the degree of component ionization. A pH gradient is usually run toward conditions of decreasing the ionization of the component, thereby inducing the component to leave the stationary phase.

11.8.7 Separation Procedures

The separation procedure can begin after all of the solvents and solutes have been properly prepared. The first step is to start the solvent running through the system. The solvent is placed in the solvent reservoir, which is usually located above the pump. A feed line connects the reservoir to the pump through an in-line filter. A priming procedure is used to fill the pump heads with solvent, and the pump is then turned ON. The pumping rate should be less than, or equal to, 1.0 mL/minute at this point.

Solvent passes out of the pump, into the column, and through the detector where the signal is monitored on the chart recorder. While some early signal noise is expected, a stable, non-fluctuating baseline should soon develop, indicating that the column has equilibrated with the solvent. Solute injections can now be made. For loop injectors, the solute solution is loaded with the injector in the "load" position. Turning the injector to "inject" places the loop in the path of the solvent flow. Depending on the chromatographic method, the components will elute from the column in 2 to 30 minutes, nominally.

Gradients are started following the injection. If all of the equipment is working correctly, and the proper separation conditions used, typical bell-shaped, well-separated, peaks should be observed. However, if this does not occur, the user must carefully troubleshoot the system for the source of the problem.

11.9 QUALITATIVE AND QUANTITATIVE ANALYSIS CHROMATOGRAM

11.9.1 Qualitative Analysis

Initially, identification must be made of the solute components. This is generally accomplished by comparison of retention times of the solute components with the retention times of pure standards. Although retention time comparisons using a single group of HPLC parameters are not an absolute method of identification, using a second group of parameters (e.g. different column, solvents and flow) will prove adequate for many situations. Since other compounds can have the same retention time, the chromatographer should be constantly alert to the presence of possible interfering substances.

Another method of identifying a particular peak is by the technique called "spiking". This technique can be used when the identity of some specific peak is questioned, and pure compounds are available. As an example, consider a vitamin separation for which the location of vitamin B_1 and vitamin B_6 have become confused. By spiking the original sample with additional vitamin B_6, the proper peak can be located by noting the peak of increased area (or height), as seen in Fig. 11.8.

Once again, it must be stated that this technique is not definitive due to other components having similar retention times.

11.9.2 Computation

Before beginning our discussion on the quantitative aspects of HPLC, it is important that two important terms be defined: accuracy and precision.

Accuracy – a measure of the exactness of an answer to a known or actual value.

Precision – a measure of the degree to which several runs are reproducible.

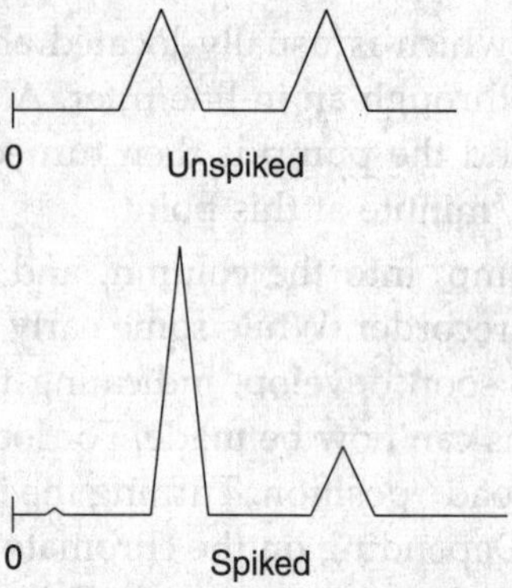

Fig. 11.8 *Peak identification by "spiking".*

The mathematical calculation used to represent precision is called "standard deviation". It is calculated in the following steps, where X = different values obtained in the measurement processes:

1. Sum (σ) the values of x, written as:
   Σx
2. Divide σx by n, the number of differing values, which gives
   $$\frac{\Sigma x}{n} = \bar{x}$$
   (X mean or average value of x)
3. Subtract each value for x from the mean. (Some values will be + others will be –):
   $$\bar{X} - x$$
4. Square the differences of the values of x from the mean.
   (All resulting numbers will be +):
   $$(\bar{X} - x)^2$$
5. Sum the squared differences for each x:
   $$\Sigma(\bar{x} - x)^2$$
6. Divide the sum of these squared differences by the total number of values, less one:
   $$\frac{\Sigma(\bar{x} - x)^2}{n - 1}$$
7. The square root of term #6 is equal to the standard deviation:
   $$\sqrt{\frac{\Sigma(\bar{X} - X)^2}{N - 1}} = \text{S.D.}$$

The Relative Standard Deviation or Coefficient of Variation, is obtained by dividing the Standard Deviation, by the mean, X; and multiplying by 100:

$$\text{Coefficient of variation} = \frac{\text{S.D.}}{X}\ X\ 100$$

The Coefficient of Variation is the measure of overall precision of the work completed. The smaller the resulting number, the better the quality of work.

11.9.3 Quantitative Analysis

Chromatographic quantitation involves the comparison of standards and samples, their areas (or heights), and retention times. This quantitation is based upon two requirements:

1. Reproducible chromatography
2. Linear response of the detector for the compounds of interest.

When these criteria have been met, and the chromatography produced, quantitation of the peaks may proceed. This process can be accomplished in any of the following ways.

11.9.4 Peak Height

Measurements are made for each peak of interest, from the baseline to the peak maximum. The drawbacks to peak height usage are related to the effects of chromatographic conditions, such as flow rate, on the peak height sensitivity. Errors may also arise from poor measurements due to insufficient resolution, drifting baseline, and peaks which are extremely small.

11.9.5 Peak Areas

(a) **Planimetry:** This method of determining area is one of the original techniques, and it requires a great deal of skill in operating the planimeter. (A planimeter is a device used to measure the area by tracing the peak. A digital counter displays the area measured under the peak.) This method is seldom used today.

(b) **Cut and Weigh:** This technique requires cutting the peak out of the chart paper and weighing it on an analytical balance. Cut and weigh is highly dependent on the operator's skill. Errors can occur from details as simple as humidity changes, or oil from the operator's hands.

(c) **Height Times Width at Half Height of Peak:** These measurements are based on the triangular shape of the Gaussian peak:

$$A = H \text{ X } W_{(\%)}$$

where H equals peak height and W(II,) equals peak width at half height.

Basically, a simple calculation, it can only be used with symmetrical peaks.

(d) **Disc® Integrators:** Good accuracy is provided with this instrument, independent of peak shape. However, it is essential that the peaks remain on scale. At one time very popular, these integrators have been replaced by digital integrators and microcomputers.

(e) **Digital Integrators:** Computing integrators automatically determine the peak areas. The user simply enters the known concentration for each component peak, and the integrator will then compute the appropriate response factors, placing these factors into memory for future use. When an unknown sample is run, the integrator measures the peak areas electronically, applies the appropriate response factor then calculates and prints out the concentration of the components — all automatically.

(f) **Microcomputers:** A microcomputer (as used in LC) differs from a microprocessor in that the computer has a user accessible language built in — usually BASIC. The microcomputer performs data acquisition and reduction in much the same fashion as a computing integrator. The same sort of parameters and commands as used on the integrator would be used to instruct the microcomputer for data handling. The big performance difference between the two devices is that the microcomputer can process the data — do calculations much more extensively than the computing integrator. For example, programs written in BASIC can do linear regressions, standard deviation, print tables of data from a series of runs, or just about any other type of calculation the user may care to enter.

Microcomputers also enable control and interfacing of peripheral equipment not possible with other systems. Imagination is generally the limiting factor on utilization of the microcomputer. Increased cost and technical demands on the operator are the major drawbacks of this type of integrator.

Once the peaks have been quantitated, the user now has numbers representing peak heights or areas. What one needs to know, however, is what these mean in real or absolute measurement terms. There are four different mathematical manipulations which can be used to translate these numbers (raw data):

- Normalization
- Internal Standard
- External Standard
- Standard Addition

11.9.6 Normalization

There are two basic forms of normalization:

1. Area normalization for compounds with similar detector response and for which 100% of the components of the solute elute. For four components, *A*, *B*, *C* & *D*, a simple normalization calculation would be made as follows:

$$\%A = \frac{\text{Area of peak } A}{\text{Total peak areas}} \times 100$$

$$(A + B + C + D)$$

2. Area normalization using response factors to correct for variations in detector response for the different components.

To calculate these response factors, a calibration curve (with a minimum of 3 concentrations) for each of the components *A*, *B*, *C*, & *D* must be made. Sample concentrations (weight) vs. peak area are plotted and the best straight line drawn through the points. The slope of this curve, X/Y, is the response factor for that particular component.

The corrected area of each component is thus calculated by dividing its peak area by the response factor.

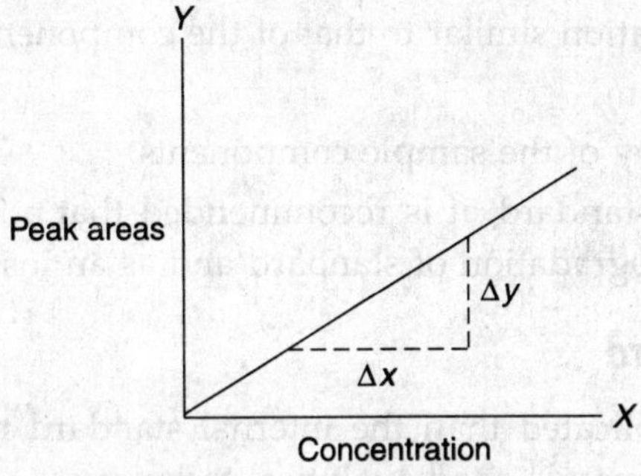

Fig. 11.9 *Calibration plot.*

$$\frac{\text{Area peak } B}{\text{Response factor component } B} = \frac{\text{Corrected area}}{\text{Component } B}$$

1.9.7 Internal Standard

This technique requires the addition to the sample of a known component of known concentration for comparative determinations. Since the internal standard is of a consistent volume and known concentration, compensations for variations in sample injection size and manipulation can be made. The mechanism for this compensatory action is based upon comparison of the peak height (or area) of the internal standard with the peak height (or area) of the component(s) of interest. With this technique it is not necessary for all components present in the solute to be eluted from the column, only those of interest and the internal standard.

As with normalization, a response factor can be calculated by mixing a known concentration of pure internal standard with various concentrations of pure component standard. A chromatogram is then run for each of the standards, and measurements of peak height (or areas) made for both component and internal standard peaks. The ratio of internal standard to standard component is made for each of the runs, and plotted vs. concentration.

The point or the curve where the peak height ratio equals one is noted and the concentration at that point determined from the graph. This number then becomes the response factor (R). Concentration of unknown samples containing components of interest is determined as follows:

$$\frac{\text{Peak height (or area) of component}}{\text{Peak height (or area) of internal std.}} \times R$$

It should be noted that the internal standard must be linear with the standard compounds in the concentration range of interest.

Internal standards must also meet the following criteria:

- Never be found as a normal component of the solute
- Completely resolved from sample component peaks

- Added at a concentration similar to that of the component(s) of interest
- Available in pure form
- Non-reactive with any of the sample components.

When using an internal standard, it is recommended that a 3-point calibration curve be run routinely to check for degradation of standard and as an instrument check.

11.9.8 External Standard

This technique is less complicated than the internal standard method, but it is extremely operator-dependent, and demands good analytical technique.

Calibration curves of pure standard of varying concentrations are first prepared. It is essential to the accuracy of the measurement that injection volumes always be identical for the components of interest, as well as the pure standards. The curve is plotted with peak height (area) vs. concentration, and the unknown concentration then read from the curve.

11.9.9 Standard Addition

This method of quantitation is a combination of internal and external standard methods. This method is generally used when working in a non-linear portion of the detector response curve.

11.9.10 Interferences

Since more than one component can have the identical retention time of the sample (peak) of interest, it is important to know which other compounds could cause interferences. Compounds which are generally found in conjunction with the component of interest, as well as naturally occurring substances in the sample matrix, should be chromatographed and retention times noted. If a common substance found in the sample matrix or compound has an identical retention time when compared to the component of interest, another separation mechanism should be used. Alternatively, a method for extracting the component of interest from the interfering substance must be found.

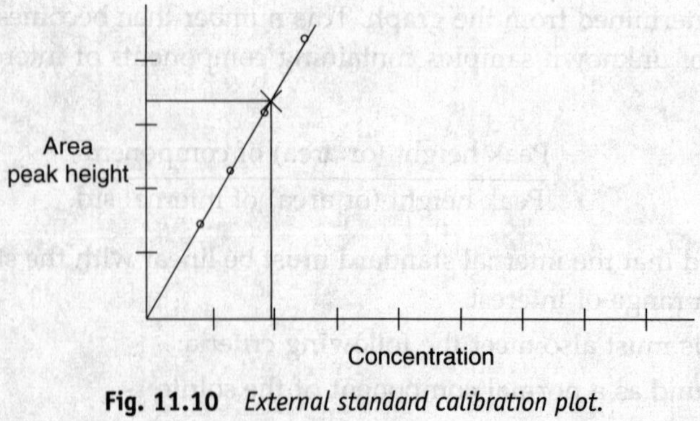

Fig. 11.10 *External standard calibration plot.*

A simplistic example of an interference study is the examination of an analgesic preparation containing caffeine, phenacetin, aspirin and acetaminophen. Should the chromatogram reveal the phenacetin and caffeine to separate as a single peak, the parameters used — perhaps the method itself— would not be satisfactory.

A more complicated example is the determination of lidocaine (an antiarrhythmic) levels in serum. There are many areas of interference which must be investigated. Blank serums need to be extracted and chromatographed in the same manner as the lidocaine sample — to look for naturally interfering serum components. After the establishment of a method for which there are no interfering serum components, serum samples spiked with many commonly prescribed arugs (and over-the-counter preparations) must be extracted and chromatographed. This will establish whether or not interferences exist.

11.9.11 Recovery

The percentage recovery of the component of interest will vary from method to method. It is important to know, therefore, what that percentage is before interpreting results. By extracting pure standards in accordance with the methodology to be used for the sample, and comparing the results with pure standards (chromatographed without following the extraction scheme), a percentage recovery of pure standard can be calculated from the extraction scheme. This calculation is made by dividing the concentration of the extracted standard by the concentration of the non-extracted sample multiplied by 100:

$$\frac{\text{standard}_{\text{ext}}}{\text{sampe}_{\text{nontext}}} \times 100 = \%\ \text{recovery}$$

1. Adsorption Chromatography – separations which occur due to differences in adsorption characteristics of the components of of interest.
2. Affinity Chromatography – a specialized type of chromatography using bonded phases which have an affinity for (or an ability to adsorb) only certain types of molecules.
3. Band – the area (or volume) in the column in which the component(s) is located at any given time.
4. Band Spreading – the diffusion of the band into a greater area (or volume) due to column effects or fittings.
5. Bonded Phase – chemically bonded functional groups to a stationary phase base such as silica.
6. Chromatography – a separation technique based upon the ability of components to adsorb to a particulate stationary phase as a constant flow of gas or liquid moves across the stationary phase.
7. Column Capacity Factor – number of moles of solute in the stationary phase divided by the number of moles of solute

in the mobile phase — retention of the component compared to the solvent front (or dead volume).

8. Dead Volume – any volume between the injector and detector where separation does not occur.
9. Distribution Coefficient (partition coefficient) – same as capacity factor.
10. Eluate – the mobile phase and components which exit the end of the column.
11. Eluent – solvent (or mobile phase) which passes through the column and removes the sample components from the stationary phase.
12. Eluotropic Series – solvent group ranked in order of eluting strength.
13. Elution Chromatography – the removal of the components of interest from the column with an eluent.
14. Elutriation – separation of particles by size using sedimentation.
15. Exclusion Chromatography – a mode of chromatography which separates components dependent upon molecular size.
16. Exclusion Limit. – the molecular size limitation for a particular exclusion method or column.
17. Flow Rate – milliliters of mobile phase which are pumped during a given time frame.
18. Flow Programming – altering the flow rate during the elution process.
19. Fronting – lack of symmetry on the upside of the peak (opposite of tailing).
20. Gel Filtration – a form of size exclusion chromatography which soft gels as the stationary phase.
21. Gel Permeation – see exclusion chromatography.
22. Gradient Program – altering the composition of the mobile phase at a specified rate per time during the chromatographic process.
23. Ion-Exchange Chromatography – a mode of chromatography based upon ionic charge affinities.
24. Linear Capacity – sample load for which the capacity factor k' does not fluctuate by more than approximately 10%.
25. Linear Velocity – is the ratio of the length of the column to the time it takes for an unretained component to elute.

26. Low Dead Volume Connection – a fitting which allows two pieces of tubing to be joined in such a manner that there is minimal volume between their ends.
27. Mass Transfer – interaction between solute components and between mobile phase and stationary phase which produces equilibrium.
28. Mobile Phase – Solvents which are pumped through the system coursing over the column and eluting components (also called eluent).
29. Molecular Exclusion – See exclusion chromatography.
30. Paired Ion Chromatography – the technique by which the functionality of the mobile phase is altered by adding ionic reagents to the mobile phase.
31. Partition Chromatography – chromatographic technique based upon the variation of solubilities between the solute and the stationary phase.
32. Pellicular Packings – packing material with a solid core such as a glass bead enveloped with a porous material.
33. Permeation Chromatography – see gel filtration.
34. Plate Height – HETP, height equivalent to a theoretical plate — measures the efficiency of the column and should be very small. HETP indicates the length of a plate in which the mobile phase and stationary phase are in equilibrium.
35. Plate Number – measurement of the number of plates in a column. It is desirable that this number be large. The larger the number of plates per meter the greater the column efficiency (N).
36. Pulse Dampener – a device used to smooth flow pulsations produced by the pump.
37. Resolution – the degree of separation of a solute mixture into its individual components (R).
38. Reverse(d) Phase – a mode of chromatography where the stationary phase is nonpolar (hydrocarbon-like) and the mobile phase is relatively polar (non-hydrocarbon-like).
39. Selectivity – the measure of the ability of two components (peaks) to separate, from one another. For proper separation

to occur this value must be greater than 1 *(a)*.

40. Solvent – the reagents which comprise the mobile phase.
41. Solvent Demixing – the phenomena through which one component of the mobile phase is being slightly adsorbed, producing a state of non-equilibrium.
42. Standard Deviation (relative) – the measure of precision (also known as coefficient of variation).
43. Stationary Phase – the immobile packing material in the column.
44. Stepwise Gradient – changing the mobile phase in a step-wise increment.
45. Tailing – asymmetry of the down-side of the peak.

11.10 APPLICATIONS

HPLC is a form of column chromatography used frequently in biochemistry and analytical chemistry to separate, identify, and quantify compounds based on their idiosyncratic polarities and interactions with the column's stationary phase. It utilizes different types of stationary phase (typically, hydrophobic saturated carbon chains), a pump that moves the mobile phase(s) and analyte through the column, and a detector that provides a characteristic retention time for the analyte. The detector may also provide other characteristic information (i.e., UV/VIS spectroscopic data for analyte if so equipped). Analyte retention time varies depending on the strength of its interactions with the stationary phase, the ratio/composition of solvent(s) used, and the flow rate of the mobile phase. With HPLC, a pump (rather than gravity) provides the higher pressure required to propel the mobile phase and analyte through the densely packed column. The increased density arises from smaller particle sizes. This allows for a better separation on columns of shorter length when compared to ordinary column chromatography.

In general, HPLC can be used whenever the sample is a liquid, or can be dissolved in a liquid. Given a particular problem, the extreme versatility of the HPLC technique often makes it the separation technique of first choice. Some of the more common uses are listed in Table 11.4.

Table 11.4 *Common HPLC uses*

Life Science	*Industrial*	*Pharmaceutical*
Drug Monitoring	Polymers	Drug analysis
Toxicology	Pesticides	Formulation testing
Special Chemistries	Oil products	Trace contaminants
Amino acids	Antioxidants	Raw products
Proteins	Surfactants	Quality control
Lipids	Complex formulation	Drug screening
Carbohydrates	Reaction mechanics	

One of the most attractive uses of HPLC is in the analysis of trace components. There are a variety of sophisticated approaches, ranging from preconcentration to selective detection, that aid in trace analysis. Two applications of HPLC which have benefited significantly from advances in this area are clinical chemistry and trace, pollution analysis. The clinical measurement of both endogenous and; exogenous compounds in body fluids is becoming increasingly practical using HPLC. The inherent sensitivity of the HPLC technique allows detection of extremely low levels of carcinogenic pollutants.

Considering its versatility, HPLC's limitations are quite minor. Some limitations are equipment related, while others are due to the stationary phases and chemical interactions. For example, the most common stationary phase is silica gel which dissolves under basic pH conditions. Since some separations are best performed with basic additives, there may be stationary phase degradation. Another example is the use of certain size exclusion chromatography stationary phases with asphalt products. In particular, a stationary phase composed of polystyrene-divinylbenzene (PSDVB) will irreversibly absorb certain components of crude oils. In time, this will degrade column performance. Of course, these situations are atypical, and because more than one way exists to solve a problem, these limitations do not really detract from the overall usefulness of the HPLC technique.

11.11 REFERENCES

1. P.A. Bristow, Liquid Chromatography in Practice, DCD (Pharmaceuticals Division), Macclesfield, U.K., 1976.
2. E.L. Johnson, R Stevenson, Basic Liquid Chromatography, Varian Associates, Inc., Palo Alto, CA, 1978.
3. J.J. Kirkland, L.R Snyder, Introduction to Modern Liquid Chromatography, John Wiley & Sons, NY, 1974.
4. H. Engelhardt, High Performance Liquid Chromatography, John Wiley & Sons, NY, 1979.
5. C.F. Simpson, Edit., Practical High Performance Liquid Chromatography, Heyden & Son Ltd., 1978.
6. Y. Shen and R.D. Smith, Electrophoresis 23 (2002) pp. 3106-3124.
7. D.A. Skoog and D.M. West, Fundamentals of Analytical Chemistry, 3rd edn., pp. 643–649.
8. D. Sievers, M.E. Swartz and B.J. Murphy, G.I.T. Laboratory Journal 5 (2004) pp. 43-45. (2004) pp. 503-504
9. L.A. Colon, J.M. Cintron, J.A. Anspach, A.M. Fermier and K.A. Swinney, The Analyst 129.
10. L.R. Snyder, J.J. Kirkland and J.L. Glajch, Practical HPLC Method Development, 2nd edn., pp. 41-43.

11 J.C. Giddings, Dynamics of Chromatography: Part I Principles and Theory (Marcel Dekker, Inc., New York: 1965) pp. 25-26, 229-230.

12. J.M. Cintron and L.A. Colon, The Analyst 127 (2002) pp. 701-704. A.D. Jerkovich, J.S. Mellors and J.W. Jorgenson, LCGC 21 (July 2003).

13. Instrumental Methods of Analysis" by Willard, Merritt and Dean, 4th edn., D. Van Nostrand Co. Inc., 1965.

CHAPTER 12

High Pressure Thin Layer Chromatography

12.1 INTRODUCTION

Thin layer chromatography (TLC) is a simple, rapid, versatile, sensitive, inexpensive analytical technique for the separation of substances. In this, the mobile phase is a liquid, containing a single solvent or a mixture of solvents and the stationary phase is an active solid, known as sorbent, wherein a mobile phase passes over a stationary phase in such as way that a mixture of substances is separated into its components. The term "thin-layer chromatography", introduced by E. Stahl in 1956, means a chromatographic separation process in which the stationary phase consists of a thin layer applied to a solid "substrate" or "support". For some years, TLC has also been referred to as planar chromatography. TLC can be used both as an analytical and a preparative technique. TLC has many advantages over other chromatographic methods for separation of simple mixtures.

TLC is used if

- the substances are non-volatile or of low volatility.
- the substances are strongly polar, of medium polarity, nonpolar or ionic.
- a large number of samples must be analyzed simultaneously, cost-effectively, and within a limited period of time.
- the samples to be analyzed would damage or destroy the columns of LC (liquid chromatography) or GC (gas chromatography).
- the solvents used would attack the sorbents in LC column packings.
- the substances in the material being analyzed cannot be detected by the methods of LC or GC or only with great difficulty.
- after the chromatography, all the components of the samples have to be detectable (remain at the start or migrate with the front).
- the components of a mixture of substances after separation have to be detected individually or have to be subjected to various detection methods one after the other (e.g., in drug screening).
- no source of electricity is available.

The modern use of TLC has seen a strong move in the direction of plate scanning and video imaging as a means of providing sensitive and reliably accurate results and more permanent record of the chromatogram. This is in addition to its obvious labor saving aspect and chemically "clean" approach. Thin layer chromatography has been successfully hyphenated with high performance liquid chromatography (HPLC), mass spectroscopy (MS), Fourier transform infrared (FTIR), and Raman spectroscopy to give far more detailed analytical data on separated compounds. Even the UV/Visible diode array technique has been utilized in TLC to determine peak purity or the presence of unresolved analytes.

12.2 HISTORY OF TLC

Although column chromatography can be traced to its discoverer, the Russian botanist, Tswett in 1903, it was not until 1938 that separations on thin layers were achieved when Izmailov and Shraiber, looking for a simpler technique, which required less sample and sorbent, separated plant extracts using aluminium oxide spread on a glass plate. The sorbent was applied to a microscope slide as a slurry, giving a layer about 2 mm thick. The sample (plant extracts) was applied as droplets to the layer. The solvent (methanol) was then added drop-wise from above on to the applied spots and a series of circular rings were obtained of differing colors on the layer. Circular TLC was born, and Izmailov and Shraiber named this new technique "drop chromatography".

In 1949, Meinhard and Hall used a starch binder to give some firmness to the layer, in order to separate inorganic ions, which they described as "surface chromatography". Further advances were made in 1951 by Kirchner et al., who used the now conventional ascending method, with a sorbent composed of silicic acid, for the separation of terpene derivatives, describing the plates used as "chromatostrips". In 1954, Reitsema used much broader plates and was able to separate several mixtures in one run. Surprisingly it was some time before the advantages of this development were recognized.

However, from 1956 a series of papers from Stahl appeared introducing "thin-layer chromatography" as an analytical procedure, describing the equipment and characterization of sorbents for plate preparation. Silica gel "nach Stahl" or "according to Stahl" became well known, with plaster of Paris (calcium sulphate) being used as a binder and TLC began to be widely used. In 1962, Kurt Randerath's book on TLC was published, followed by those of Stahl and co-workers, entitled 'Thin Layer Chromatography – A Laboratory Handbook' (1965), and Kirchner's, 'Thin Layer Chromatography' (1967). Then, in 1969 the 2nd edition of Stahl's book appeared which was greatly expanded. These authors showed the wide versatility of TLC and its applicability to a large spectrum of separation problems and also illustrated how quickly the technique had gained acceptance throughout the world. (By 1965 Stahl could quote over 4500 publications.) With Stahl's publication the importance of factors such as controlling the layer thickness, layer uniformity, the binder level and the standardisation of the sorbents as regards pore size and volume, the specific surface area and particle size, were recognised as crucial to obtaining highly reproducible, quality separations.

The next major advance was the advent of HPTLC (High performance thin-layer chromatography). In 1973, Halpaap was one of the first to recognize the advantage of using a smaller average particle size of silica gel (about 5–6 μm) in the preparation of TLC plates. He compared the effect of particle size on development time, R_f values and plate height. By the mid-1970s it was recognized that HPTLC added a new dimension to TLC as it was demonstrated that precision could be improved tenfold, analysis time could be reduced by a similar factor, less mobile phase was required, and the development distances on the layers could be reduced. The technique could now be made fully instrumental to give accuracy comparable with HPLC. Commercially the plates were first called "nano-TLC" plates by the manufacturer, (Merck), but this was soon changed to the designation "HPTLC". In 1977, the first major HPTLC publication appeared, simply called "HPTLC high performance thin-layer chromatography" edited by Zlatkis and Kaiser. In this volume Halpaap and Ripphahn described their comparative results with the new 5 × 5 cm HPTLC plates versus conventional TLC for a series of lipophilic dyes. Bonded phases then followed in quick succession.

Reversed-phase HPTLC was reported in 1980 by Halpaap et al. and this soon became commercially available as pre-coated plates. In 1982, Jost and Hauck reported an amino (NH_2^-) modified HPTLC plate, which was soon followed by cyano-bonded (1985) and diol-bonded (1987) phases. The 1980s also saw improvements in spectrodensitometric scanners with full computer control becoming possible, including options for peak purity and the measurement of full UV/visible spectra for all separated components. Automated multiple development (AMD) made its appearance in 1984 due to the pioneering work of Burger. This improvement enabled a marked increase in the number and resolution of the separated components.

In recent years TLC/HPTLC research has entered the chiral separation field using a number of chiral selectors and chiral stationary phases. Only one type of chiral pre-coated plate commercially available, which is based on a ligand exchange principle and is produced commercially either as a TLC or HPTLC plate. Gunther has reported results with amino-acids and derivatives on the TLC plate and Mack and Hauck similarly with their HPTLC equivalent.

Now all the steps of the TLC process can be computer controlled. The use of highly sensitive charge coupled device (CCD) cameras have enabled the chromatographer to electronically store images of chromatograms for future use (identity or stability testing) and for direct entry into reports at a later date. Commercially available HPTLC plates coated with specially pure 4–5 μm spherical silica gel have added further capabilities to the technique. Background interference has been reduced, and resolution further improved, which has enabled TLC to be hyphenated effectively with Raman spectroscopy.

12.3 BASIC PRINCIPLES OF HPTLC

Thin layer chromatography (TLC) and high performance thin layer, chromatography (HPTLC) also called planar chromatography – are, like all chromatographic techniques, based on a multistage distribution processes. This involves a suitable adsorbent (the stationary phase), solvents or solvent mixtures (the mobile phase or element), and the sample molecules, for thin layer chromatography the adsorbent is coated as a thin layer into a suitable support (e.g., glass plate, polyester or aluminium sheet), on this layer the substance mixture is separated by elution with a suitable solvent. Solute and solvent compete for "sites" on the adsorbent, to be adsorbed, the solute molecule must first displace a solvent molecule. Molecules with polar functional groups or those capable of hydrogen bonding will have a strong affinity for the adsorbent surface and will be strongly retained. The adsorbent surface consists of discrete adsorption sites. In the case of silica, these are hydroxyl (ooh) groups, the following SI-OH is known as silanol group. The ability of a solute to distribute itself between the phases according to the separation coefficient is:

$$K = Cs/Cm.$$

If the process involves the exchange of negatively charged ions, it is known as anion exchange. The complementary process is known as cation-exchange. The retention of solute takes place on the basis of size or shape. The applicability of thin layer chromatography has been enhanced by the development of new adsorbents and supports.

12.3.1 Main Difference of HPTLC and TLC: Particle and Pore Size of Sorbents

The other differences are

	HPTLC	TLC
Layer of Sorbent	• 100 μm	• 250 μm
Efficiency	• High due to smaller particle size generated	• Less
Separations	• 3 - 5 cm	• 10-15 cm
Analysis Time	• Shorter migration distance and the analysis time is greatly reduced	• Slower
Solid support	• Wide choice of stationary phases like silica gel for normal phase and C8 , C18 for reversed phase modes	• Silica gel, Alumina & Kiesulguhr
Development chamber	• New type that requires less amount of mobile phase	• More amount
Sample spotting	• Auto sampler	• Manual spotting
Scanning	• Use of UV/ Visible/ Fluorescence scanner scans the entire chromatogram qualitatively and quantitatively and the scanner is an advanced type of densitometer	• Not possible

12.3.2 Features of HPTLC

1. Simultaneous processing of the sample and the standard — better analytical precision and accuracy, less need for internal standard
2. Several analysts work simultaneously
3. Less analysis time and low cost per analysis
4. Low maintenance cost
5. Simple sample preparation — handle samples of divergent nature
6. No prior treatment for solvents like filtration and degassing
7. Low mobile phase consumption per sample
8. No interference from previous analysis — fresh stationary and mobile phases for each analysis — no contamination
9. Visual detection possible — open system
10. Non-UV absorbing compounds detected by post-chromatographic derivatization

12.4 STEPS INVOLVED IN HPTLC

1. Selection of chromatographic layer
2. Sample and standard preparation
3. Layer pre-washing
4. Layer pre-conditioning
5. Application of sample and standard
6. Chromatographic development
7. Detection of spots
8. Scanning
9. Documentation of chromatic plate

Selection of chromatographic layer

- Precoated plates, different support materials, different sorbents available
- 80% of analysis, silica gel GF. Basic substances, alkaloids and steroids

Aluminum oxide

- Amino acids, dipeptides, sugars and alkaloids, cellulose
- Nonpolar substances, fatty acids, carotenoids, cholesterol, RP2, RP8 and RP18
- Preservatives, barbiturates, analgesic and phenothiazines, hybrid plates-RPWF254s

Sample and Standard Preparation

- Avoid interference from impurities and water vapors
- Use low signal to noise ratio, Straight base line- Improvement of LOD

Solvents used are Methanol, Chloroform: Methanol (1:1), Ethyl acetate: Methanol (1:1), Chloroform: Methanol: Ammonia (90:!0:1), Methylene chloride : Methanol (1:1), 1% Ammonia or 1% Acetic acid

Dry the plates and store in dust-free atmosphere.

Activation of pre-coated plates

- Freshly open box of plates do not require activation.
- Plates exposed to high humidity or kept on hand for long time to be activated.
- By placing in an oven at 110-120ºC for 30 minutes prior to spotting aluminum sheets should be kept in between two glass plates and placing in oven at 110-120ºC for 15 minutes.

Application of sample and standard

- Usual concentration range is 0.1-1 µg / µL
- Above this causes poor separation
- Linomat IV (automatic applicator); nitrogen gas sprays sample and standard from syringe on TLC plates as bands
- Band-wise application; better separation; high response to densitometer

Selection of mobile phase

- Trial and error
- One's own experience and literature
- Normal phase
- Stationary phase is polar
- Mobile phase is nonpolar
- Nonpolar compounds eluted first because of lower affinity with stationary phase
- Polar compounds retained because of higher affinity with the stationary phasereversed phase
- Stationary phase is nonpolar
- Mobile phase is polar
- Polar compounds eluted first because of lower affinity with stationary phase
- Nonpolar compounds retained because of higher affinity with the stationary phase
- 3 - 4 component mobile phase should be avoided
- Multi-component mobile phase once used not recommended for further use and solvent composition is expressed by volumes (v/v) and sum of volumes is usually 100
- Twin trough chambers are used only 10-15 mL of mobile phase is required
- Components of mobile phase should be mixed introduced into the twin trough chamber

Pre-conditioning (Chamber saturation)

- Unsaturated chamber causes high Rf values
- Saturated chamber by lining with filter paper for 30 minutes prior to development - uniform distribution of solvent vapors - less solvent for the sample to travel - lower Rf values.

Chromatographic development and drying

- After development, remove the plate and mobile phase is removed from the plate to avoid contamination of lab atmosphere
- Dry in vacuum desiccator, avoid hair drier, essential oil components may evaporate

Detection and visualization

- Detection under UV light is first choice; non-destructive
- Spots of fluorescent compounds can be seen at 254 nm (short wavelength) or at 366 nm (long wavelength)
- Spots of non-fluorescent compounds can be seen; fluorescent stationary phase is used; silica gel GF
- Non-UV absorbing compounds like ethambutol, dicylomine etc., dipping the plates in 0.1% iodine solution
- When individual component does not respond to UV; derivatisation required for detection

Quantification

- Sample and standard should be chromatographed on same plate; after development chromatogram is scanned
- Camag TLC scanner III scans the chromatogram in reflectance or in transmittance mode by absorbance or by fluorescent mode; scanning speed is selectable up to 100 mm/s; spectra recording is fast; 36 tracks with up to 100 peak windows can be evaluated
- Calibration of single and multiple levels with linear or non-linear regressions are possible· When target values are to be verified such as stability testing and dissolution profile single level calibration is suitable
- Statistics such as RSD or CI report automatically
- Concentration of analyte in the sample is calculated by considering the sample initially taken and dilution factors

Documentation

E-Merck introduced plates with imprinted identification code – supplier name. Item number, batch number and individual plate number. Avoid manipulation of data at any stage – coding automatically recorded during photo documentation

12.5 THIN LAYER CHROMATOGRAPHY TO PREPARATIVE CHROMATOGRAPHY

One of the first steps in scale-up of preparative liquid chromatography separations is selection of an appropriate mobile phase. Two methods are commonly used to determine the proper mobile phase composition: Thin Layer Chromatography (TLC) or High Performance Liquid Chromatography (HPLC). The use of TLC will be discussed here to deal with the successful correlation between the TLC separation to the preparative silica column. TLC is a liquid-solid adsorption technique where the mobile phase ascends the thin layer of stationary phase coated onto a backing support such as glass by capillary action. There is a similar relationship to column chromatography where the solvent travels down through the column's adsorbent. The similar relationship allows TLC to be a rapid method for determining solvent composition for preparative separations.

Steps for Method Development

Choose stationary phase

Choose a scalable TLC plate, preferably that has an identical media as the preparative column. Choose between normal and reverse phase based on sample polarity and solubility.

Choose a Mobile Phase

Criteria for choosing a preparative solvent

- Solubility
- Affinity
- Resolution

1. **Solubility:** Many solvent systems provide the minimal solubility for the sample, but to elute a sample from a column the mobile phase must have a greater solubility for the sample, as the sample concentration is usually very high. When possible, it is best to dissolve the sample in the mobile phase. The first step in solvent selection is determination of the solubility of the sample. The desired mobile phase would provide the greatest solubility, while providing affinity for the sample on the stationary phase.

Solvent Solubility Screening Table

Solvent	
Water	Increasing Polarity
Methanol	
Ethanol	
Acetone	
Diethyl Ether	
Ethyl Acetate	
Dichloromethane	
Toluene	
Chloroform	
Cyclohexane	
Petroleum Ether	
Hexane	

2. **Affinity:** To achieve a separation, the sample must have a relatively equal affinity for the solvent and the packing material. If the sample has a higher affinity for the stationary phase than the solvent, the sample will remain at the origin (R_f value will be too low).

3. **Resolution:** Resolution is improved by optimizing the affinity between sample, solvent, and support. The optimum solvent for separating two or more compounds will maximize the difference in the compounds. Most TLC and preparative mobile phase systems contain a polar solvent and a chromatographically dissimilar less polar solvent. As a guide for method development, a substitution in the polar solvent often results in a change in resolution, while a change in the less-polar solvent results primarily in a change in R_f of the sample components. The table below shows some common tendencies of various functional groups to adsorb onto the silica.

Affinity of Functional Groups for Silica Gel

- NH_2 Amine
- COOH Carboxylic Acid
- COH Alcohol
- $CONH_2$ Amide
- C=O Carbonyl
- $C{=}{-}CO_2R$ Ester
- C-O-C Ether
- C1 Halocarbons
- CC- Hydrocarbons

Select Visualization Technique

Once a mobile phase is selected, visualization techniques will need to be determined. Common techniques include SWUV, I_2 /SWUV, I_2 /KI for nitrogen containing compounds, H_2SO_4/LWUV, H_2SO_4/PMA for non-nitrogen containing compounds.

Perform TLC Analysis

Look up the affinity for the type of compound as well as the solvent strengths to find a starting point for method development or look up a reference from a similar structure, then adjust the mobile phase composition to adjust the R_f. It is common to try 3–6 solvent systems for the first round of method development. Review the results after visualization and adjust the R_f if necessary, increase the separation and evaluate visualization techniques to make sure you are seeing all necessary compounds.

Optimizing TLC Separations for Preparative Separations

The optimum separation of compounds by TLC is usually achieved when R_f values are between 0.3–0.5.

$$R_f = \frac{\text{Distance from origin to center of spot}}{\text{Distance from origin to solvent front}}$$

Generally, adjusting the compound's R_f between 0.3–0.5 is done first for a TLC separation. For scale-up to preparative separations, the TLC solvent system's polarity must be decreased to lower the R_f between 0.15–0.35. This R_f range is optimal for a preparative separation, in terms of sample load, resolution, residence time, and solvent usage.

Determination of Column Volumes (C.V.)

The equation C.V. = 1/ R_f relates the TLC values and the preparative LC column volumes to elute each component. This equation is only a guideline and the relationship between the R_f values and the column volume will vary in use. Generally, the LC column volumes will be equal to or less than the calculated values. The elution volume will also be dependent upon the sample load and solvent used to solubilize the sample.

12.6 SAMPLE APPLICATION

Samples should be placed on a TLC plate as a spot that is as small as possible. Sample volumes of 1 to 2 mL are common but when using the so called high performance TLC plates (HPTLC plates) coated with particles 5-7 mm in diameter a maximum loading of about 100-200 nL can only be tolerated. Ideally, the sample area should be circular and not greater than 1 mm in diameter, on any HPTLC plate. As this requires the use of samples a few nanoliters in volume, most will need to be concentrated. Manually, the sample can be applied with a micro-pipette and the solvent is then allowed to evaporate. This procedure is repeated until sufficient sample is placed on the plate. Micro-syringes can also be employed to place a sample on to the plate. With care, and a little local heating, sample concentration can also be accomplished. The contents of the syringe are slowly but continuously discharged onto the plate and at the same time the solvent is progressively evaporated. This procedure can be automated, using computer controlled syringes and, in this way, samples can be focused onto the plate and constrained to a very small area.

There are various devices that are commercially available that will apply samples to a TLC plate either individually or as multiple groups. An example of a device for automatically concentrating a sample and then placing it on a plate is shown in Fig. 12.1.

12.7 QUALITY STANDARDIZATION BY HPTLC

12.7.1 Occurrence of Curcuminoids in *Curcuma longa*

Paramasivam, M. et.al. (2008) have developed a simple high performance thin layer chromatographic (HPTLC) method for the simultaneous determination of the pharmacologically important active curcuminoids, viz. curcumin, demethoxycurcumin and bis-demethoxycurcumin in *Curcuma longa* L. The assay combines the separation and quantification of the analytes on silica gel 60 GF254 HPTLC plates with visualization under UV and scanning at 425 nm. Using this technique, the alkaloidal content of different parts of the title plant has been determined.

Methods: The methods involve:

Plant material: Rhizomes of *C. longa* were collected from the Experimental Farm for the Development of Medicinal and Aromatic Plants (BCKV, Mohanpur, India). The plant specimens were authenticated and a voucher specimen is deposited in the herbarium.

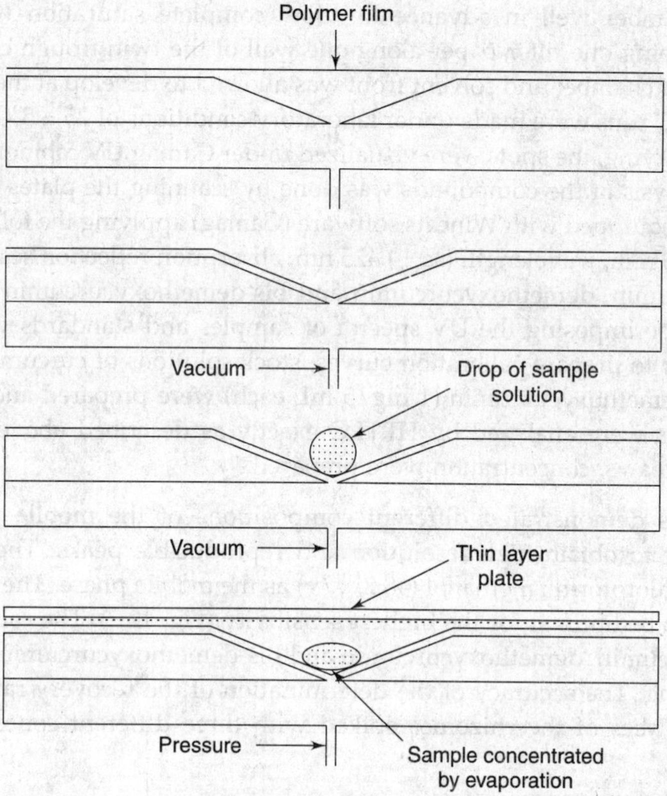

Fig. 12.1 *TLC sampling device incorporating sample concentration.*

Extraction of plant material for analysis: Air dried (35- 50°C) rhizomes of seven germplasm of *C. longa* and market turmeric powder samples (1 g each) were ultrasonically extracted separately in 20 mL HPLC grade methanol for 15 min (3 times) and filtered through Whatman No. 42 filter paper after each extraction. Extracts were concentrated under vacuum and finally made up to 20 mL with HPLC grade methanol and ready for HPTLC analysis.

Chromatographic conditions: Chromatography was performed on glass-backed silica gel 60 GF254 HPTLC layers (20 × 20 cm, 300 μm layer thickness) prepared using a Camag (Multenz, Switzerland) TLC plate auto-coater. Methanolic solutions of samples and standard compounds curcumin, demethoxycurcumin and bis-demethoxycurcumin of known concentrations were applied to the layers as 7 mm wide bands positioned 15 mm from the bottom and 20 mm from the side of the plate, using a Camag Linomat 5 automated TLC applicator with the nitrogen flow providing a delivery speed of 150 nL/s from the syringe. These parameters were kept constant throughout the analysis of samples.

Detection and quantification of the alkaloids: After sample application plates were developed in a Camag twin trough glass tank pre-saturated with the mobile phase chloroform: methanol (48:2, v/v) for one hour. It was then poured in twin trough glass solvent development chamber well in advance to allow complete saturation which was further enhanced by keeping one filter paper along one wall of the twin trough chamber. The plate was then kept in a chamber and solvent front was allowed to develop at the height of 7 cm on the plate. The TLC runs were made under laboratory conditions of 25 ± 5°C and 50 % relative humidity. After drying, the spots were visualized under Camag UV cabinet (254 and 366 nm). Quantitative analysis of the compounds was done by scanning the plates using Camag TLC scanner model 3 equipped with Wincats software (Camag) applying the following conditions: slit width 6 × 0.45 mm, wavelength (λ_{max}) 425 nm, absorption reflection scan mode. The identification of curcumin, demethoxycurcumin and bis-demethoxycurcumin in rhizomes were confirmed by superimposing the UV spectra of samples and standards within the same R_f window. In order to prepare calibration curves, stock solutions of curcumin, demethoxycurcumin and bis-demethoxycurcumin (1 mg/5 mL each) were prepared and various volumes of these solutions were analyzed by HPTLC exactly as described above. Then calibration curves of peak area vs. concentration were prepared.

Results: Results demonstrated different compositions of the mobile phase for HPTLC analysis in order to obtain high resolution and reproducible peaks. The desired aim was achieved using chloroform : methanol (48:2, v/v) as the mobile phase. The wavelength of 425 nm was found to be optimal for the highest sensitivity (Fig. 12.2). The calibration curves for the alkaloids curcumin, demethoxycurcumin and bis-demethoxycurcumin were linear in the range 100-1,000 ng. The accuracy of the determination of the recovery rate was determined by triplicate analyses of the rhizomes spiked with three different concentrations of stock

Fig. 12.2 *HPTLC chromatogram of the standard curcuminoids.*

solution of curcumin, demethoxycurcumin and bis-demethoxycurcumin. The recovery rates were 97.3%, 92.9% and 95.4% for curcumin, demethoxycurcumin and bis-demethoxycurcumin, respectively.

For the quantitative determination of curcumin, demethoxycurcumin and bis-demethoxycurcumin, the analyses of turmeric rhizome specimens of *C. longa* were repeated three times. It is clear that three alkaloids were present in two cultivars, viz. Kalimpong and PTS-43 having their maximum concentrations in the Kalimpong cultivar.

In summary, the HPTLC method for the simultaneous analysis of curcumin, demethoxycurcumin and bis-demethoxycurcumin from *C. longa* reported here is very simple, sensitive, economic and suitable for rapid screening of large number of plant samples. Moreover, this analysis can be performed without any special sample pre-treatment and fifteen samples can be analyzed on a single TLC laver (20 × 20 cm).

12.7.2 Determination of Caffeine, Ergotamine and Metamizol by HPTLC/UV/FLD and HPTLC/ESI-MS

Mario Aranda and Gertrud Morlock has developed simultaneous Determination of Caffeine, Ergotamine and Metamizol by HPTLC/UV/FLD and HPTLC/ESI-MS.

The objective of this work was to develop a high throughput analytical method to detect simultaneously caffeine, ergotamine tartrate and metamizol in headache tablets by HPTLC-UV/FLD with mass confirmation via HPTLC-ESI/MS. IN methodology the compounds were dissolved with 100 mL of methanol-water mixture (7:3 v/v) and to complete the dissolution process, the flask was shaken in a mechanical shaker for 20 minutes and sonicated for 10 minutes in an ultrasonic bath. Chromatography was carried out on 20 × 10 cm HPTLC plate silica gel 60. Solutions of samples and the standard mixture were applied with Automatic TLC Sampler IV (ATS IV) from CAMAG (Muttenz, Switzerland). Detection was performed by multi-wavelength scanning using the scanner 3 (CAMAG).

Calibration plots showed regressions with determination coefficients (r^2) > 0.999. Recoveries in commercially available pharmaceutical products were 102.8% ± 2.8% for ergotamine, 106.6% ± 3.2% for caffeine and 104.7% ± 2.2% for metamizol (Fig. 12.3).

HPTLC/ESI-MS

The compounds mass spectra were obtained using a plunger-based extraction device for HPTLC/ESI-MS which employment was recently demonstrated in food and trace analysis (Fig. 12.4).

This new HPTLC method offers a reliable and low-cost high throughput alternative (Fig. 12.5).

The compounds online identification via HPTLC/MD gives a reliable approach to ensure correct peak identification and is also helpful for deciphering the type of impurities present.

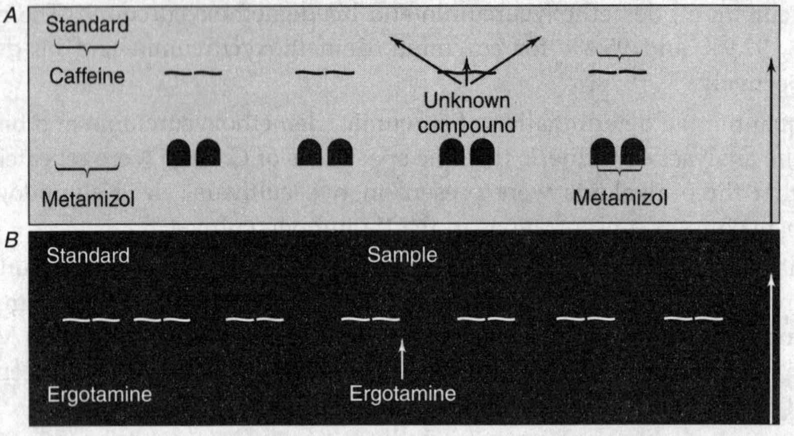

Fig. 12.3 *HPTLC plate silica gel 60 illuminated at UV 254 nm (A) and 366-400 nm (B).*

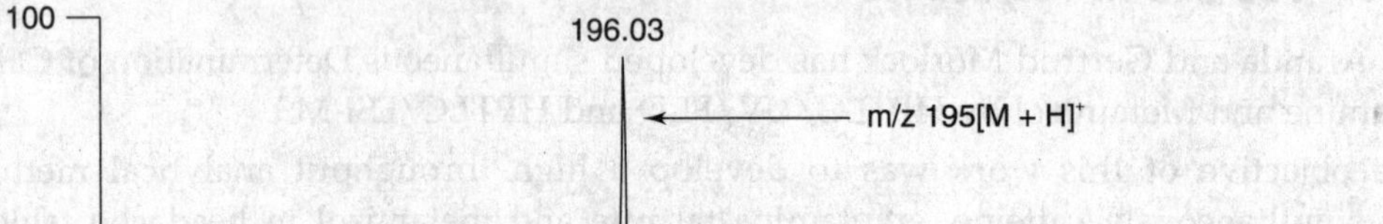

Fig. 12.4 *Caffeine mass spectrum.*

12.8 TLC/HPTLC APPLICATIONS

Pharmaceuticals and Drugs

Identification, purity testing and determination of the concentration of active ingredients, auxiliary substances and preservatives in drugs and drug preparations, process control in synthetic manufacturing processes.

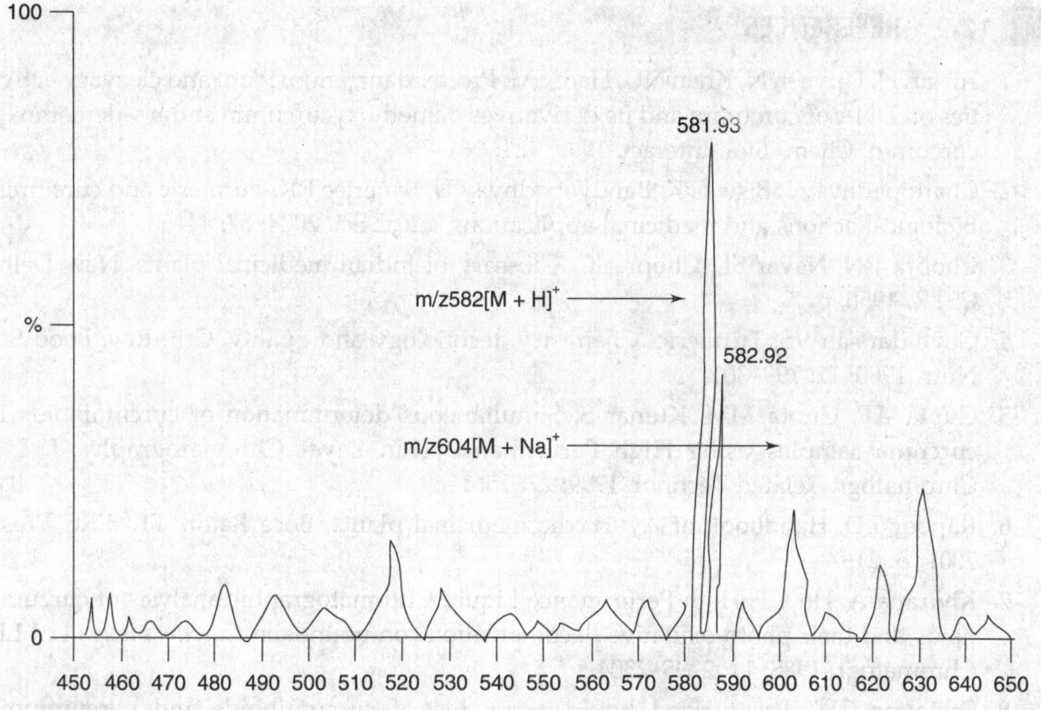

Fig. 12.5 *Ergotamine mass spectrum examples of mass spectra obtained directly from the compound band.*

Clinical Chemistry, Forensic Chemistry and Biochemistry

Determination of active substances and their metabolites in biological matrices, diagnosis of metabolic disorders such as PKU (phenylketonuria), cystinuria and maple syrup disease in babies.

Cosmetology

Dye raw materials and end products, preservatives, surfactants, fatty acids, constituents of perfumes.

Food Analysis

Determination of pesticides and fungicides in drinking water, residues in vegetables, salads and meat, vitamins in soft drinks and margarine, banned additives in Germany (e.g., sandalwood extract in fish and meat products), compliance with limit values (e.g., polycyclic compounds in drinking water, aflatoxins in milk and milk products).

Environmental Analysis

Groundwater analysis, determination of pollutants from abandoned armaments in soils and surface waters, decomposition products from azo dyes used in textiles.

Analysis of Inorganic Substances

Determination of inorganic ions (metals).

Other Areas

Electrolytic technology (*meta*-nitrobenzoic acid in nickel plating baths).

12.9 REFERENCES

1. Ahsan H, Parveen N, Khan NU, Hadi SM. Pro-oxidant, antioxidant and cleavage activities on DNA of curcumin and its derivatives demethoxycurcumin and bis-demethoxycurcumin. Chem. Biol. Interact. 1999; 121: 161–75.
2. Chattopadhyay I, Biswas K, Bandyopadhyay U, Banerjee RK. Turmeric and curcumin: Biological actions and medicinal applications. Curr. Sci. 2004; 87: 44–50.
3. Chopra RN, Nayar SL, Chopra IC. Glossary of Indian medicinal plants. New Delhi, CSTR, 1956, p. 7.
4. Govindarajan VS. Turmeric- Chemistry, technology and quality. Crit. Rev. Food Sci. Nutr. 1980; 12: 199–301.
5. Gupta AP, Gupta MM, Kumar S. Simultaneous determination of curcuminoids in curcuma samples using High Performance Thin Layer Chromatography. J. Liq. Chromatogr. Related Technol. 1999; 22: 1561–69.
6. Kapoor LD. Handbook of Ayurvedic medicinal plants. Boca Raton, FL, CRC Press, 2001, p. 216.
7. Khurana A, Ho CT. High Performance Liquid Chromatographic analysis of curcuminoids and their photo-oxidative decomposition compounds in *Curcuma longa* L. J Liq Chromatogr. 1988; 11: 2295–2304.
8. Schieffer GW. Pressurized liquid extraction of curcuminoids and curcuminoid degradation products from turmeric (*Curcuma longa*) with subsequent HPLC assays. J. Liq. Chromatogr. Related Technol. 2002; 25: 3033–44.
9. Taylor SJ, McDowell IJ. Determination of the curcuminoid pigments in turmeric (*Curcuma domestica* Val.) by reversed phase High Performance Liquid Chromatography. Chromatographia 1992; 34: 73–77.
10. J. O'Sullivan and J.T. Mecabe, AANAJ. 2006, 74:61–69
11. H.J. Luftmann, Anal. Bioanal, Chem. 2004, 380:964–968
12. M. Aranda and G. Morlock J. Chromatogr. A. (Submitted)
13. G. Morlock, W. Schwack, Anal. Bioanal, Chem. 2006, 385:586–595.

CHAPTER 13

Gas Chromatography – Mass Spectrometry

13.1 INTRODUCTION

Gas Chromatography and mass spectrometry are in many ways highly compatible techniques. In both the techniques, the sample is in the vapor phase and both techniques deal with about the same amount of sample. Both Gas Chromatography (GC) and Mass Spectrometry (MS), bring something to their union. GC can separate volatile and semivolatile compounds with great resolution, but it cannot identify them. MS can provide detailed structural information on most compounds such that they can be exactly identified, but it cannot readily separate them. Therefore, the combination of the two techniques was suggested after the development of GC in the mid-1950s.

Through mass spectroscopy (MS) one can determine the molecular weight, molecular formula and the functional groups present on a compound. The molecular weight can be determined by identifying the molecular ion peak. Looking at the mass to charge ratio ($M/2$) and the distribution of isotopes one is able to determine an average mass for a compound. The mass contribution of each isotope is determined by each isotope's mass and relative abundance. Adding all the isotopes, mass contributions will then result in an average mass for the compound. The molecular formula for a compound can then be ascertained from the relative peak heights of the isotopes. Peak areas and heights can also be useful because they can indicate the concentration of various components present in the sample. When looking at mass spectra, the majority of peaks that appear are attributed to fragments; these fragments are pieces of the original compound that have separated from the compound. The appearance of these fragments makes interpretation slightly more difficult that is why the use of a library can become advantageous. By comparing the sample to knowns, a positive identification is possible.

The main advantage of interfacing a gas chromatograph with a mass spectrometer is its efficiency. With combined instrumentation, only one sample needs to be prepared. This saves on both time and material. The physical bench space taken up is minimized as well. The GC-MS/MS allows one to obtain more detailed information about unknown compounds faster and more easily than the instruments separated. Combining these technologies allows one to analyze more complex compounds faster than GC, MS or GC-MS.

13.2 HISTORICAL DEVELOPMENT

Eugen Goldstain, in 1986, observed rays in gas discharges under low pressure that traveled away from the anode and through channels in a perforated cathode, opposite to the direction of negatively charged cathode rays (which travel from cathode to anode). Goldstein called these positively charged anode rays "Kanalstrahlen", the standard translation of this term into English is "canal rays". Wilhelm Wein found that strong electric or magnetic fields deflected the canal rays and, in 1899 constructed a device with parallel electric and magnetic field that separated the positive rays according to their charge-to-mass ratio (Q/m). Wein found that the charge-to-mass ratio depended on the nature of the gas in the discharge tube. English scientist J.J. Thomson later improved on the work of Wein by reducing the pressure to create a mass spectrograph.

The first application of mass spectrometry to the analysis of amino acids and peptides was reported in 1958. Carl-One Anderson highlighted the main fragment ions observed in the ionization of methyl esters.

Some of the modern techniques of mass spectrometry were devised by Arthur Jeffrey Dempster and F.W. Aston in 1918 and 1919 respectively. In 1989, half of the Nobel Prize in Physics was awarded to Hans Dehmelt and Wolfgang Paul for the development of the ion trap technique in 1950s and 1960s. In 2002, the Nobel Prize in Chemistry was awarded to John Bennett Fenn for the development of electrospray ionization (ESI) and Koichi Tanaka for the development of Soft Laser Dispersion (SLD) and their application to the ionization of biological macromolecules, especially proteins.

The development of affordable and miniatured computers has helped in the simplification of the use of this instrument, as well as allowed great improvements in the amount of time it takes to analyze a sample. In 1996, the top-of-the-line high-speed GC-MS units completed analysis in 90 seconds, whereas first generation GC-MS would have required 15 minutes. This has led to their widespread adoption in a number of fields.

13.3 PRINCIPLE

Gas chromatography-mass spectroscopy (GC-MS) is an analytical technique for identification of a wide array of unknown organic substance in sample matrices. It is also used for elucidating the chemical structures of molecules and other chemical compounds. The MS principle consists of ionizing chemical compounds to generate charged molecules or molecule fragments and measurement of their mass-to-charge ratio.

The method is based on the principle of chromatographic separation of components of a mixture on a GC column, followed by their identification from their mass spectra. The compounds are separated on a suitable GC column following which the components eluted from the column are subjected to electron-impact or chemical ionization. The fragmented and molecular ions are identified from their characteristic mass spectra. Thus, the substances present in the sample are determined from their characteristic primary and secondary ions and also from their retention times.

MS instruments consists of three molecules : an **ion source**, which can convert gas phase sample molecules into ions; a **mass analyzer**, which sorts the ions by their masses by applying electromagnetic fields, and a **detector**, which measures the value of an indicator quantity and thus provides data for calculating the abundances of each ion present.

In mass spectroscopy working – upon passing through the column the particles continue on to the mass spectrometer where the sample is "bombarded by high energy electrons". These electrons collide with the atoms in the sample and as a result, knock an electron of the sample's atoms, inducing positively charged particles. The particles are then separated by a variety of means depending upon the type of mass spectrometer used. For example, in an ion trap mass spectrometer the particles are deflected depending upon their mass/charge ratio. A detector determines the particles' "relative positions and abundances" which can then be used to determine the identity of the compound. The largest peak in the spectrum is called the base peak. The heights of all the additional peaks in the spectrum are then measured as percent relative to the base peak.

The technique has both qualitative and quantitative uses. These include identifying unknown compounds, determining the isotopic composition of elements in a molecule, and determining the structure of a compound by observing its fragmentation. Other uses include quantifying the amount of a compound in a sample or studying the fundamental of gas phase ion chemistry. MS is now in very common use in analytical laboratories that study physical, chemical or biological properties of a great variety of compounds.

13.4 INSTRUMENTATION : GAS CHROMATOGRAPHY MASS SPECTROMETRY

The capabilities of integrated gas chromatography mass spectrometry are almost unique in meeting the requirements for analytical methods which are exquisitely sensitive but also specific and reliable in providing information about the biological effects of specific compounds as a function of their concentrations.

Gas chromatography (Fig. 13.1) is generally recognized as one of the most powerful separative methods available, and is widely employed in biomedical and environmental research. Although several selective detectors have been described, the identity of the separated components which elute must generally be inferred from their retention times relative to known standards. Mass spectrometry is a complementary technique in that it can provide a great deal of information about the identity of a compound provided that a pure sample is available, and may be sufficient to identify it rigorously. The combined system (GC/MS) can utilize both instruments to maximum advantage and, under favorable conditions can identify every major component of a complex mixture.

The GC portion of this system provides high resolution separation of volatile organic solutes in a mixture in the gas phase. As each solute exits the GC column, it is diverted into a mass spectrometer which is capable of both monitoring the amount of and identifying the chemical nature of the solute. In this way, both quantitative and qualitative information about the mixture can be obtained.

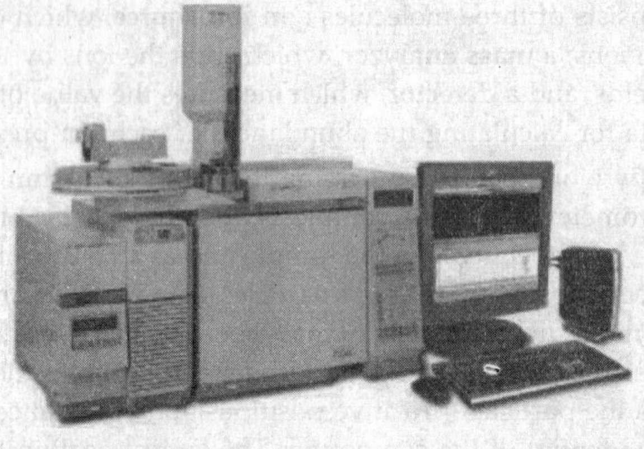

Fig. 13.1 *Gas chromatography.*

The MS portion of the system takes each gaseous solute exiting the GC and ionizes it in an electron beam. The ions formed by a specific solute will depend on the nature of the bonds in the molecule, and both ionized molecules and ion fragments of the molecule are possible. The ions are then directed down a separator which isolates and counts the ions according to mass. The sequence and relative intensity of the mass peaks give information about the chemical identity of the solute. The absolute intensity of the peaks provides information about the amount of substance present.

Figure 13.2 demonstrates components of a GC-MS instrument:

1. Sample introduction through the gas chromatograph.
2. The mass spectrograph components are kept under vacuum the (i) ion source (ii) mass analyzer, and (iii) detector information processed by a computer.
3. Data output.

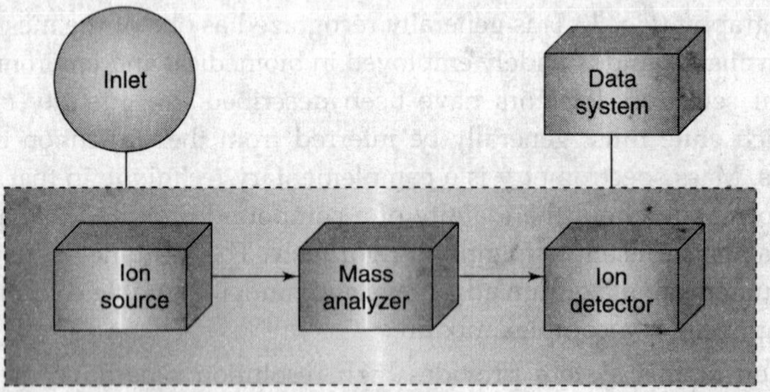

Fig. 13.2 *GCMS – instrumentation.*

13.4.1 Instrumentation

13.4.1.1 Ion source technologies

The ion source is the part of the mass spectrometer that ionizes the material under analysis (the analyte). The ions are then transported by magnetic or electric fields to the mass analyzer. Techniques for ionization have been key to determining what type of samples can be analyzed by mass spectrometry. The different ion source technologies include:

- **Chemical Ionization (CI):** CI uses a reagent ion to react with the analyte molecules to form ions by either a proton or hydride transfer.

  $$MH + C_2H_5^+ \rightarrow MH_2^+ + C_2H_4$$

  $$MH + C_2H_5^+ \rightarrow M^+ + C_2H_6$$

  The reagent ions are produced by introducing a large excess of methane (relative to the analyte) into an electron impact (EI) ion source. Electron collisons produce CH_4^+ and CH_3^+ which further react with methane to form CH_5^+ and $C_2H_5^+$

  $$CH_4^+ + CH_4 \rightarrow C_2H_5^+ + CH_3$$

  $$CH_3^+ + CH_4 \rightarrow C_2H_5^+ + H_2$$

- **Plasma and Glow Discharge:** Plasma is a hot, partially ionized gas that effectively excites and ionizes atoms. The most common plasma source is inductively-coupled plasma. A flow discharge is low-pressure plasma maintained between two electrodes. It is particularly effective at sputtering and ionizing material solid surfaces.
- **Electron Impact (EI):** EI source uses an electron beam, usually generated from a tunsten filament, to ionize gas phase atoms or molecules. An electron from the beam knocks an electron off of the analyte atoms or molecules to create ions.
- **Electrospray Ionization (ESI):** ESI source consists of a very fine needle and a series of skimmers. A sample solution is sprayed into the source chamber to form droplets. The droplets carry charge when one exits the capillary and, as the solvent evaporates, the droplets disappear leaving highly charged analyte molecules. ESI is particularly useful for large biological molecules that are difficult to vaporize or ionize.
- **Fast-atom bombard (FAB):** In FAB a high-energy beam of neutral atom, typically Xe of Ar, strikes a solid or low-vapor-pressure liquid sample causing desorption and ionization. It is used for large biological molecules that are difficult to get into the gas phase. The sample is usually dispersed in a matrix such as glycerol. FAB causes little fragmentation and usually gives a large molecular ion peak, making it useful for molecular weight determination. The atomic beam is produced by accelerating ions from an ion source through a charge exchange all. The ions pick up an electron in collisions with neutral atoms to form a beam of high energy atoms.
- **Field ionization:** Molecules can lose an electron when placed in a very high electric field. High fields can be created in an ion source by applying a high voltage between a

cathode and an anode called a field emitter. A field emitter consists of a wire covered with microscopic carbon dendrites, which greatly amplify the effective field at the carbon points.

- **Laser ionization (LIMS):** A laser pulse ablates material from the surface of a sample, and creates microplasma that ionizes source of the sample constituents. The laser pulse accomplished both vaporization and ionization of the sample.
- **Matrix-Assisted Laser Desorption Ionization (MALDI):** MALDI is a LIMS method of vaporizing and ionizing large biological molecules such as proteins or DNA fragments. The biological molecules are dispersed in a solid matrix such as nicotinic acid or dihydroxybenzoic acid. A UV laser pulse ablates the matrix which carries some of the larger molecules into the gas phase in an ionized form so they can be extracted into a mass spectrometer.
- **Plasma-Desorption Ionization (PD):** Decay of 252 CF produces two fisison fragments that travel in opposite directions. One fragment strikes the sample knocking out 1-10 analyte ions. The other fragment strikes a detector and triggers the start of data acquisition. This ionization method is especially useful for large biological molecules.
- **Resonance Ionization (RIMS):** One or more laser beams are tuned in resonance to transistions of a gas-phase atom or more to promote it in a stepwise fashion above its ionization potential to create an ion solid samples must be vaporized by heating, sputtering or laser ablation.
- **Spark Source:** A spark source ionizes analytes in solid samples by pulsing an electric current across two cathodes. If the sample is a metal it can serve as one of the electrodes otherwise it can be mixed with graphite and placed in a cup-shaped electrode.
- **Thermal ionization (TIMS):** Thermal ionization is used for elemental or refractory materials. A sample is deposited on a metal ribbon, such as Pt or Pe, and an electric current heats the metal to a high temperature. The ribbon is often coated with graphite to provide a reducing effect.

Ion attachment ionization is a newer soft ionization technique that allows for fragmentation free anlysis.

13.4.2 Mass Analyser Technologies

Mass analyzers separate the ions according to their mass-to-charge ratio. The following two laws govern the dynamics of charged particles in electric and magnetic fields in vacuum.

$$F = Q\,(E + V \times B) \text{ (Lorentz Force Law)}$$

(Newton's second law of motion in non-realistic case, i.e., valid only at ion velocitry much lower than the speed of light.)

Here F is the force applied to the ion, m is the mass of the ion, a is the acceleration, Q is the ion charge, E is the electrid field, and $V \times B$ is the vector cross product of the ion velocity and the magnetic field.

Equating the above expression for the force applied to the ion yields:

$$(m/Q)\, a = E + V \times B$$

The differential equation is the classic equation at motion for charged particles. Together with the particle's initial conditions, it completley determines the particle's motion in space and time in terms of m/Q. Thus, mass spectrometer could be thought of as "mass-to-charge spectrometers", when presenting data, it is common to use the dimension less m/z, where z is the number of elementary charges (e) on the ion ($z = Q/e$). This quantity, although it is internally called the mass-to-charge ratio, more accurately speaking, represents the ratio of the mass number and the charge number, z.

There are many types of mass analyzers, using both static or dynamic fields, and magnetic or electric fields, but all operate according to the above differential equation. Each analyzer type has its strengths and weaknesses. Many mass spectrometers use two or more mass analyzers for tandem mass spectrometry (MS/MS).

13.4.3 Detector

The final element of the mass spectrometer is the detector. The detector records either the charge induced or the current produced when an ion passes on or hits a surface. In a scanning instrument the signal produced in the detector during the course of the scan versus where the instrument is in the scan (at what m/Q) will produce a mass spectrum, a record of ions as a function of m/Q.

Detectors are of the following types:

- **Channeltron:** A channeltron is a horn-shaped continuous dynode structure that is coated on the side with a electron emissive material. An ion striking the channeltron creates secondary electrons that have an avalanche effect to create more secondary electrons and finally a current pulse.
- **Daly Detector:** A daly detector consists of a metal that emits secondary electrons when struck by an ion. The secondary electrons are accelerated onto a scintillator that produced light that is then detected by a photomultiplier tube.
- **Electron Multiplier Tube (EMT):** Electron multiplier tubes are similar in design to photomultiplier tubes. They consist of a series of biased dynodes that eject secondary electrons when they are struck by an ion. They therefore multiply the ion current and can be used in analog or digital mode.
- **Faraday Cup:** A Faraday, Cup is a metal cup that is placed in the path of the ion beam. It is attached to an electrometer which measures the ion-beam current. Since Faraday cup can only be used in an anolog mode it is less sensitive than other detectors that are capable of operating in pulse-counting mode.

- **Microchannel Plate:** A microchannel plate (MCP) consists of an array of glass capillaries (10 – 25 in inner diameter) that are coated on the inside with a electron-emissive material. The capillaries are biased at a high voltage and like the channeltron, an ion that strikes the inside wall one of the capillaries creates an avalanche of secondary electrons. This cascading effect creates a gain of 10^3 to 10^4 and produces a current pulse at the output.

13.5 GC- MS WORKING

The GC-MS instrument is made up of two parts. The gas chromatography (GC) portion separates the chemical mixture into pulses of pure chemicals and the mass spectrometer (MS) identifies and quantifies the chemicals.

The GC separates chemicals based on their volatility, or ease with which they evaporate into a gas. It is similar to a running race where a group of people begin at the starting line, but as the race proceeds, the runners separate based on their speed. The chemicals in the mixture separate based on their volatility. In general, small molecules travel more quickly than larger molecules.

The MS is used to identify chemicals based on their structure. Let's say after completing a puzzle, one can accidentally drop it on the floor. Some parts of the puzzle remain attached together and some individual pieces break off completely. By looking at these various pieces, you are still able to get an idea of what the original puzzle looked like. This is very similar to the way that the mass spectrometer works.

1. **Gas chromatography (GC)**
   - **Injection port** – One **microliter** (1 µl, or 0.000001 L) of solvent containing the mixture of molecules is injected into the GC and the sample is carried by inert (non-reactive) gas through the instrument, usually helium. The inject port is heated to 300°C to cause the chemicals to become gases.
   - **Oven** – The outer part of the GC is a very specialized **oven**. The column is heated to move the molecules through the column. Typical oven temperatures range from 40°C to 320°C.
   - **Column** – Inside the oven is the **column** which is a 30 meter thin tube with a special polymer coating on the inside. Chemical mixtures are separated based on their votality and are carried through the column by helium. Chemicals with high volatility travel through the column more quickly than chemicals with low votality.
2. **Mass spectrometer (MS)**
   - **Ion Source** – After passing through the GC, the chemical pulses continue to the MS. The molecules are blasted with electrons, which cause them to break into pieces and turn into positively charged particles called **ions**. This is important because the particles must be charged to pass through the filter.

- **Filter** – As the ions continue through the MS, they travel through an electromagnetic field that filters the ions based on mass. The scientist using the instrument chooses what range of masses should be allowed through the filter. The filter continuously scans through the range of masses as the stream of ions come from the ion source.
- **Detector** – A detector counts the number of ions with a specific mass. This information is sent to a computer and a **mass spectrum** is created. The mass spectrum is a graph of the number of ions with different masses that traveled through the filter.

3. **Computer**

   The data from the mass spectrometer is sent to a computer and plotted on a graph called **mass spectrum**.

13.6 ACHIEVING MAXIMUM SENSITIVITY IN GC-MS

The combined technique of gas chromatography-mass spectrometry (GC-MS) provides a powerful tool for separation, identification and quantification of compounds in complex mixtures. Increasingly sensitive instruments are invaluable to experts in a variety of fields who were previously unable to identify trace compounds in complicated or difficult samples.

Achieving Sensitivity

There are a number of ways to optimize GC-MS sensitivity. When sensitivity is defined in terms of signal-to-noise ratio, or S/N, increasing sensitivity is achieved by increasing signal, decreasing noise, or a combination of both. Because S/N is a simple ratio, factors affecting signal or noise have equivalent influence on sensitivity (S/N).

Intensity

Optimizing the transfer of analyte from the injector through all the components of the GC-MS system to the detector can increase the signal. Optimizing the response of the detector to obtain adequate signal without excessive noise is a complementary approach to increasing signal.

Many components of the GC-MS system can contribute to background, or chemical, noise. In addition, noise can originate from the sample itself, from the injection technique, or from the type of chromatographic column. Significant noise can also emanate from the electrical power source. Alternate data acquisition modes (SIM and recently introduced Scan/SIM techniques) have dramatic effects on GC-MS sensitivity. In a similar manner, alternate ionization techniques (chemical ionization, negative ion chemical ionization) can enhance sensitivity, especially for selected analytes. The present discussion of sensitivity is limited to the full-scan electron impact mode of operation and applies primarily to single-quadruple instruments.

Increasing Signal

Optimizing the signal can be considered in three phases: maximizing the injection, or transfer, to the GC column; optimizing the GC column for maximum sensitivity; and adjusting the detector (MS) conditions.

GC Injection

Efficient sample transfer to the GC column is dependent upon optimizing the injector conditions. Minor inefficiencies in injector operation become major obstacles when trace quantities of analyte are injected. The selection of injection-port liner, split- and septum-purge flow rates, syringe needle penetration, column positioning in the injector and other parameters affect the efficient transfer of analyte to the GC column. Using alternate injector types can enhance analyte transfer to the GC column.

GC Column Selection

For most GC-MS applications, the GC typically employs relatively long narrow-bore capillary columns (0.18 to 0.320-mm OD; 15 to 75 M long). Sensitivity increases as the GC column diameter decreases because narrower columns give taller, narrower peaks. If noise remains constant, S/N is greater with tall, narrow chromatographic peaks.

Optimizing the MS Signal

Signal generation in the MS comprises three processes:

- Ion creation (from the ion source)
- Ion transmission (through the lenses and mass filter)
- Ion detection (with the electron multiplier)

The combined efficiency of these processes results in optimizing signal intensity. In some cases, inefficiencies in these processes also affect noise. These efficiencies are affected by instrument design and condition (cleanliness, age), as well as operating parameters.

Ionization and Ion Transmission

Sensitivity is affected considerably by ion source condition (cleanliness) and optimum tuning. Ion source operating parameters have a significant effect on sensitivity because these parameters control ion production.

Various design improvements in mass spectrometer sources in recent years have resulted in significant improvements in sensitivity compared to previous designs. Filament construction and placement is one such design improvement that results in enhanced sensitivity. Filament shielding can ensure efficient ion transport to the detector while protecting thermally labile compounds. Thus, a high-efficiency ion source provides more uniform temperature control for increased sensitivity.

The vacuum system significantly affects ion transmission. Differentially pumped vacuum systems, in some mass spectrometer designs, maintain a higher level of vacuum in the mass analyzer region. An increased vacuum results in longer mean-free paths for ions, which allows for more efficient ion transmission through the mass filter. This, in turn, enables greater sensitivity.

Detectors

Electron multipliers detect and convert ions to a signal. When impacted by a charged particle (ion), the surface of the multiplier emits several electrons in a process called secondary emission. This process repeats several times to give up to a million electrons for each ion impact on the electron multiplier surface. The gain, or signal amplification, is determined by the voltage applied across the entire electron-several hundred to several thousand volts.

Most GC-MS instruments employ two types of electron multipliers: continuous dynode types and discrete dynode types. Discrete dynode electron multipliers typically operate at lower voltages, so they show slightly less noise than continuous dynode types. Typically, the electron multiplier's sensitivity can be optimized to obtain a maximum signal with minimum noise.

Decreasing Noise

Challenges and solutions in the GC that affect the S/N ratio directly affect the sensitivity of the GC-MS system as a whole. Lowering the background signal originating from the GC and improving the chromatographic resolution can improve GC sensitivity. As the quantity of a specific analyte to be detected is decreased, the effects of minor interferences on the ability to detect the specific analyte become an increasingly significant problem.

Carrier Gas

Achieving the highest possible purity in the carrier gas helps to decrease noise, specifically "chemical noise." For example, carbon dioxide gives a background signal at m/z 44 that increases chemical noise. Oxygen degrades column phase, which results in increased bleed (i.e., background signal at m/z 207). Using pure carrier gas and eliminating contamination and leaks in carrier gas lines are important factors affecting instrument sensitivity as well as overall chromatographic performance. Use of high-purity gases with appropriate filters can significantly lower the chromatographic background signal and therefore increase sensitivity.

Injection Ports

Chromatographers face the continuing challenge of assuring complete sample volatilization and transfer of sample to the GC. Using appropriate injection-port operating parameters and maintenance procedures ensures optimum sample transfer to the GC column.

Background noise often arises from siloxanes from the GC septa, glass wool used in injection port liners, deactivation of injection port liners and other sources. These peaks are commonly seen in normal GC-MS backgrounds. Use of low-bleed septa, preconditioning of septa and liners, and use of injection-port septum purge can minimize this source of contamination.

GC Columns

Another common source of background noise originates from the GC column stationary phases. Traces of oxygen in the carrier gas can degrade the liquid phase. To minimize this problem, the GC column, injection port, and transfer line temperatures should never exceed the maximum-rated temperature of the liquid phase of the capillary column. Thin liquid phases result in lower column bleed and decreased chemical noise. In addition, several proprietary low-bleed column phases, designed specifically for GC-MS use, have been introduced in recent years.

Electronic and Vibrational Noise

Electronic noise is determined largely by the instrument design and manufacture. Minimizing overall electronic noise is a major consideration in instrument design and selection of electronic components. Noise is frequently minimized by supplying "clean" or conditioned electrical power for instrument operation. In addition, minimizing vibration from motors and other devices, such as mechanical pumps, is an important consideration in minimizing overall instrument noise. (Source: www.ssi.shimandzu.com/sens).

13.7 ANALYSIS OF ORGANIC COMPOUNDS

The organic compounds in samples are analyzed based on two categories: (i) volatile organics and (ii) semivolatile organics. This classification is based on the analytical technique used rather than the chemical structures of compounds.

13.7.1 Volatile Organics – Purge and Trap Method

The techniques may be applied to transfer the volatile analytes from the sample matrices into the GC column. Purse and trap technique and thermal desorption. In the purge and trap method, an aliquot of aqueous sample (usually 5 mL for wastewaters and 25 mL for drinking waters) is bubbled through by helium or nitrogen for 11 min. The analytes are purged out from the sample and carried over with the purging gas onto a trap consisting of activated charcoal, tenax and silica gel adsorptions where they are adsorbed. The trap is then heated at 180°C for 4 minutes. The analytes are desorbed and transported by the carrier gas onto the GC column.

For soil or solid waste samples, an accurately weighted amount of sample is treated with methanol. A small portion of the methanol extract, usually between 10 and 100 mL, depending on the expected levels of analytes concentrations in the sample and the presence of matrix interference substances, is injected onto 5 mL of laboratory reagent – grade water taken in the purging vessel which is then purged with an inert gas. Analytes adsorbed over the trap are desorbed by heating and transported to the GC column.

13.7.2 Semivolatile Organics – Tuning Method

In GC/MS technique, the principle of analysis of the semivolatile organics is not so distinctly different from that of the volatile organics. Only, the method of extraction of analytes from the sample matrices and the sample concentration steps for the semivolatile organic compounds differs from the volatile organics. Tuning method is described below:

GC/MS instrument is equipped with a tuning substance – perfluorotributylamine (PFTBA). The characteristics mass ions of this compound or any other tuning compound and the relative area response corresponding to these mass ions have been well established and set as guidelines to check the performance of the instrument called tuning. All instruments must meet such tuning criteria to perform general organic analysis satisfactorily. The use of additional tuning substances to set the acceptable performance criteria for analysis of organic substances by GC/MS are:

- 4 bromofluorobenzene (BFB)
- decafluorotriphenylphosphine (DFTPP)
- For LC/MS analysis tuning substance is:
- Decafluorotriphenyl phosphine oxide (DFTAPPO)

In GC/MS operation, inject 2 µL (50 mg) of BFB or DFTPP solution. Alternatively, 2 µL of BFB solution may be added to 5 mL reagent water and analyzed following purge and trap concentration. A background correction should be made prior to checking the m/z criteria of the mass spectra. If the instrument has passed the DFTPP test earlier in the day, it may not be necessary to perform a BFB test for volatiles when using the same instrument. After meeting the tuning criteria, analyze the standards, blanks and samples.

A column performance test for base/neutral and acid fractions must be performed to test the efficiency of chromatographic column for separation of the analyte semivolatile organics.

13.7.3 Compound Identification

The analytes are identified from their mass spectra and retention times. The retention time must fall within 13 sec. of the retention time of the known compound. The mass spectra of the unknown peak must have the primary and all secondary characteristic ions (molecular ion).

For calibration, quantification of an analyte is performed by three methods: (i) External Standard Method, (ii) Internal Standard Method, and (iii) Isotope Dilution Method.

13.8 MASS SPECTRUM ANALYSIS

Mass spectrum analysis is an integral part of spectroscopy and mass spectrometry dealing with the interpretation of mass spectra. Organic chemists obtain mass spectra of chemical compounds as part of structure elucidation and the analysis is part of every organic chemistry curriculum (Fig. 13.3).

- the molecular ion,
- isotope peaks
- fragmentation peaks
- metastable peaks

Mass spectra first of all display the **molecular ion** (or **parent ion**) peak which is a radical cation $M^{+\cdot}$ as a result of removing one electron from the molecule. In the spectrum for toluene for example the molecular ion peak is located at 92 m/e corresponding to its molecular mass. The molecular ion peak does not always appear or can be weak. The height of the molecular ion peak diminishes with branching and with increasing mass in a homologous series. Identifying the molecular ion can be difficult. A useful aid is the nitrogen rule: if the mass is an even number, the compound contains no nitrogen or an even number of nitrogens. Molecular ion peaks are also often preceded by a M-1 or M-2 peak resulting from loss of a hydrogen radical or dihydrogen.

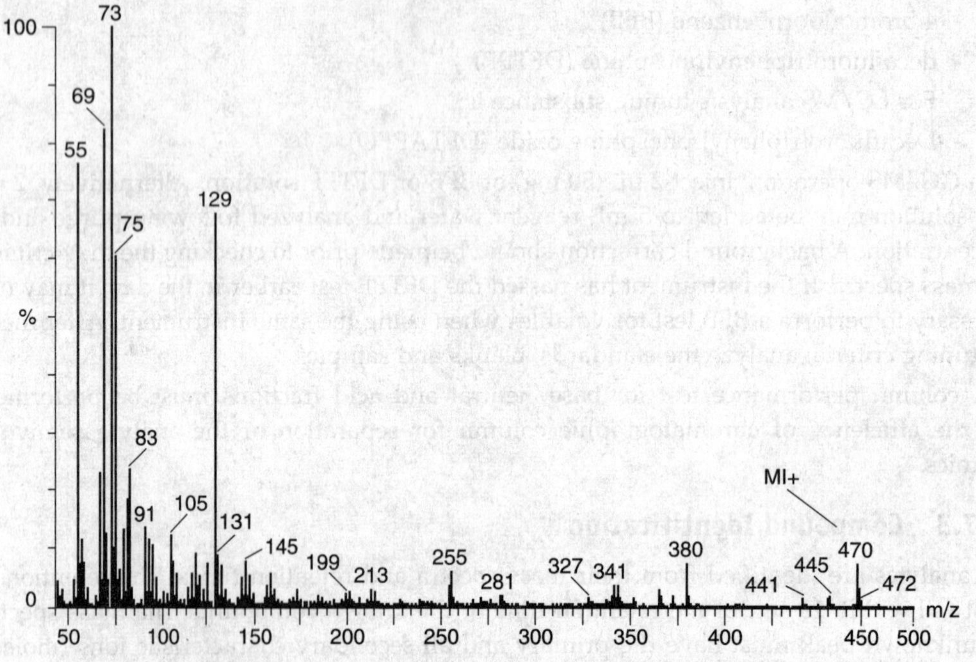

Fig. 13.3 *Mass spectra have several distinct sets of peaks.*

More peaks are visible with m/e ratios larger than the molecular ion peak due to isotope distributions. The value of 92 in the toluene example corresponds to the monoisotopic mass of a molecule of toluene entirely composed of the most abundant isotopes (^{1}H and ^{12}C). The so-called **M+1 peak** corresponds to a fraction of the molecules with one higher isotope incorporated ((^{2}H or ^{13}C) and the M+2 peak has two higher isotopes. The natural abundance of the higher isotopes is low for frequently encountered elements such as hydrogen, carbon and nitrogen and the intensity of isotope peaks subsequently low and the intensity quickly diminishes with total mass. In halogens, on the other hand, higher isotopes have a large abundance which results in a specific mass signature for halogen containing compounds.

Peaks with mass less than that the molecular ion are the result of fragmentation of the molecule. These peaks are called daughter peaks. The peak with the highest ratio is called the base peak which is not necessarily the molecular ion. Many reaction pathways exist for fragmentation but only newly formed cations will show up in the mass spectrum and not radical fragments or neutral fragments.

Metastable peaks are broad peaks at non-integer mass values. These peaks result from molecular fragments with lower kinetic energy because of fragmentations taking place ahead of the ionization chamber. They are not of analytical value.

Fragmentation

The fragmentation pattern not only allows the determination of the mass of an unknown compound but also allows guessing the molecular structure especially in combination the calculation of the degree of unsaturation from the molecular formula (when available). Neutral fragments frequently lost are carbon monoxide, ethylene, water, ammonia, and hydrogen sulfide.

Fragmentations arise from:

- Homolysis processes. An example is the cleavage of carbon-carbon bonds next to a heteroatom
- In this depiction single-electron movements are indicated by a fishhook arrow.
- Rearrangement reactions, for example, a retro Diels-Alder reaction extruding neutral ethylene;

  or the McLafferty rearrangement. As it is not always obvious where a lone electron resides in a radical cation a square bracket notation is often used.

Some General Rules:

- Cleavage occurs at alkyl substituted carbons reflecting the order generally observed in carbocations.
- Double bonds and arene fragments tend to resist fragmentation.
- Allylic cations are stable and resist fragmentation.
- the even-electron rule stipulates that even-electron species (cations but not radical ions) will not fragment into two odd-electron species but rather to another cation and a neutral molecule.

Toluene example

The mass spectrum for toluene has around 30 signals. Several peaks can be rationalized in this fragmentation pattern (Fig. 13.4)

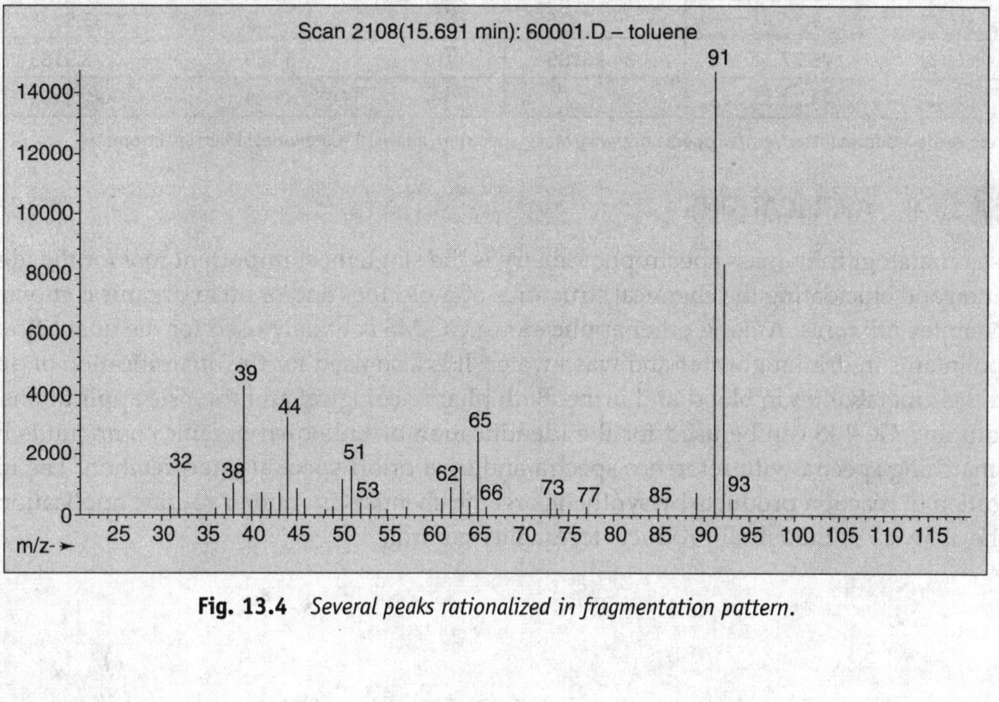

Fig. 13.4 *Several peaks rationalized in fragmentation pattern.*

Isotope effects

Isotope peaks within a spectrum can help in structure elucidation. Compounds containing halogens (especially chlorine and bromine) produce very distinct isotope peaks. The mass spectrum of methylbromide has two prominent peaks of equal intensity at 94 (M) and 96 (M+2) and then two more at 79 and 81 belonging to the bromine fragment.

Even when compounds only contain elements with less intense isotope peaks (carbon or oxygen), the distribution of these peaks can be used to assign the spectrum to the correct compound. For example, two compounds with identical mass of 150, $C_8H_{12}N_3^+$ and $C_9H_{10}O_2^+$, will have two different M+2 intensities which make it possible to distinguish between them.

Natural abundance of some elements

Table 13.1 gives the isotope distributions for some elements. Some elements like phosphorus and fluorine only exist as a single isotope, with a natural abundance of 100%.

Table 13.1 *Natural abundance of some elements*

Isotope	% Nat. Abundance	Atomic Mass	Isotope	% Nat. Abundance	Atomic Mass
^{1}H	99.985	1.007825	^{12}C	98.89	12 (definition)
^{2}H	0.015	2.0140	^{13}C	1.11	13.00335
^{16}O	99.76	15.99491	^{14}N	99.64	14.00307
^{17}O	0.04		^{15}N	0.36	15.00011
^{18}O	0.2				
^{28}Si	92.23	27.97693	^{32}S	95.0	31.97207
^{29}Si	4.67	28.97649	^{33}S	0.76	32.97146
^{30}Si	3.10	29.97376	^{34}S	4.22	33.96786
^{35}Cl	75.77	34.96885	^{79}Br	50.69	78.9183
^{37}Cl	24.23		^{81}Br	49.31	80.9163

Source: Retrieved from (http://em/wikipedia.org/wiki/Mass_spectrum_analysis) -Categories: Mass Spectrometry

13.9 APPLICATIONS

Gas chromatography-mass spectrophotometry is the singlemost important tool for the identification and elucidating the chemical structures of molecules and or other organic compounds in complex mixtures. Among other applications, GC-MS is widely used for the quantification of pollutants in drinking water and wastewater. It is also used for the quantification of drugs and their metabolites in blood and urine. Both pharmacological and forensic applications are significant. GC-MS can be used for the identification of unknown organic compounds both by matching spectra with reference spectra and by a priori spectral interpretation. The identification of reaction products by synthetic organic chemists is another routine application, as is the analysis of industrial products for quality control.

In GC-MS use, the organic compounds must be in solution for injection into the gas chromatography. The solvent must be volatile and organic (for example, hexane ore dichloromethane). Depending on the ionization method, analytical sensitivities of 1 to 100 pg per component are routine. Sample preparation can range from simply dissolving some of the sample in a suitable solvent to extensive cleanup procedure using various forms of liquid chromatography. In addition to the sample preparation time, the instrumental analysis time is usually fixed by the duration of the gas chromatography run, typically between 20 and 100 min. Data analysis can take another 1 to 20 hr (or more) depending on the level of detail necessary.

GS-MS is used for the quantitative measurement of individual components and the qualitative identification. The quantitative and qualitative data analysis strategies are discussed.

13.9.1 Quantitative

GC-MS can also be used to measure the concentration of one or more analytes in a complex mixture. Quantitation can be based on peak areas from mass chromatograms or from selected ion monitoring. With the selected ion monitoring technique, the mass spectrometer is not scanned over all masses, instead the instrument jumps from one selected mass to another. The advantage of this approach is that the mass spectrometer spends much more time at the given mass, the signal-to-noise ratio at that mass improves, and the overall sensitivity of the experiment increases by a factor of 100 to 1000. For example, the mass spectrometer might integrate for 500 msec at mass m1, jump to mass m2 in 10 msec, integrate for 500 msec at that mass, and jump backm to m1 in 10 msec, repeating this cycle for the duration of the GC run. In essence, the intensities of the two masses are recorded as a function of GC retention time, with measurement made every 1.02 sec. In practice, rather than only two masses, 5 to 10 masses are usually monitored simultaneously for 100 to 200 msec each. In this manner, the GC-MS response of 5 to 10 compounds, depending on the specificity of the selected masses can be measured.

The difference between mass chromatogram and selected ion monitoring is significant. With the latter technique, the responses from only a few preselected masses are recorded. With mass chromatogram all of the masses are scanned; thus no preselection is required. This is the necessary trade-off for higher sensitivity. Clearly, the mass spectrum of the analyte must be known so that the masses that uniquely characterize it can be selected. This information can be obtained from the literature or from the laboratory. Each selected set of masses can be monitored for the duration of the complete GC run or for only selected GC retention times. Selected ion monitoring is almost fully software driven, thus, it is very flexible. Different gas of masses, different time windows, and different integration times can be easily set up.

The peak areas must be calibrated, to convert the peak areas to mass of analyte, whether from mass chromatograms or from selected ion monitoring. The two main strategies are based on external and internal standards. With external standards, the areas of one or more mass chromatogram are calibrated with a known amount of the analyte injected into the GC-MS in a different experiment. Detection limits of a few nanograms can be achieved with this technique. However, the strategy that gives the most accurate quantitative results is the use of internal standards, which are known amounts of compounds added to the sample before isolation of the analytes begins. After sample extraction and cleanup, only trhe ratio of response between the analyte and the internal standard must be measured. This ratio multiplied by the amount of the internal standard gives the amount of the analyte injected into the GC-MS system. This can be converted to concentration using the correct dilution factors. Depending on the relationship of the internal standard to the analyte, the precision and accuracy of most analyses are improved.

Quantitative Determination of volatile organics

Analyte concentration in the solid matrix can be calculated as given below:

For example: Two grams of soil was treated with 10 mL methanol. A 100 mL aliquot of methanol extract was injected into 5 mL reagent grade water for purge and trap analysis of a volatile organic compound. The concentration of the compound was found to be 22 mg/L. Determine its concentration in the soil. The solution to this problem can be solved as given below:

1. Mass Calculation

$$22 \text{ mg compound } L = 22 \text{ mg} \times \frac{5 \text{ mL water}}{1000 \text{ mL water}}$$

$$= 0.11 \text{ mg compound in 5 mL water}$$

This means 0.11 mg compound was sound to be present in 5 mL water in the purging vessel spiked with methanol extract. This amount must have come from 100 μL in the purging vessel, i.e., 0.11 mg compound/100 mL methanol extract.

$$\text{Extract} = 0.11 \text{ mg} \times \frac{10 \text{ mL (or 10,000 μL) extract}}{100 \text{ μL extract}}$$

$$= 0.11 \text{ mg compound in 10 mL methanol extract,}$$

which would have come from 2 g sample.

Thus, 2 g sample contains 11 mg compound or 1000 g sample would contain:

$11 \text{ mg} \times \frac{1000\text{g}}{2\text{g}}$ or 5500 mg of compound, i.e.

the concentration of the compound in the soil = 5500 mg/kg.

2. Dimensional Analysis

$$\frac{22 \text{ mg compound}}{1 \text{ L water}} \times \frac{1 \text{ mL water}}{1000 \text{ mL}} \times \frac{5 \text{ mL water}}{100 \text{ μL}} \times \frac{1000 \text{ μL methanol}}{1 \text{ mL method}}$$

$$\times \frac{1 \text{ mL water}}{2 \text{ g soil}} \times \frac{1000 \text{ g soil}}{1 \text{ kg soil}}$$

$$= \frac{22 \times 5 \times 1000 \times 10 \times 1000}{1000 \times 1000 \times 2} \text{ μL compound/kg soil}$$

$$= 5500 \text{ μg compound per kg soil}$$

13.9.2 Qualitative

There are three ways of examining GC-MS data. First, the analyst can go through the gas chromatogram as reproduced by the mass spectrometer and look at the mass spectra scanned at each GC peak maximum. This has the advantage of being relatively quick but the disadvantage of missing components of the mixture that are not fully resolved by the GC column. The second approach is to look at each mass spectrum in turn, in essence stacking up the mass spectra one behind the other and examining them individually. This has the advantage of completeness but the disadvantage of tedium. The third approach is to look at the intensity of one particular mass as a function of time. This third approach makes use of the three-dimensional nature of GC-MS data. Two of these dimensions are the mass versus intensity of the normal mass spectrum; the third dimension is the GC retention time over which the mass spectral data are acquired.

Mass chromatograms are also useful for determining whether a given mass belongs in a given mass spectrum. To ensure qualitative identification of an organic compound using GC-MS, several criteria should be met: first the mass spectra of the unknown compound and of the authentic compound must agree over the entire mass range of the spectra. It is particularly important to compare the pattern within narrow mass ranges (for example, from m/2 to 50 to 60); these patterns should agree almost exactly. In this case, the spectrum of the authentic compound might come from a library of reference spectra or from the actual compound itself. In the latter case, the compound could be purchased or synthesized. Second, the GC retention time of the unknown compound and of the authentic compound must agree within about $\pm$ 1 to 2 sec. It is often convenient to do this experiment by coinjecting the unknown mixture and the authentic compound. The GC peak in question should increase in size by the correct factor.

GS-MS has a few limitations. Only components with vapour pressures exceeding about 10^{-10} torr can be analysed by GC-MS. Many compounds that have lower pressure can be analyzed if they are chemically derivatized (for example, as trimethylsilyl ethers). Determining positional substitution on aromatic rings is often difficult. Certain isomeric compounds cannot be distinguished by mass spectrometry (for example, naphthalene versus azulene), but they can often be separated chromatographically. Quantitative accuracy is controlled by the overall analytical method calibration. Using isotopic internal standards, accuracy of $\pm$ 20% relative standard deviaiton is typical.

13.10 CHROMATOGRAPHIC TECHNIQUES – COMBINED WITH MASS SPECTROMETRY

13.10.1 Gas Chromatography (GC)

Gas-liquid chromatography (GLC), or simply gas chromatography (GC), is a type of chromatography in which the mobile phase is a carrier gas, usually an inert gas such as helium or an unreactive gas such as nitrogen, and the stationary phase is a microscopic layer of liquid or polymer on an inert solid support, inside glass or metal tubing, called a column. The instrument used to perform gas chromatographic separations is called a gas chromatograph (also: aerograph, gas separator).

Gas chromatography is different from other forms of chromatography (HPLC, TLC, etc.) because the solutions travel through the column in a gas state. The interactions of these gaseous analytes with the walls of the column (coated by different stationary phases) causes different compounds to elute at different times called retention time. The comparison of these retention times is the analytical power to GC. This makes it very similar to HPLC.

13.10.2 Gas Chromatography-Mass Spectrometry (GC-MS)

Gas chromatography-mass spectrometry (GC-MS) is a method that combines the features of gas-liquid chromatography and mass spectrometry to identify different substances within a test sample. Applications of GC-MS include drug detection, fire investigation, environmental analysis, explosives investigation, and identification of unknown samples. GC-MS can also be used in airport security to detect substances in luggage or on human beings. Additionally, it can identify trace elements in materials that were previously thought to have disintegrated beyond identification.

The GC-MS has been widely heralded as a "gold standard" for substance identification because it is used to perform a specific test. A specific test positively identifies the actual presence of a particular substance in a given sample. A non-specific test, however, merely indicates that a substance falls into a category of substances. Although a non-specific test could statistically suggest the identity of the substance, this could lead to false positive identification.

13.10.3 Gas Chromatography-Mass Spectrometry-Mass Spectrometry (GC-MS-MS)

This GC-MS-MS instrument is the same as the gas chromatography-mass spectrometry (GC-MS), except a second phase of mass fragmentation is added. For example, using a second quadrupole in a quadrupole instrument, it is called MS-MS or Tandem MS. For organic analysis of environmental samples, including many major classes of pesticides, the GC-MS-MS is very sensitive and effective. (GC-MS-MS instruments are also referred to as GC-MS-MS, GC-MS-MS, and GC MS MS among other variations.)

13.10.4 Liquid Chromatography-Mass Spectrometry (LC-MS)

Liquid chromatography-mass spectrometry (LC-MS) is an analytical chemistry technique that combines the physical separation capabilities of liquid chromatography (aka HPLC) with the mass analysis capabilities of mass spectrometry. LC-MS is a powerful technique used for many applications which has very high sensitivity and specificity. This technique is used to analyze compounds that are too large, too polar, or too thermally labile for the GC technique. Generally, its application is oriented towards the specific detection and potential identification of chemicals in the presence of other chemicals (in a complex mixture).

13.10.5 Liquid Chromatography-Mass Spectrometry (LC-MS-MS)

This LC-MS-MS instrument is the same as the LC-MS, but has the advantage of a tandem MS-MS.

13.10.6 Inductively Coupled Plasma-Mass Spectroscopy (ICP-MS)

ICP-MS, short for inductively coupled plasma mass spectrometry, is a type of mass spectrometry which is very sensitive and capable of determining of a range of metals and several non-metals at concentrations below one part in 10^{12}. The ICP-MS allows detection of trace metals to parts per trillion. The instrument is based on coupling together inductively coupled plasma as a method of producing ions (ionization) with a mass spectrometer as a method of separating and detecting the ions. ICP-MS is also capable of monitoring isotopic speciation for the ions of choice.

13.11 REFERENCES

1. Sparkman, O. David (2000). Mass spectrometry desk reference. Pittsburgh: Global View Pub. ISBN 0-9660813-2-3.
2. "Definition of spectrograph. (http://dev.m-w.com/dictionary/spectrograph0 " Merriam Webster. Accessed 13 June 2008.
3. KM Downard (2007). "William Aston - the man behind the mass spectrograph". European Journal of Mass Spectrometry **13** (3): 177–190. doi:10.1255/ejms.878.(http://dx.doi.org/10.1255%2Fejms.878) PMID 17881785. (http://www.ncbi.nlm.nih.gov/pubmed/17881785)
4. Harper, Douglas. "Spectrum. (http://www.etymonline.com/index.php?search=spectrum&searchmode=none) " Online Etymology Dictionary. Nov. 2001. Accessed 07-12-2007. **Note**: This part of the article only makes descriptive claims about the information found in the primary source, the accuracy and applicability of which is easily verifiable by any reasonable, educated person without specialist knowledge. (See WP:PSTS)
5. Squires, Gordon (1998). "Francis Aston and the mass spectrograph". Dalton Transactions: 3893–3900. doi:10.1039/a804629h (http://dx.doi.org/10.1039%2Fa804629h) .
6. KM Downard (2007). "Francis William Aston - the man behind the mass spectrograph". European Journal of Mass Spectrometry **13** (3): 177–190. doi:10.1255/ejms.878. (http://dx.doi.org/10.1255%ejms.878) PMID 17881785. (http://www.ncbi.nlm.nih.gov/pubmed/17881785)
7. Thomson, J.J. (1913). Rays of Positive Electricity and Their Application to Chemical Analysis. (http://www.archive.org/details/RaysOfPositiveElectricity) London: Longman's Green and Company. http://www.archive.org/details/RaysOfPositiveElectricity.

8. Siri, William (1947). "Mass spectroscope for analysis in the low-mass range". Review of Scientific Instruments **18** (8): 540–545. doi:10.1063/1.1740998. (http://dx.doi.org/10.1063%2F1.1740998)
9. Price, Phil (1991). "Standard definitions of terms relating to mass spectrometry. A report from the Committee on Measurements and Standards of the American Society for Mass Spectrometry". Journal of the American Society for Mass Spectrometry **2** (4): 336–348. doi:10.1016/1044-0305(91)80025-3. (http://dx.doi.org/10.1016%2f1044-0305%2891%80025-3)
10. J.J., Thomson (1913). Rays of Positive Electricity and Their Application to Chemical Analysis. (http://www.archive.org/details/RaysOfPositiveElectricity) London: Longman's Green and Company. (http://www.archive.org/details/RaysOfPositiveElectricity.
11. Carl-Ove Andersson, Acta. Chem. Scand. **1958**, 12, 1353
12. Mass Spec Anniversary, Chemical & Engineering News, **87**, 6 (9 Feb. 2009), p. 4
13. "The Nobel Prize in Chemistry 2002: Information for the Public". (http://nobelprize.org/nobel_prizes/chemistry/laureates/2002/public.html) The Nobel Foundation. 9 October 2002. http://nobelprize.org/nobel_prizes/chemistry/laureates/2002/public.html. Retrieved 2007-08-29.
14. Michael A. Grayson, ed (2002). Measuring Mass: From Positive Rays to Proteins. (http://books.google.com/books?id=qQ7EE9MXVx0C&printsec=frontcover) Philadelphia, Pa.: Chemical Heritage Press. ISBN 0-941901-31-9. http://books.google.com/books?id=qQ7EE9MXVx0C&printsec=frontcover.
15. Karas, M.; Bachmann, D.; Hillenkamp, F. (1985). "Influence of the Wavelength in High-Irradiance Ultraviolet Laser Desorption Mass Spectrometry of Organic Molecules". Anal. Chem. **57**: 2935–9. doi:10.1021%2Fac00291a042. (http://dx.doi.org/10.1021%ac00291a042)
16. Karas M, Hillenkamp F (1988). "Laser desorption ionization of proteins with molecular masses exceeding 10,000 daltons". Anal. Chem. **60** (20): 2299–301. doi:10.1021/AC00171a028. (http://dx.doi.org/10.1021%2Fac00171a028) PMID 3239801. (http://www.ncbi.nlm.nih.gov/pubmed/3239801)
17. Tanaka, Koichi; Hiroaki Waki, Yutaka Ido, Satoshi Akita, Yoshikazu Yoshida, Tamio Yoshida, T. Matsuo (1988). "Protein and polymer analyses up to m/z 100 000 by laser ionization time-of-flight mass spectrometry". Rapid Communications in Mass Spectrometry **2** (8): 151–153. doi:10.1002/rcm.1290020802. (http://dx.doi.org/10.1002%2Frcm.1290020802)
18. A. P. Bruins (1991). "Mass spectrometry with ion sources operating at atmospheric pressure". Mass Spectrometry Reviews **10** (1): 53–77. doi:10.1002/mas.1280100104. (http://dx.doi.org/10.1002%2Fmas.1280100104)
19. John S Cottrell, Roger J Greathead (1986). "Extending the Mass Range of a Sector Mass Spectrometer". Mass Spectrometry Reviews **5**: 215–247. doi:10.1002/mas.1280050302. (http://dx.doi.org/10.1002%2Fmas.1280050302)

20. Wollnik, H. (1993). "Time-of-flight mass analyzers". Mass Spectrometry Reviews **12**: 89. doi:10.1002/mas.1280120202. (http://dx.doi.org/10.1002%2Fmas.1280120202)
21. Paul W., Steinwedel H. (1953). "Ein neues Massenspektrometer ohne Magnetfeld". Zeitschrift für Naturforschung A **8** (7): 448-450. Bibcode: 1953ZNatA...8..448P. (http://adsabs.harvard.edu/abs/1953ZNatA...8..448P)
22. R. E. March (2000). "Quadrupole ion trap mass spectrometry: a view at the turn of the century". International Journal of Mass Spectrometry **200** (1-3): 285–312. doi:10.1016/S1387-3806(00)00345-6. (http://dx.doi.org/10.1016%2FS1387-3806%2800%2900345-6)
23. Schwartz, Jae C.; Michael W. Senko and John E. P. Syka (2002). "A two-dimensional quadrupole ion trap mass spectrometer". Journal of the American Society for Mass Spectrometry **13** (6): 659–669. doi:10.1016/S1044-0305(02)00384-7. (http://dx.doi.org/10.1016%2FS1044-0305%2802%2900384-7) PMID 12056566. (http://www.ncbi.nlm.nih.gov/pubmed/12056566)
24. M. B. Comisarow and A. G. Marshall (1974). "Fourier transform ion cyclotron resonance spectroscopy". Chemical Physics Letters **25** (2): 282–283. doi:10.1016/0009-2614-(74)89137-2. (http://dx.doi.org/10.1016%2F0009-2614%2874%2989137-2)
25. Marshall, A. G.; Hendrickson, C. L.; Jackson, G. S. (1998). "Fourier transform ion cyclotron resonance mass spectrometry: a primer". Mass Spectrometry Reviews **17** (1): 1–34. doi:10.1002/(SICI)1098-2787(1998)17:1<1::AID-MAS1>3.0.CO;2-K. (http://dx.doi.org/10.1002%F%28SICI%291098-2787%281998%2917%3C1%3A1%3AAID-MAS1%3E3.0.CO%3B2-K)
26. Q. Hu, R. J. Noll, H. Li, A. Makarov, M. Hardman and R. G. Cooks (2005). "The Orbitrap: a new mass spectrometer". Journal of Mass Spectrometry **40** (4): 430–443. doi:10.1002/jms.856. (http://dx.doi.org/10.1002%2Fjms.856)
27. F. Dubois, R. Knochenmuss, R. Zenobi, A. Brunelle, C. Deprun and Y. L. Beyec (1999). "A comparison between ion-to-photon and microchannel plate detectors". Rapid Communications in Mass Spectrometry **13** (9): 786–791. doi:10.1002/(SICI)1097-0231(19990515)13:9<786::AID-RCM566>3.0.CO;2-3. (http://dx.doi.org/10.1002%2F%28SICI%291097-0231%2819990515%2913%3A9%3C786%3A%3AAID-RCM566%3E3.0.CO%3B2-3
28. M. A. Park, J. H. Callahan and A. Vertes (1994). "An inductive detector for time-of-flight mass spectrometry". Rapid Communications in Mass Spectrometry **8** (4): 317–322. doi:10.1002/rcm.1290080407. (http://dx.doi.org/10.1002%2Frcm.1290080407)
29. Robert K. Boyd (1994). "Linked-scan techniques for MS/MS using tandem-in-space instruments". Mass Spectrometry Reviews **13** (5-6): 359–410. doi:10.1002/mas.1280130502. (http://dx.doi.org/10.1002%2Fmas.1280130502)
30. Eiceman, G.A. (2000). Gas Chromatography. In R.A. Meyers (Ed.), Encyclopedia of Analytical Chemistry: Applications, Theory, and Instrumentation, pp. 10627. Chichester: Wiley. ISBN 0-471-97670-9

31. Verbeck, GF and Ruotolo, BT and Sawyer, HA and Gillig, KJ and Russell, DH, G; Ruotolo, B; Sawyer, H; Gillig, K; Russell, D (2002). "A fundamental introduction to ion mobility mass spectrometry applied to the analysis of biomolecules". (http://jbt.highwire.org/cgi/content/abstract/13/2/56) J Biomol Tech **13** (2): 56-61. PMID 19498967. (http://www.ncbi.nlm.nih.gov/pubmed/19498967) PMC 2279851. (http://www.pubmedcentral.gov/articlerender.fcgi?tool=pmcentrez&artid=2279851) (http://jbt.highwire.org/cgi/content/abstract/13/2/56.

32. L. M. Matz, G. R. Asbury and H. H. Hill (2002). "Two-dimensional separations with electrospray ionization ambient pressure high-resolution ion mobility spectrometry/quadrupole mass spectrometry". Rapid Communications in Mass Spectrometry **16** (7): 670–675. doi:10.1002/rcm.623. (http://dx.doi.org/10.1002%2Frcm.623)

33. Rena A. Sowell, Stormy L. Koeniger, Stephen J. Valentine, Myeong Hee Moon and David E. Clemmer (2004). "Nanoflow LC/IMS-MS and LC/IMS-CID/MS of Protein Mixtures". Journal of the American Society for Mass Spectrometry **15** (9): 1341–1353. doi:10.1016/j.jasms.2004.06.014. (http://dx.doi.org/10.1016%2Fjasms.2004.06.014)

34. Tureček, František; McLafferty, Fred W. (1993). Interpretation of mass spectra. (http://books.google.com/books?id=xQWk5WQfMQAC&printsec=frontcover) Sausalito: University Science Books. ISBN 0-935702-25-3. http://books.google.com/books?id=xQWk5WQfMQAC&printsec=frontcover

35. Mistrik, R.(2004). A New Concept for the Interpretation of Mass Spectra Based on a Combination of a Fragmentation Mechanism Database and a Computer Expert System. (http://www.highchem.com/publications/a-new-concept-for-the-interpretation-of-mass-spectra.html) in Ashcroft, A.E., Brenton, G., Monaghan,J.J. (Eds.), Advances in Mass Spectrometry, Elsevier, Amsterdam, Vol. 16, pp. 821.

36. Hsieh, Yunsheng; Korfmacher, WA (2006). "Systems for Drug Metabolism and Pharmacokinetic Screening, Y. Hsieh and W.A. Korfmacher, Current Drug Metabolism". Current Drug Metabolism **7** (5): 479–489. doi:10.2174/138920006777697963. (http://dx.doi.org/10.2174%2F138920006777697963) PMID 16787157. (http://www.ncbi.nlm.nih.gov/pubmed/16787157)

37. Covey, T.R.; Lee, E.D.; Henion, J.D. (1986). "Mass Spectrometry for the Determination of Drugs in Biological Samples". Anal. Chem. **58**: 2453–2460. doi:10.1021/ac00125a022. (http://dx.doi.org/10.1021%2Fac00125a022)

38. Covey, Tom R.; Crowther, Jonathan B.; Dewey, Elizabeth A.; Henion, Jack D. (1985). "Mass Spectrometry Determination of Drugs and Their Metabolites in Biological Fluids". Anal. Chem. **57** (2): 474–81. doi:10.1021/ac50001a036. (http://dx.doi.org/10.1021%2Fac50001a036)

39. S. Petrie and D. K. Bohme (2007). "Ions in space". Mass Spectrometry Reviews **26** (2): 258–280. doi:10.1002/mas.20114. (http://dx.doi.org/10.1002%Fmas.20114)

40. "Cassini Plasma Spectrometer". (http://caps.space.swri.edu/) Southwest Research Institute. http://caps.space.swri.edu/. Retrieved 2008-01-04.

41. Riker JB, Haberman B (1976). "Expired gas monitoring by mass spectrometry in a respiratory intensive care unit". Crit. Care Med. **4** (5): 223–9. doi:10.1097/00003246-197609000-00002. (http://dx.doi.org/10.1097%2F00003246-197609000-00002) PMID 975846. (http://www.ncbi.nlm.nih.gov/pubmed/975846)

42. J. W.W. Gothard, C.M. Busst, M.A. Branthwaite, N.J.H. Davies and D.M. Denison (1980). "Applications of respiratory mass spectrometry to intensive care". Anaesthesia **35** (9): 890–895. doi:10.1111/j.1365-2044.1980.tb03950.x. (http://dx.doi.org/10.1111%2Fj.1365-2044.1980.tb03950.x)

43. Spectrometric identification of organic compounds Silverstein, Bassler, Morrill 4th edn.

44. Organic spectroscopy William Kemp 2nd edn.ISBN 033342171

45. Lide, D. R., Ed. (2002), CRC Handbook of Chemistry and Physics (83rd edn.), Boca Raton, FL: CRC Press, ISBN 0-8493-0483-0

CHAPTER

14 NMR Spectroscopy

14.1 INTRODUCTION

Nuclear magnetic resonance spectroscopy, most commonly known as NMR spectroscopy, is the name given to a technique which exploits the magnetic properties of certain nuclei. Nuclear Magnetic Resonance (NMR) is a branch of spectroscopy in which radio frequency waves induce transitions between magnetic energy levels of nuclei of a molecule. The magnetic energy levels are created by keeping the nuclei in a magnetic field. Without the magnetic field the spin states of nuclei are degenerated, i.e., possesses the same energy, and energy level transition is not possible. When a magnetic field is applied, the separate levels and radio frequency radiation can cause transitions between these energy levels. The most important applications for the organic chemist are proton NMR and carbon-13 NMR spectroscopy. In principle, NMR is applicable to any nucleus possessing spin. The impact of NMR spectroscopy on the natural sciences has been substantial. It can, among other things, be used to study mixtures of analytes, to understand dynamic effects such as change in temperature and reaction mechanisms, and is an invaluable tool in understanding protein and nucleic acid structure and function. It can be applied to a wide variety of samples, both in the solution and the solid state.

14.1.1 Basis of NMR

NMR spectra arise from a property that some nuclei have, usually called spin. This is a quantum effect; a model of it imagines the nucleus spinning on its axis, this spin being quantised, i.e., having only two directions, up or down. The nuclear spin is a characteristic property of the nucleus.

The spinning nucleus generates a magnetic field, since it is a moving charge. If an external magnetic field is applied to the nucleus, then the spin state where the nuclear magnetic field is aligned with the external field has a different, lower, energy from that spin state that gives rise to an opposing field. The fields are shown by arrows on the diagram below (Fig. 14.1).

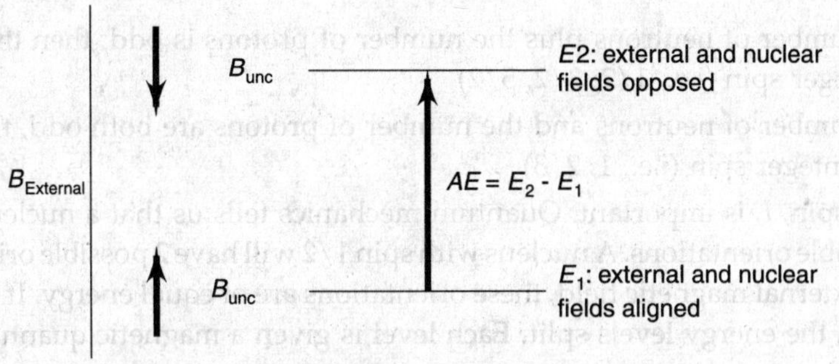

Fig. 14.1 *External and nuclear fields: opposed and aligned.*

Since the two states have different energies, it is possible to convert a nucleus from the lower-energy state E_1 to the higher state E_2 by the input of suitable energy. The energy difference *DE* is such that radio frequency waves will perform this switch. The NMR spectrum arises because nuclei in different parts of the molecule experience different local magnetic fields according to the molecular structure, and so have different frequencies at which they absorb. This difference is called the chemical shift, dealt with on the next page. The radios frequency used depends on the strength of the magnetic field; early machines used magnetic field strengths of around 1.5 Tesla (15000 Gauss), the proton magnetic resonances being around 60 MHz. The ability to achieve higher magnetic fields of around 5 T (50 kG) using superconducting magnets has moved the resonances to 200 MHz. In practice the radio frequency is kept constant, and the magnetic field is then swept over a narrow range of field strength. The absorptions are plotted on a graph where the frequency differences (in reality magnetic field differences) are plotted relative to some standard compound which defines the zero; for proton magnetic resonance this compound is TMS, tetramethylsilane, $Si(CH_3)_4$.

14.2 NMR SPECTROSCOPY

Nuclear Magnetic Resonance spectroscopy is a powerful and theoretically complex analytical tool. It is important to remember that, with NMR, experiments are performed on the **nuclei** of atoms, not the electrons. The chemical environment of specific nuclei is deduced from information obtained about the nuclei.

14.2.1 Nuclear Spin and the Splitting of Energy Levels in a Magnetic Field

Subatomic particles (electrons, protons and neutrons) can be imagined as spinning on their axes. In many atoms (such as ^{12}C) these spins are paired against each other, such that the nucleus of the atom has no overall spin. However, in some atoms (such as ^{1}H and ^{13}C) the nucleus does possess an overall spin. The rules for determining the net spin of a nucleus are as follows:

1. If the number of neutrons and the number of protons are both even, then the nucleus has NO spin.

2. If the number of neutrons plus the number of protons is odd, then the nucleus has a half-integer spin (i.e., 1/2, 3/2, 5/2)
3. If the number of neutrons and the number of protons are both odd, then the nucleus has an integer spin (i.e., 1, 2, 3)

The overall spin, *I*, is important. Quantum mechanics tells us that a nucleus of spin *I* will have $2I + 1$ possible orientations. A nucleus with spin 1/2 will have 2 possible orientations. In the absence of an external magnetic field, these orientations are of equal energy. If a magnetic field is applied, then the energy levels split. Each level is given a magnetic quantum number, *m*.

Energy levels for a nucleus with spin quantum number 1/2

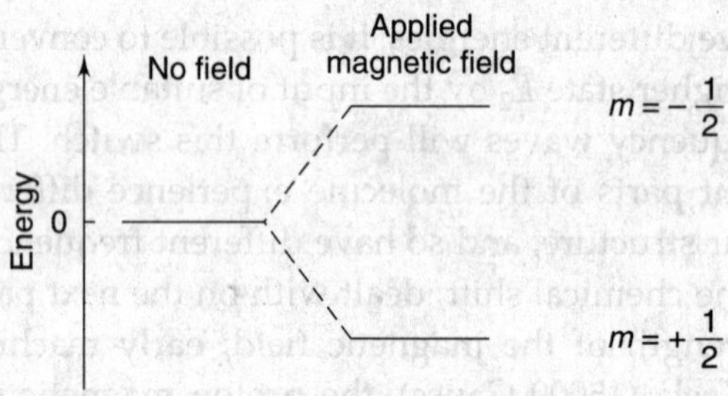

When the nucleus is in a magnetic field, the initial populations of the energy levels are determined by thermodynamics, as described by the Boltzmann distribution. This is very important, and it means that the lower energy level will contain slightly more nuclei than the higher level. It is possible to excite these nuclei into the higher level with electromagnetic radiation. The frequency of radiation needed is determined by the difference in energy between the energy levels.

14.2.2 Calculating Transition Energy

The nucleus has a positive charge and is spinning. This generates a small magnetic field. The nucleus therefore possesses a magnetic moment, *u*, which is proportional to its spin, *I*.

$$\mu = \frac{\gamma I h}{2\pi}$$

The constant is called the *magnetogyric ratio* and is a fundamental nuclear constant which has a different value for every nucleus. *h* is Plancks constant.

The energy of a particular energy level is given by;

$$E = -\frac{\gamma h}{2\pi} mB$$

where, *B* is the strength of the magnetic field at the nucleus.

The difference in energy between levels (the transition energy) can be found from

$$\Delta E = \frac{\gamma h B}{2\pi}$$

This means that if the magnetic field, B, is increased, so is ΔE. It also means that if a nucleus has a relatively large magnetogyric ratio, then ΔE is correspondingly large.

14.2.3 The Absorption of Radiation by a Nucleus in a Magnetic Field

A "classical" view of the behavior of the nucleus, that is, the behavior of a charged particle in a magnetic field will be described.

Imagine a nucleus (of spin 1/2) in a magnetic field. This nucleus is in the lower energy level (i.e., its magnetic moment does not oppose the applied field). The nucleus is spinning on its axis. In the presence of a magnetic field, this axis of rotation will *precess* around the magnetic field (Fig. 14.2).

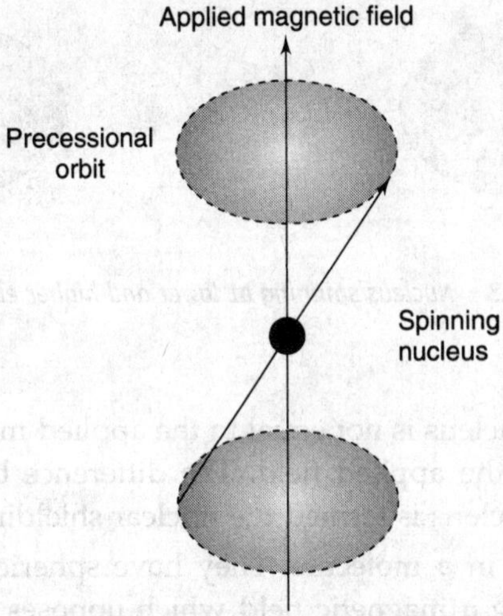

Fig. 14.2 *Nucleus spinning.*

The frequency of precession is termed the *Larmor frequency*, which is identical to the transition frequency.

The potential energy of the precessing nucleus is given by

$$E = -\mu B \cos \theta$$

where θ is the angle between the direction of the applied field and the axis of nuclear rotation.

If energy is absorbed by the nucleus, then the angle of precession, θ, will change. For a nucleus of spin 1/2, absorption of radiation "flips" the magnetic moment so that it opposes the applied field (the higher energy state).

It is important to realize that only a small proportion of "target" nuclei are in the lower energy state (and can absorb radiation). There is possibility that by exciting these nuclei, the populations of the higher and lower energy levels will become equal (Fig. 14.3). If this occurs, then there will be no further absorption of radiation. The spin system is *saturated*. The possibility of saturation means that we must be aware of the relaxation processes which return nuclei to the lower energy state.

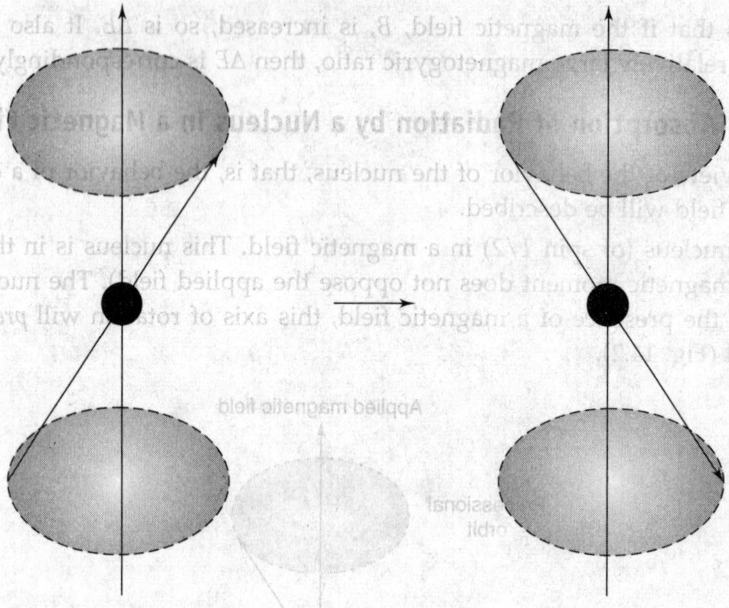

Fig. 14.3 *Nucleus spinning at lower and higher energy state.*

14.2.4 Chemical Shift

The magnetic field at the nucleus is not equal to the applied magnetic field; electrons around the nucleus shield it from the applied field. The difference between the applied magnetic field and the field at the nucleus is termed the nuclear shielding.

Consider the s-electrons in a molecule. They have spherical symmetry and circulate in the applied field, producing a magnetic field which opposes the applied field. This means that the applied field strength must be increased for the nucleus to absorb at its transition frequency. This up field shift is also termed diamagnetic shift.

Electrons in p-orbitals have no spherical symmetry. They produce comparatively large magnetic fields at the nucleus, which give a low field shift. This "deshielding" is termed paramagnetic shift (Fig. 14.4).

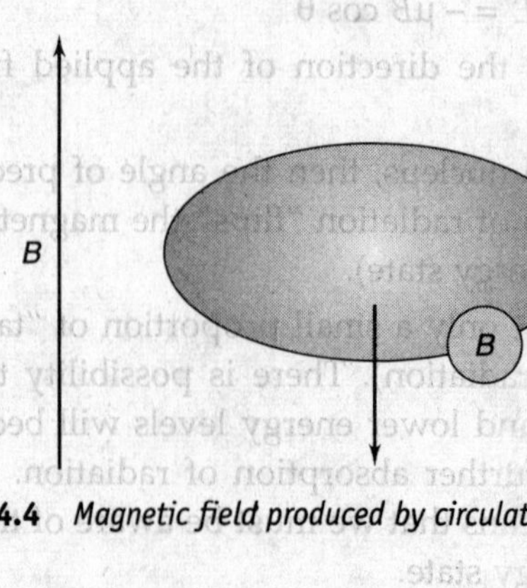

Fig. 14.4 *Magnetic field produced by circulating electron.*

In proton (^1H) NMR, p-orbitals play no part (there aren't any), that is why only a small range of chemical shift (10 ppm) is observed. One can easily see the effect of s-electrons on the chemical shift by looking at substituted methanes, CH_3X. As X becomes increasingly electronegative, so the electron density around the protons decreases, and they resonate at lower field strengths (increasing d_H values).

Chemical shift is defined as nuclear shielding/applied magnetic field. Chemical shift is a function of the nucleus and its environment. It is measured relative to a reference compound. For ^1H NMR, the reference is usually tetramethylsilane, Si $(CH_3)_4$.

14.2.5 Spin-Spin Coupling

Consider the structure of ethanol

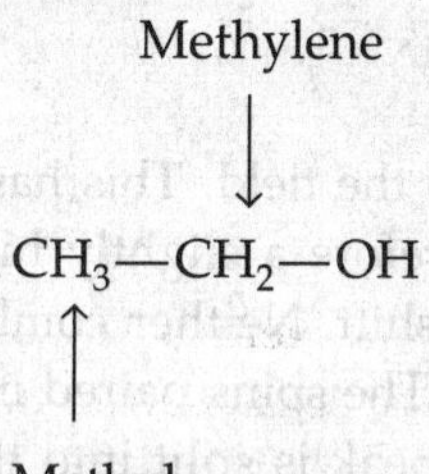

The ^1H NMR spectrum of ethanol (below) shows the methyl peak has been split into three peaks (a *triplet*) and the methylene peak has been split into four peaks (a *quartet*). This occurs because there is a small interaction (*coupling*) between the two groups of protons. The spacings between the peaks of the methyl triplet are equal to the spacings between the peaks of the methylene quartet. This spacing is measured in Hertz and is called the *coupling constant*, *J*. (Fig. 14.5).

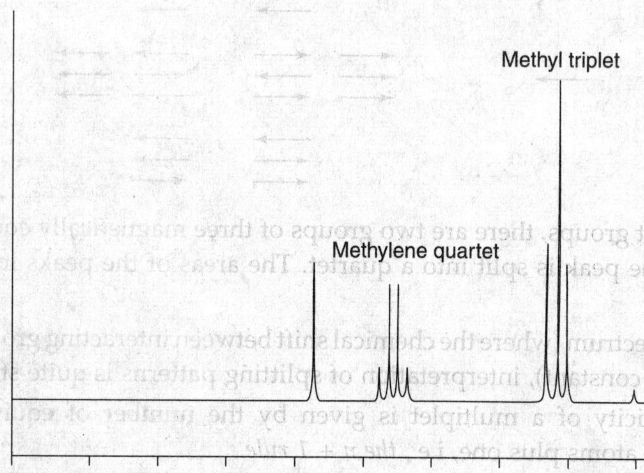

Fig. 14.5 *Coupling constant: methyl triplet and methyl quartet.*

To see why the methyl peak is split into a triplet, let's look at the **methylene** protons. There are two of them, and each can have one of the two possible orientations (aligned with or opposed against the applied field). This gives a total of four possible states:

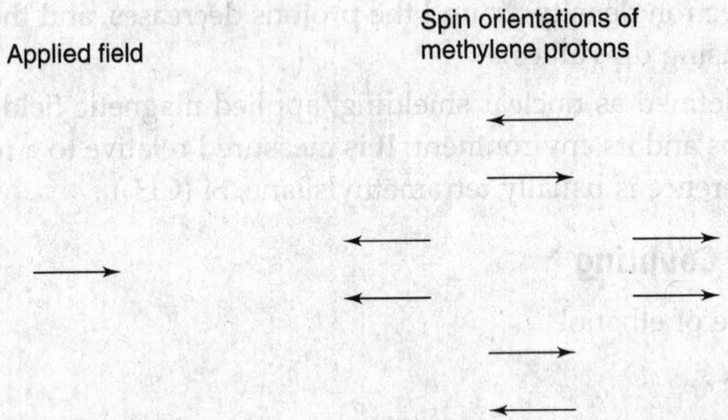

In the first possible combination, spins are paired and opposed to the field. This has the effect of reducing the field experienced by the methyl protons; therefore a slightly higher field is needed to bring them to resonance, resulting in an up field shift. Neither combination of spins opposed to each other has an effect on the methyl peak. The spins paired in the direction of the field produce a downfield shift. Hence, the methyl peak is split into three, with the ratio of areas 1:2:1.

Similarly, the effect of the methyl protons on the methylene protons is such that there are eight possible spin combinations for the three methyl protons:

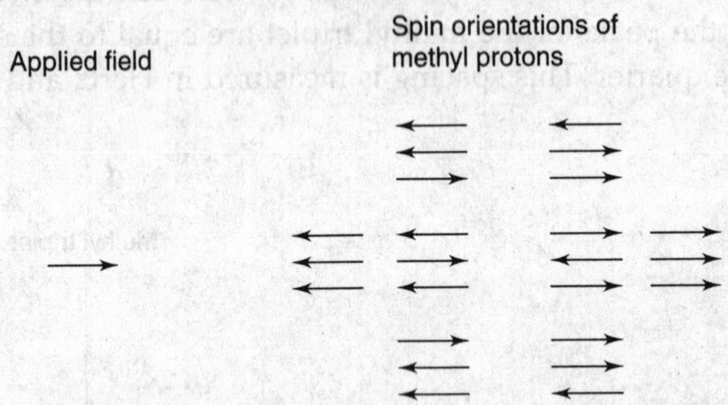

Out of these eight groups, there are two groups of three magnetically equivalent combinations. The methylene peak is split into a quartet. The areas of the peaks in the quartet have the ratio 1:3:3:1.

In a first-order spectrum (where the chemical shift between interacting groups is much larger than their coupling constant), interpretation of splitting patterns is quite straightforward:

- The multiplicity of a multiplet is given by the number of equivalent protons in neighboring atoms plus one, i.e., *the n + 1 rule.*
- Equivalent nuclei do not interact with each other. The three methyl protons in ethanol cause splitting of the neighboring methylene protons; they do not cause splitting among themselves.

- The coupling constant is not dependent on the applied field. Multiplets can be easily distinguished from closely spaced chemical shift peaks.

14.3 PRINCIPLE OF NMR

The nuclei of all elements carry a charge. When the spins of the protons and neutrons comprising these nuclei are not paired, the overall spin of the charged nucleus generates a magnetic dipole along the spin axis, and the intrinsic magnitude of this dipole is a fundamental nuclear property called the nuclear magnetic moment, μ. The symmetry of the charge distribution in the nucleus is a function of its internal structure and if this is spherical (i.e., analogous to the symmetry of a 1s hydrogen orbital), it is said to have a corresponding spin angular momentum number of $I = 1/2$, of which examples are ^1H, ^{13}C, ^{15}N, ^{19}F, ^{31}P, etc. Nuclei which have a non-spherical charge distribution (analogous to, e.g., a hydrogen 3d orbital) have higher spin numbers (e.g., ^{10}B, ^{14}N, etc).

In quantum mechanical terms, the nuclear magnetic moment of a nucleus can align with an externally applied magnetic field of strength **Bo** in only $2I + 1$ ways, either reinforcing or opposing **Bo**. The energetically preferred orientation has the magnetic moment aligned parallel with the applied field (spin +1/2) and is often given the notation Alfa, whereas the higher energy anti-parallel orientation (spin – 1/2) is referred to as Beta. The rotational axis of the spinning nucleus cannot be orientated exactly parallel (or anti-parallel) with the direction of the applied field **Bo** (defined in our coordinate system as about the *z*-axis) but must precess about this field at an angle (for protons about 54) with an angular velocity given by the expression;

$$\omega_o = \gamma \mathbf{Bo} \qquad \text{(the Larmor frequency, in Hz)} \qquad ...(1)$$

The constant γ is called the magnetogyric ratio and relates the magnetic moment *u* and the spin number *I* for any specific nucleus:

$$\gamma = 2\pi\mu/hI \qquad (h \text{ is Planck's constant}) \qquad ...(2)$$

For a single nucleus with $I = 1/2$ and positive γ, only one transition is possible ($\Delta I = 1$, a single quantum transition) between the two energy levels.

NMR is all about how to interpret such transitions in terms of chemical structure. We will first consider the energy of a typical NMR transition. If angular velocity is related to frequency by $\omega_o = 2...\nu$, then

$$\nu = \gamma \mathbf{Bo}/21/4 \qquad ...(3)$$

It follows that proton NMR transitions ($\Delta I = 1$) have the following energy;

$$h\nu = \Delta E = h\gamma \mathbf{Bo}./2\pi \qquad ...(4)$$

For a proton $\gamma = 26.75 \times 10^7$ rad T^{-1} s^{-1} and **Bo** ~ 2*T*, $\Delta E = \mathbf{6 \times 10^{-26}}$ J. The relative populations of the higher (n^2) and lower (n^1) energy levels at room temperature are given by the Boltzmann law:

$$n^2/n^1 = e{-}^{\Delta}E/kT \sim 0.99999. \qquad ...(5)$$

For NMR, this means that the probability of observing a transition from n^1 to n^2 is only slightly greater than that for a downward transition, i.e., the overall probability of observing absorption of energy is quite small. This relationship also explains why a larger **Bo** favors sensitivity in NMR measurements, increasing as it does the difference between the two Boltzmann levels, and why NMR becomes more sensitive at lower temperatures.

14.4 INSTRUMENTATION

14.4.1 Instrument Details

Model: Mercury Plus 300 MHz NMR SPECTROMETER

Specifications: 5 mm autoswitchable probe with PFG (1H/ 13C/ 31P/ 19F) 5 mm Dual Broadband probe with PFG for Multinuclear NMR (13C, 15N, 27Al, 31P, 29Si, 77Se, 119Sn, 125Te, 199Hg, 51V, 7Li etc.)

14.4.2 Main Features of the Instrument

MODEL: Mercury Plus 300 NMR spectrometer.

Specification: 5 mm multinuclear probes for solution studies. Multinuclear CP-MAS probe for solids application.

VT Accessory: Variable temperature facility is available from –80 degree to +130 degree with suitable solvent.

The NMR Probe is switchable between high frequency range (1H, 19F) and broadband frequency range (13C, 15N, 27Al, 31P,29Si, 77Se, 119Sn, 125Te etc.). In case of solids, only broadband frequency range is available and the sample can be spun up to a maximum of 5KHz at the magic angle.

High power decoupling in case of solids is limited to a power of 100 W. Total spectral width is limited to 100 kHz.

Nuclei with non-zero spins, when placed in a strong magnetic field precess at specific orientations with respect to the applied magnetic field. When appropriate energy is supplied in the form of radio frequency, these nuclei flip to a higher energy state. The energy absorbed during this transition is a function of nucleus type and its chemical environment in the molecule. The excited nuclei are allowed to precess freely and come back to their equilibrium positions. During this process an electric signal is induced in a suitably placed RF coil. This signal which is monitored with respect to time is called free induction decay (FID). The FID, which is in time domain gives its equivalent frequency domain signal on Fourier transformation. A plot of the absorption frequency versus the intensity of the absorption constitutes the NMR spectrum (Fig. 14.6).

14.4.3 Sample Holder

Sample should be held in a holder which should be chemically inert, durable and transparent to rf radiation. Generally, glass tubes are employed which are sturdy, practical and cheap. They are generally about 7.5 cm long and approximately 0.3 cm in diameter.

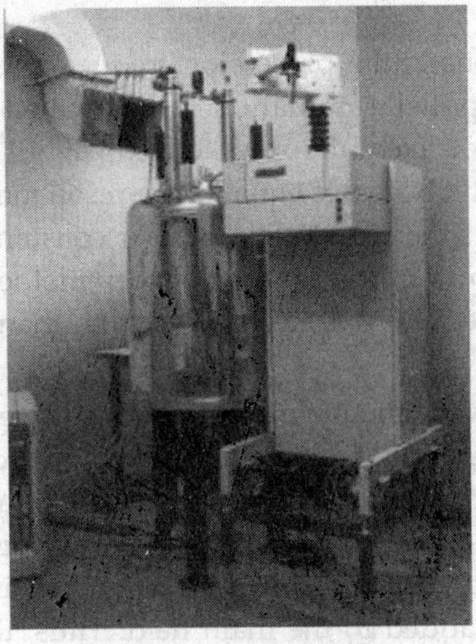

Fig. 14.6 *NMR instrument.*

14.4.4 Magnet

Permanent magnet or electromagnet can be used in a nuclear magnetic resonance instrument. The important feature of the magnet is that it should give homogenous magnetic field, i.e., the strength and direction of the magnetic field should not change from point to point. Moreover, the strength of the field should be very high, i.e., at least 20000 gauss (*G*), because the chemical shifts are proportional to the field strength. Thus, the factors which are important in the design of the magnets for NMR spectroscopy are the homogeneity or uniformity of the field, the constancy of the field, strength and maximum obtainable strength of the field.

Permanent magnets can provide fields which are sufficiently constant provided that these are thermostatted very carefully. Conventional electromagnets are less stable but the desired stability can be produced by producing ancillary stabilizing devices. Recently electromagnets consisting of superconducting solenoid operating in liquid helium crystals have been developed and are highly stable. Electromagnets interact much less with external magnetic effect as compared to the permanent magnets.

Permanent magnets as well as electromagnets can give highly homogeneous fields by carefully machining and alignment of the pole faces. Further, homogeneity can be obtained with the help of current shims and consists of coils and are located at the pole faces. The current shims are used to produce field gradients in various directions so as to cancel out gradient inherent in the main field.

Permanent and conventional electromagnets are generally used in spectrometers operating up to 100 MHz. Superconducting magnets have been found to realize fields corresponding to 230 MHz.

14.4.5 Sweep Generator

In order for a nucleus to resonate, the precession frequency of the nucleus should become equal to the frequency of the applied rf radiation. If the applied magnetic field H_o is kept constant, the precession frequency is fixed. In order to bring about resonance, the frequency of rf field should be changed so that it becomes equal to the resonance frequency (frequency sweep method). On the other hand, if the rf radiation is kept constant, the resonance frequency of the nucleus must be changed by varying H_o (field sweep method). Generally, the field sweep method is regarded as better because it is easier to vary H_o than the rf radiation so as to bring about resonance in nuclei.

Practically, it is not easy to vary the magnetic field of a large stable magnet. This technical problem is solved by superimposing a small variable magnetic field on the main field. This can be done by fixing a pair of Helmholtz coils in the pole faces of the main magnet. The magnetic field induced by these coils can be varied by varying the current flowing through them. The small magnetic field generated by Helmholtz coils is in the same direction as the main field and therefore, it is added to the main field. Thus, the sample is exposed to both fields and they appear as one field to the molecule (Fig. 14.7).

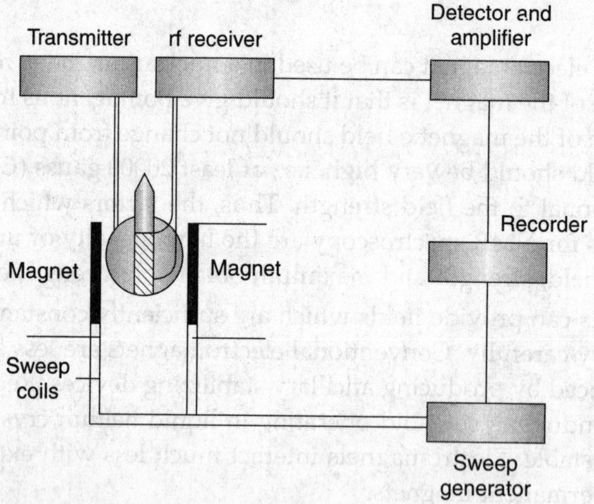

Fig. 14.7 *Schematic diagram of a NMR spectrometer.*

14.4.6 Radio Frequency Generator

Radio frequency oscillator is used to generate radio frequency radiation. To achieve the maximum interaction of the *rf* radiation with the sample, the coil of oscillator is wound around the sample container. The oscillator irradiates the sample with rf radiation.

The oscillator coil is wound perpendicular to the applied magnetic field. This is done so that the applied rf field should not change the effective magnetic field in the process of irradiation.

14.4.7 Radio Frequency Receiver

When the radio frequency radiation is passed through the magnetized sample, the two phenomena, namely, absorption and dispersion, may occur. The line shapes associated with the absorption and dispersion are shown in Figs. 14.8 and 14.9. The observation of either dispersion or absorption will enable the resonance frequency to be determined. It is found that the interpretation of absorption spectrum is easier as compared to dispersion spectrum.

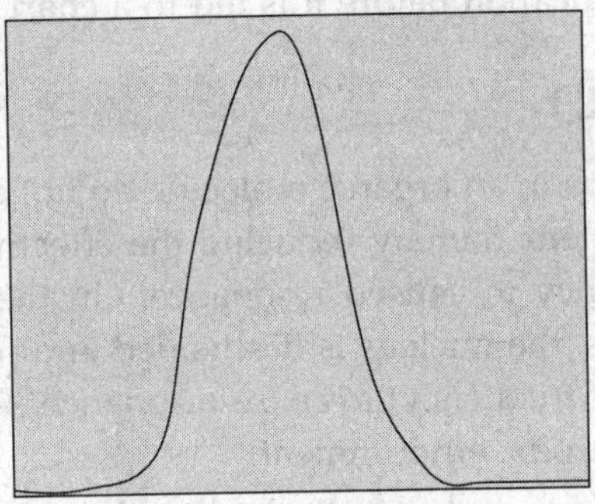

Fig. 14.8 *Absorption signal.*

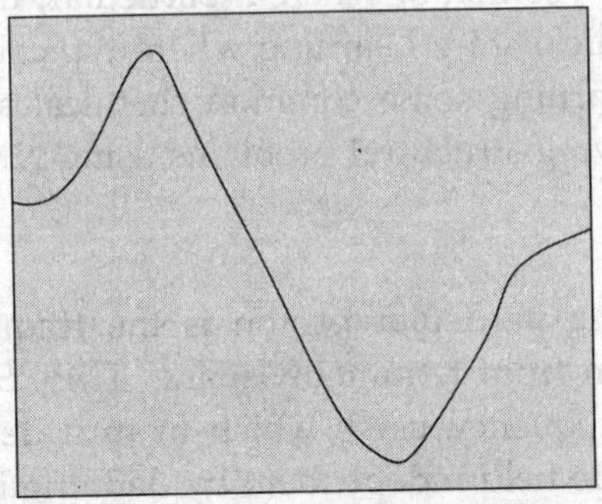

Fig. 14.9 *Dispersion signal.*

The detector should be capable of separating absorption signal from dispersion signal and from that of the rf oscillator. There are two main methods of detection. These are:

(i) The first method uses a radio frequency bridge. Its network balances the transmitter signal and allows the absorption and dispersion signals to appear as an out-of-e.m.f across the bridge. In this method, the coil used for surrounding the samples serves as both a transmitter and a receiver coil.

(ii) The second method employs a separate receiver coil. This method is some times called the crossed coil or nuclear induction method. If the two coils are fixed at right angles to each other as well as to the direction of static magnetic field, they will not be effectively coupled. In this way, the transmitter signal is separated from the absorption and dispersion signals.

The separation of absorption and dispersion signal is achieved by knowing the fact that they differ in phase by an angle of 90°. The general method in practice is to use a phase sensitive detector which helps the operator to select the phase of the signal to be detected.

14.4.8 Readout System

The absorption signal received from radio frequency receiver is extremely weak. Therefore, it requires considerable amplification before it is fed to a chart recorder.

14.5 CHEMICAL SHIFT

Chemically different hydrogens in an organic molecule do not experience the same magnetic field. Electrons shield the nucleus thereby reducing the effective magnetic field and requiring energy of a lower frequency to caused resonance. On the other hand, when electrons are withdrawn from a nucleus, the nucleus is deshielded and feels a stronger magnetic field requiring more energy (higher frequency) to cause resonance. Thus, NMR can provide information about hydrogen's electronic environment.

Generally, hydrogens bound to carbons attached to electron withdrawing groups tend to resonate at higher frequencies (more downfield to the left) from TMS, tetramethylsilane, a common NMR standard. The position of where a particular, hydrogen atom resonates relative to TMS is called its chemical shift. Learning where typical hydrogens resonate requires experience and study, but learning some common chemical shift will provide you with a tremendous advantage at solving structural problems using NMR.

The chemical shift in NMR

The chemical shift given by a particular proton is the frequency difference between its absorption and the absorption from tetramethylsilane, TMS, $Si(CH_3)_4$. The actual chemical shift depends on the radio frequency used, which in turn depends on the magnetic field strength. The shift can be made field-independent by defining it thus:

$$D = \frac{(n - n_o)\ s\ 10^6}{n_o} = \frac{Dn \times 10^6}{n_o}$$

The value of d is quoted in parts per million – this means that for a machine operating around 200MHz the shifts correspond to frequency differences of a few hundred Hz.

The chemical shift depends on the environment of the hydrogen atom concerned. The local external magnetic field, the one with which the nucleus interacts, is not the same as the applied external field. This is because the nucleus is shielded from this field to a greater or lesser extent by the other atoms in the vicinity and their electrons. The effect can be seen in the low-resolution spectrum of ethanol, CH_3CHO, given below (Fig. 14.10).

The less the shielding experienced, the higher the chemical shift. The relatively exposed hydrogen in the aldehyde group has $d = 9.8$, whereas the more heavily shielded hydrogen atoms in the methyl group are found at $d = 2.2$. TMS hydrogen atoms are by definition found at $d = 0$.

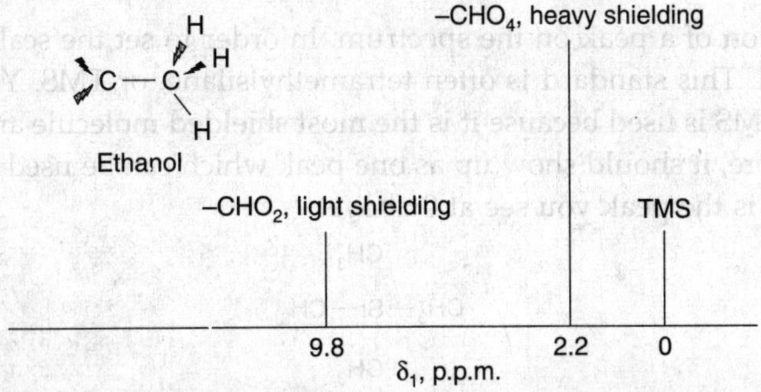

Fig 14.10 *Low resolution spectrum of ethanol and aldehyde.*

Typical ^1H chemical shifts relative to TMS = 0.

$RCOO\mathbf{H}$	carboxylic acids	+9 to +13	Can be broad and solvent-dependent
$RCON\mathbf{H}_2$	amide	+5 to +12	Broad and solvent-dependent
$RC\mathbf{H}O$	aldehyde	+8 to +10	Sharp
$C_6\mathbf{H}_6$ etc	aromatics	+6 to +10	
$R_2C{=}C\mathbf{H}R$	alkene	+4 to +8	
$RN\mathbf{H}_2$	amines	+1 to +6	Broad and solvent-dependent
$RO\mathbf{H}$	alcohols	+0.5 to +8	Broad and solvent-dependent
$RC\mathbf{H}_2R$	methylene	+1.5 to +4.5	
$RC\mathbf{H}_3$	methyl	0 to +4	

14.5.1 Chemical Shifts

There are many different nuclei which can be observed by NMR spectroscopy: ^1H (proton), ^{13}C (carbon 13), ^{15}N (nitrogen 15), ^{19}F (fluorine 19), and many others. ^1H, and ^{13}C are used the most often. Greg Brust described ^1H because it is representative of NMR spectroscopy.

All nuclei have a positive charge and are constantly spinning like a top. We know from physics that a charge in motion creates a magnetic field. In NMR, when we hit the nucleus with radio frequency (rf) radiation, it causes the nucleus and its magnetic field to flip (or it makes the nuclear magnet resonate, hence the name NMR). In the NMR spectrometer there is a coil of wire which surrounds the sample. We also know that a magnet moving in a wire coil causes a charge to flow in the wire. So when the magnetic field of the nucleus flips, it causes a current in the wire which the computer can detect.

The pattern seen here is called *free induction decay*, or FID. It looks like it does because when the nuclei are pulsed, the spins of like nuclei group together and after the pulse, the spins slowly move apart, or decay. This is visible in the FID.

Before you can determine which peaks correspond to specific nuclei, you must set the scale for the chemical shifts of the spectrum (Fig. 14.11 a & b). *Chemical shift* is a fancy scientific name for the position of a peak on the spectrum. In order to set the scale of the spectrum, a standard is needed. This standard is often tetramethylsilane, or TMS. You can see its structure on the right. TMS is used because it is the most shielded molecule and all its protons are equivalent. Therefore, it should show up as one peak which can be used to set the 0 mark on the spectrum. This is the peak you see at 0 above.

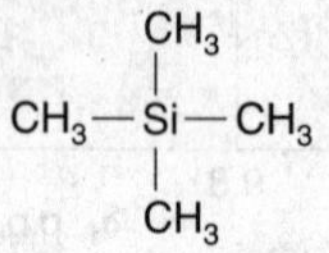

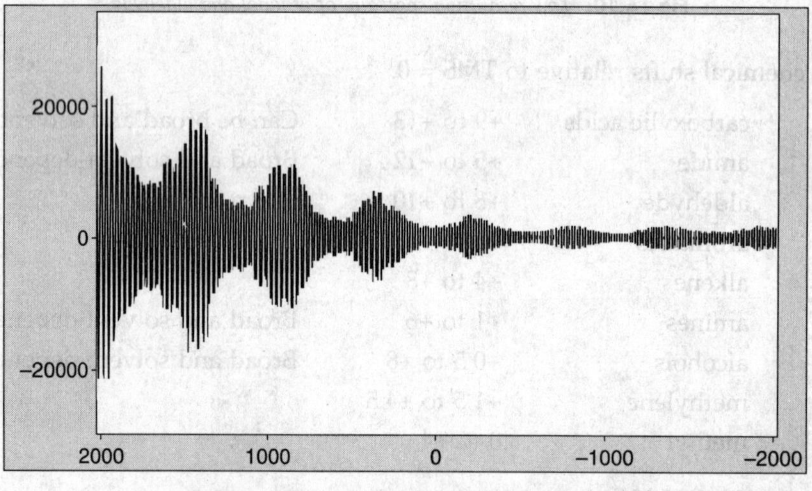

Fig. 14.11(a) *Chemical shift spectrum.*

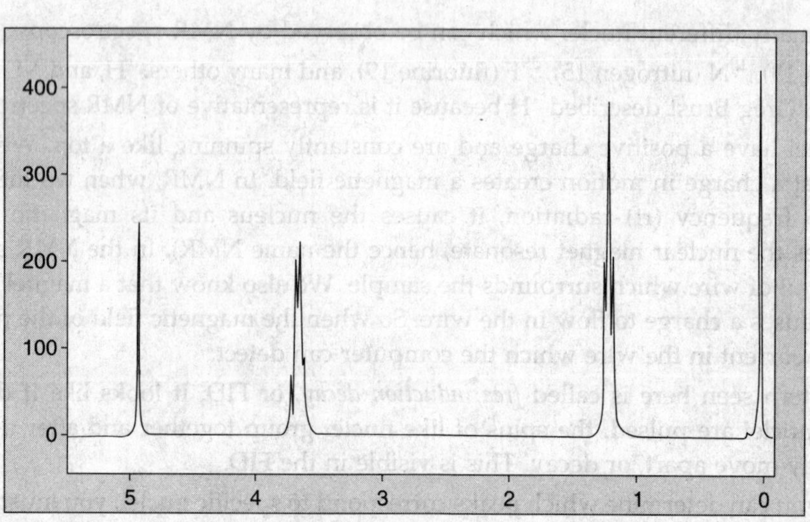

Fig 14.11(b) *Chemical shift spectrums.*

The first thing you might notice when looking at the spectrum is that there are not just a few peaks on the spectrum, but rather some single peaks, some double peaks, and even larger groups of peaks. These groups of peaks are each due to one type of nuclei, not many as one might assume. The reason there are a group of peaks instead of just one is that hydrogens on one carbon are *coupled* to, or influenced by the magnetic fields of hydrogens on adjacent nuclei. This coupling "splits" the signal into the multiple peaks seen in the spectrum. This splitting follows what is known as the "*N* plus one rule," which states that the number of peaks seen for each type of hydrogen is equal to the number of hydrogens on adjacent nuclei (N) plus one. For example, the spectrum above is of ethyl alcohol, the structure of which is $H_3C–CH_2–OH$. The peak between 1 and 2 corresponds to the hydrogens on the CH_3 group. It is split into three peaks by the hydrogens on the CH_2 group (2 + 1 = 3). The peak between 3 and 4 is the peak for the hydrogens on the CH_2 group. It is split into four peaks by the hydrogens on the CH_3 group (3 + 1 = 4).

Below is a table of just a few of the many different chemical shifts (given in parts per million, or ppm) possible in ^1H NMR and the protons associated with those shifts, which are highlighted in red.

Proton	**chemical shift (ppm)**
ethane ($H_3C–CH_3$)	1.25
propane ($H_3C–CH_2–CH_3$)	1.17
cyclopropane (C_3H_6)	0.22
cyclopentane (C_5H_{10})	1.40
cyclohexane (C_6H_{12})	1.35
toluene ($H_3C–C_6H_5$)	2.87
styrene ($H_2C=CH–C_6H_5$)	5.09
styrene ($H_2C=CH–C_6H_5$)	6.68

Aromatics	**ortho**	**meta**	**para**
benzene (C_6H_6)	7.54	7.54	7.54
toluene (H_3C-C_6H_5)	7.16	7.16	7.16
styrene (H_2C=CH-C_6H_5)	7.24	7.24	7.24
bromobenzene (Br-C_6H_5)	7.49	7.14	7.24
phenol (HO-C_6H_5)	6.77	7.13	6.87

14.6 NMR SPECTRA – THE EFFECT OF THE PROTON ENVIRONMENT

The number of peaks in the low-resolution spectrum depends on the number of different environments that the hydrogen atoms have in the molecule.

14.6.1 One Environment

Benzene, C_6H_6 has only one sort of hydrogen atom, so that the NMR spectrum shows a single peak (the TMS peak is omitted) (Fig. 14.12):

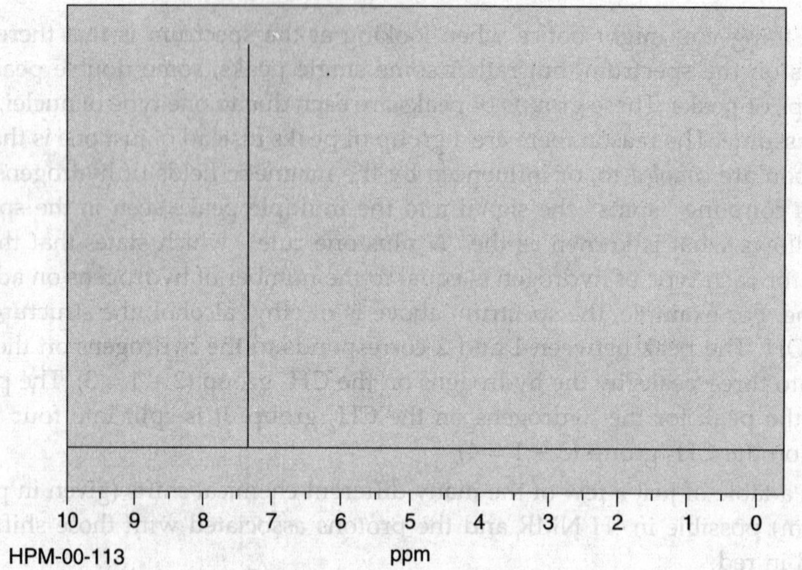

Fig. 14.12 *NMR-benzene single spectrum.*

14.6.2 Two Environments

Ethanal CH_3CHO has two sorts of hydrogen atom, those on the methyl group and that on the aldehyde group. It therefore has two peaks in its spectrum (the TMS peak is omitted) (Fig. 14.13):

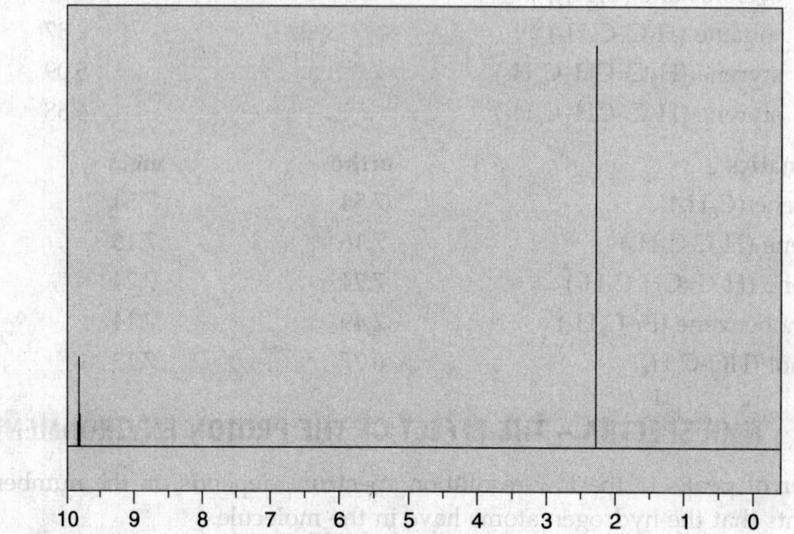

Fig. 14.13 *NMR-aldehyde two spectrum.*

14.6.3 Three Environments

Ethanol· CH_3CH_2OH has methyl hydrogen, methylene hydrogen, and hydroxyl hydrogen. It therefore has three peaks in its spectrum (the TMS peak is omitted). However, this spectrum is the high-resolution spectrum, which shows spin-spin coupling; this is dealt with below. It can be neglected as far as A level examinations are concerned, but it is an important topic and merits some consideration.

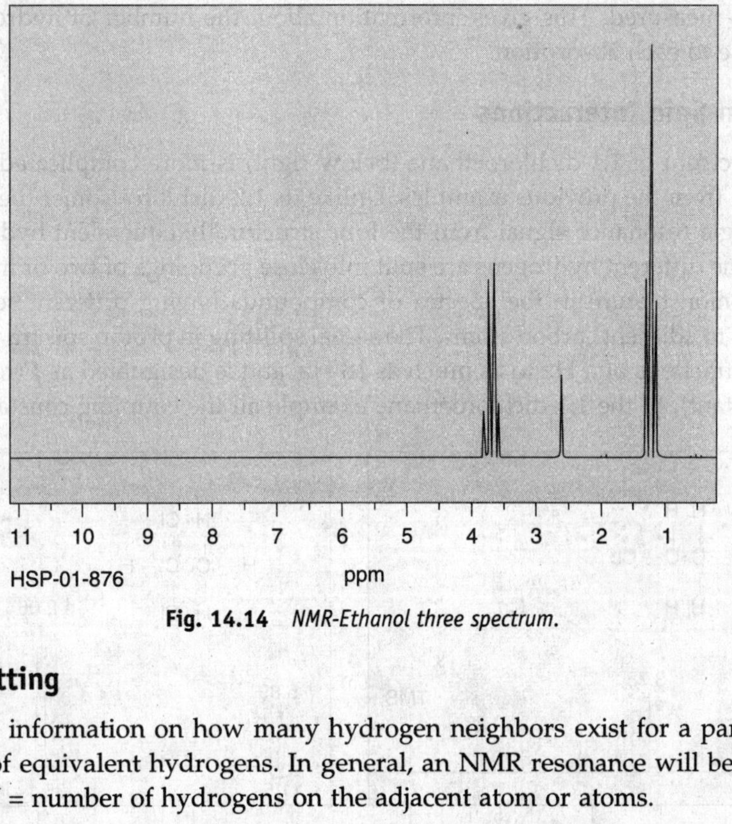

Fig. 14.14 *NMR-Ethanol three spectrum.*

14.6.4 Spiltting

NMR provides information on how many hydrogen neighbors exist for a particular hydrogen or group of equivalent hydrogens. In general, an NMR resonance will be split into N+! peaks where N = number of hydrogens on the adjacent atom or atoms.

If there are no hydrogens on the adjacent atoms, then the resonance will remain a single peak, a singlet. If there is one hydrogen on the adjacent atoms, the resonance will be split into two peaks of equal size, a doublet. Two hydrogens on the adjacent atoms will split the resonance into three peaks with an area in the ratio of 1:2:1, a triplet. If there are 3 hydrogens on the adjacent atoms, the resonance will be split into four peaks with an area in the ratio of 1:3:3:1, a quarter.

14.6.5 Spin-Spin Coupling

A given hydrogen nucleus can exist in one of the two spin states; it is the transition between these states that gives rise to the NMR spectral line. Other adjacent hydrogen atoms in the same environment, for example, all the methyl hydrogen atoms in ethanol, do not affect the absorption. However, this is no longer true when the hydrogen atoms are non-equivalent. In ethanol, for example, the hydrogen atoms on the methyl group interact with those on the methylene group – their magnetic fields *couple*. The effect of coupling on the spectrum is that the lines are split into multiplets. Most coupling occurs between hydrogen atoms on adjacent carbon atoms, so in the ethanol spectrum there is splitting of the lines due to the methyl and methylene hydrogen atoms, but not that of the hydroxyl hydrogen – it is too far away.

The spectra are commonly integrated, that is the area under each peak (or set of peaks in a multiplet) is measured. This gives information about the number of hydrogen atoms that have given rise to each absorption.

14.6.6 Spin-Spin Interactions

The NMR spectrum of 1,1-dichloroethane (below right) is more complicated than we might have expected from the previous examples. Unlike its 1,2-dichloro-isomer (below left), which displays a single resonance signal from the four structurally equivalent hydrogens, the two signals from the different hydrogens are split into close groupings of two or more resonances. This is a common feature in the spectra of compounds having different sets of hydrogen atoms bonded to adjacent carbon atoms. The signal splitting in proton spectra is usually small, ranging from fractions of a Hz to as much as 18 Hz, and is designated as *J* (referred to as the coupling constant). In the 1,1-dichloroethane example all the coupling constants are 6.0 Hz.

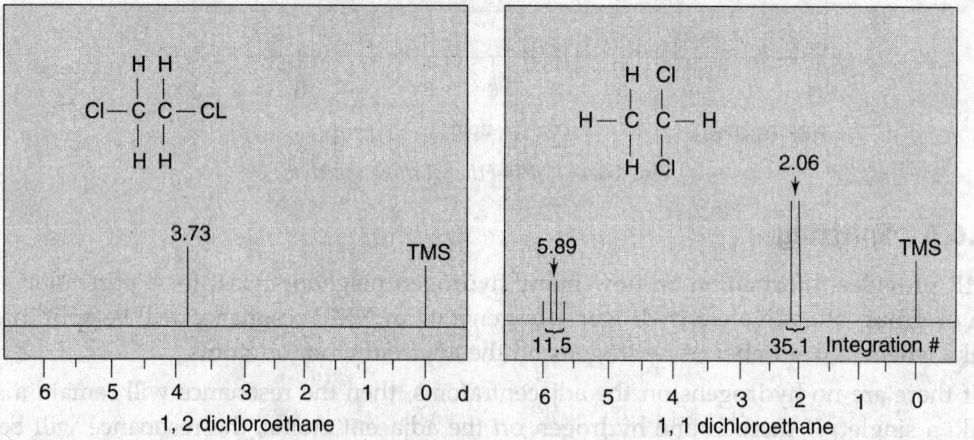

The splitting patterns found in various spectra are easily recognized, provided the chemical shifts of the different sets of hydrogen that generate the signals differ by two or more ppm. The patterns are symmetrically distributed on both sides of the proton chemical shift, and the central lines are always stronger than the outer lines. The most commonly observed patterns have been given descriptive names, such as doublet (two equal intensity signals), triplet (three signals with an intensity ratio of 1:2:1) and quartet (a set of four signals with intensities of 1:3:3:1). Four such patterns are displayed in the following illustration. The line separation is always constant within a given multiplet, and is called the coupling constant (*J*). The magnitude of *J*, usually given in units of Hz, is magnetic field independent.

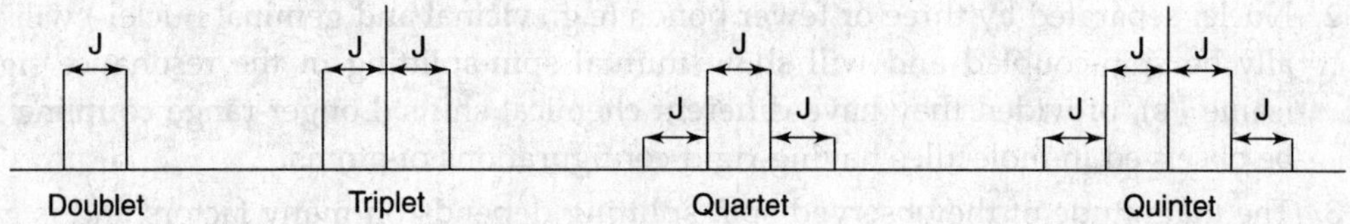

The splitting patterns shown above display the ideal or "**First-Order**" arrangement of lines. This is usually observed if the spin-coupled nuclei have very different chemical shifts (i.e., Δ is large compared to *J*). If the coupled nuclei have similar chemical shifts, the splitting patterns are distorted (second order behavior). In fact, signal splitting disappears if the chemical shifts are the same. Two examples that exhibit minor 2nd order distortion are shown below (both are taken at a frequency of 90 MHz). The ethyl acetate spectrum on the left displays the typical quartet and triplet of a substituted ethyl group. The spectrum of 1,3-dichloropropane on the right demonstrates that equivalent sets of hydrogens may combine their influence on a second, symmetrically located set.

Even though the chemical shift difference between the *A* and *B* protons in the 1,3-dichloroethane spectrum is fairly large (140 Hz) compared with the coupling constant (6.2 Hz), some distortion of the splitting patterns is evident. The line intensities closest to the chemical shift of the coupled partner are enhanced. Thus the *B* set triplet lines closest to *A* are increased, and the *A* quintet lines nearest *B* are likewise stronger. A smaller distortion of this kind is visible for the *A* and *C* couplings in the ethyl acetate spectrum.

Causes for signal splitting

If an atom under examination is perturbed or influenced by a nearby nuclear spin (or set of spins), the observed nucleus responds to such influences, and its response is manifested in its resonance signal. This spin-coupling is transmitted through the connecting bonds, and it functions in both directions. Thus, when the perturbing nucleus becomes the observed nucleus, it also exhibits signal splitting with the same *J*. For spin-coupling to be observed, the sets of interacting nuclei must be bonded in relatively close proximity (e.g., vicinal and geminal locations), or be oriented in certain optimal and rigid configurations. Some spectroscopists place a number before the symbol *J* to designate the number of bonds linking the coupled nuclei. Using this terminology, a vicinal coupling constant is ^3J and a geminal constant is ^2J.

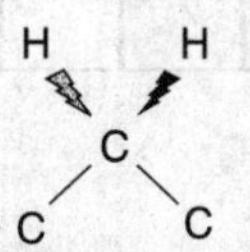

Geminal Hydrogens

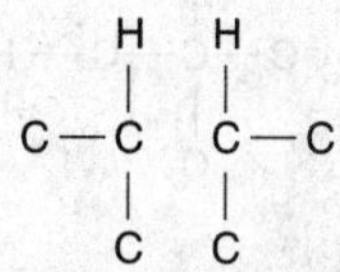

Vicinal Hydrogens

Important requirements and characteristics for spin 1/2 nuclei

1. Nuclei having the same chemical shift (called **isochronous**) do not exhibit spin-splitting. They may actually be spin-coupled, but the splitting cannot be observed directly.

2. Nuclei separated by three or fewer bonds (e.g., vicinal and geminal nuclei) will usually be spin-coupled and will show mutual spin-splitting of the resonance signals (same J's), provided they have different chemical shifts. Longer-range coupling may be observed in molecules having rigid configurations of atoms.
3. The magnitude of the observed spin-splitting depends on many factors and is given by the coupling constant J (units of Hz). J is the same for both partners in a spin-splitting interaction and is independent of the external magnetic field strength.
4. The splitting pattern of a given nucleus (or set of equivalent nuclei) can be predicted by the ***n*+1 rule**, where n is the number of neighboring spin-coupled nuclei with the same (or very similar) Js. If there are 2 neighboring, spin-coupled, nuclei the observed signal is a triplet (2+1 = 3); if there are three spin-coupled neighbors the signal is a quartet (3+1 = 4). In all cases the central line(s) of the splitting pattern are stronger than those on the periphery. The intensity ratio of these lines is given by the numbers in Pascal's triangle. Thus, a doublet has 1:1 or equal intensities, a triplet has an intensity ratio of 1:2:1, a quartet 1:3:3:1, etc. To see how the numbers in Pascal's triangle are related to the Fibonacci series.

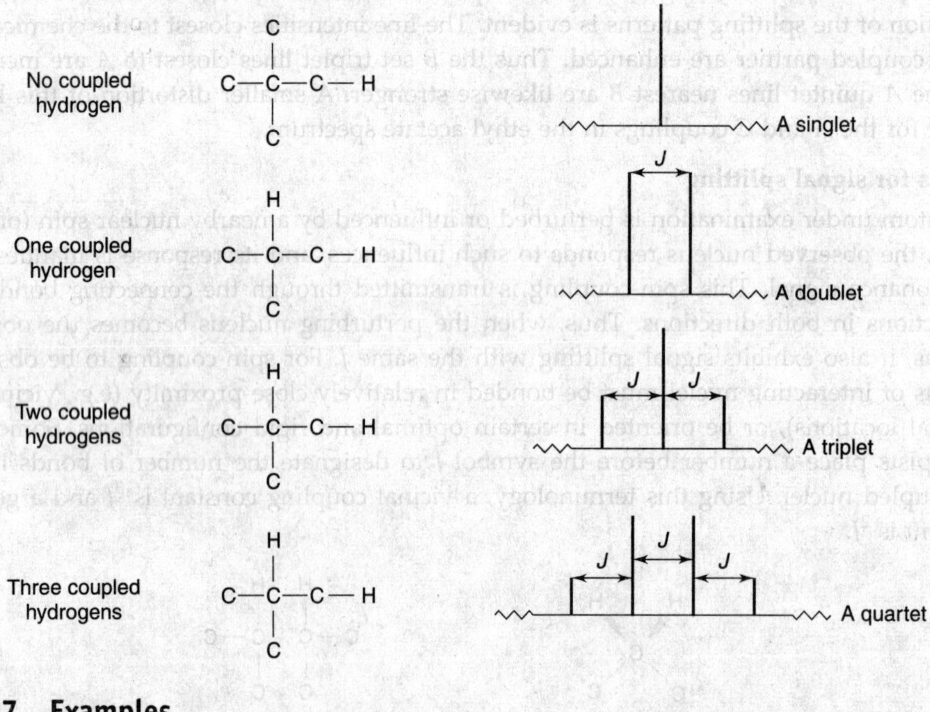

14.6.7 Examples

Test your ability to interpret ^1H NMR spectra by analyzing the examples presented below. The seven spectra may be examined in turn by clicking the "Toggle Spectra" button. Try to associate each spectrum with a plausible structural formula.

Although the first four cases are relatively simple, keep in mind that the integration values provide ratios, not absolute numbers. In two cases additional information from infrared spectroscopy is provided. When you have made an assignment you may check your answer by clicking on the spectrum itself. In the sixth example, a similar constitutional isomer cannot be ruled out by the data given.

14.6.7.1 Carbon NMR spectroscopy

The power and usefulness of ^{1}H NMR spectroscopy as a tool for structural analysis should be evident from the past discussion. Unfortunately, when significant portions of a molecule lack C-H bonds, no information is forthcoming. Examples include polychlorinated compounds such as chlordane, polycarbonyl compounds such as croconic acid, and compounds incorporating triple bonds (structures below, orange colored carbons).

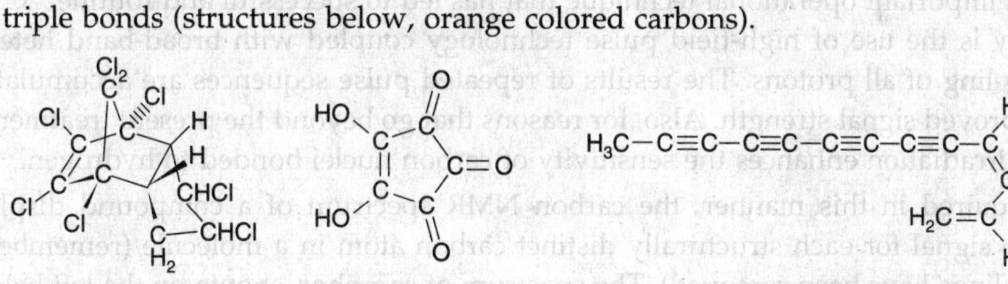

Chlordane   Croconic acid   A polyacetylene from dahlia

Even when numerous C-H groups are present, an unambiguous interpretation of a proton NMR spectrum may not be possible. The following diagram depicts three pairs of isomers (A & B) which display similar proton NMR spectra. Although a careful determination of chemical shifts should permit the first pair of compounds to be distinguished, the second and third cases might be difficult to identify by proton NMR alone.

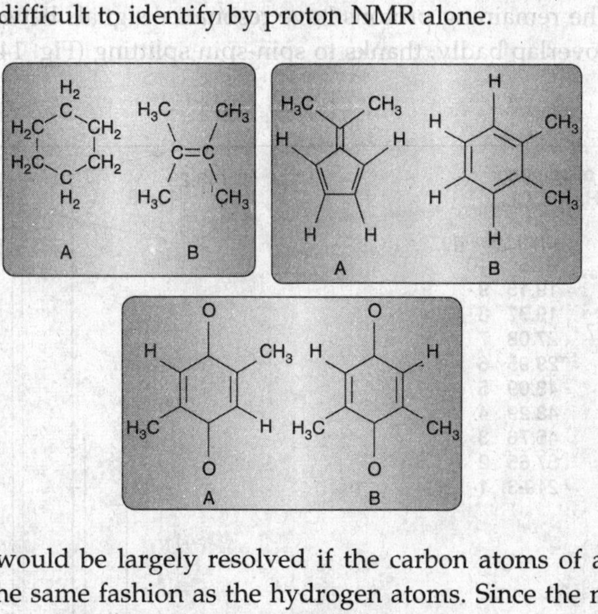

These difficulties would be largely resolved if the carbon atoms of a molecule could be probed by NMR in the same fashion as the hydrogen atoms. Since the major isotope of carbon (^{12}C) has no spin, this option seems unrealistic. Fortunately, 1.1% of elemental carbon is the ^{13}C isotope, which has a spin $I = 1/2$, so in principle it should be possible to conduct a carbon NMR experiment. It is worth noting here, that if much higher abundances of ^{13}C

were naturally present in all carbon compounds, proton NMR would become much more complicated due to large one-bond coupling of ^{13}C and ^{1}H.

Many obstacles needed to be overcome before carbon NMR emerged as a routine tool

(i) As noted, the abundance of ^{13}C in a sample is very low (1.1%), so higher sample concentrations are needed.

(ii) The ^{13}C nucleus is over fifty times less sensitive than a proton in the NMR experiment, adding to the previous difficulty.

(iii) Hydrogen atoms bonded to a ^{13}C atom split its NMR signal by 130 to 270 Hz.

The most important operational technique that has led to successful and routine ^{13}C NMR spectroscopy is the use of high-field pulse technology coupled with broad-band heteronuclear decoupling of all protons. The results of repeated pulse sequences are accumulated to provide improved signal strength. Also, for reasons that go beyond the present treatment, the decoupling irradiation enhances the sensitivity of carbon nuclei bonded to hydrogen.

When acquired in this manner, the carbon NMR spectrum of a compound displays a single sharp signal for each structurally distinct carbon atom in a molecule (remember, the proton couplings have been removed). The spectrum of camphor, shown on the left below, is typical. Furthermore, a comparison with the ^{1}H NMR spectrum on the right illustrates some of the advantageous characteristics of carbon NMR. The dispersion of ^{13}C chemical shifts is nearly twenty times greater than that for protons, and this together with the lack of signal splitting makes it more likely that every structurally distinct carbon atom will produce a separate signal. The only clearly identifiable signals in the proton spectrum are those from the methyl groups. The remaining protons have resonance signals between 1.0 and 2.8 ppm from TMS, and they overlap badly, thanks to spin-spin splitting (Fig. 14.15).

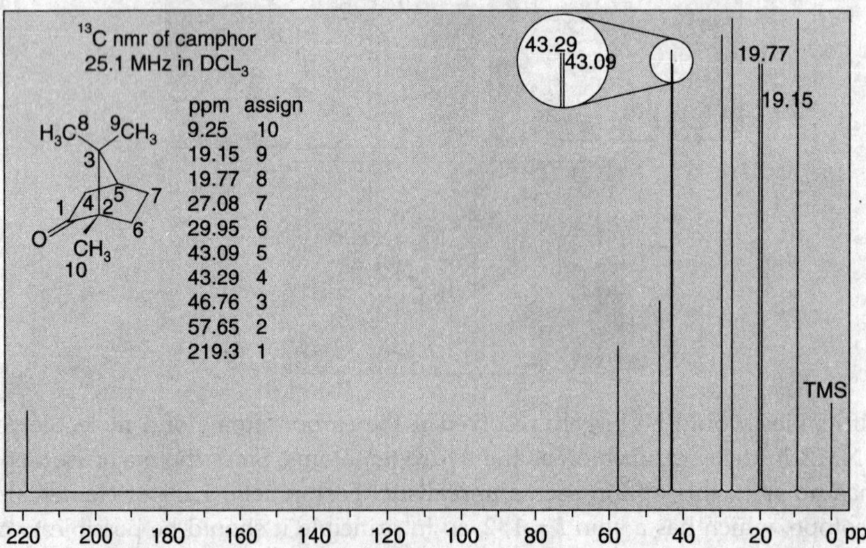

Fig. 14.15 *Spin-spin splitting.*

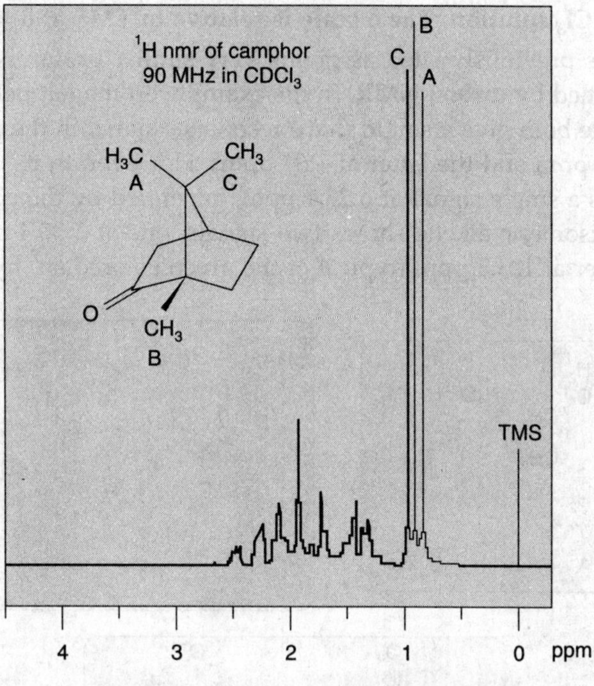

Unlike proton NMR spectroscopy, the relative strength of carbon NMR signals are not normally proportional to the number of atoms generating each one. Because of this, the number of discrete signals and their chemical shifts are the most important pieces of evidence delivered by a carbon spectrum. The general distribution of carbon chemical shifts associated with different functional groups is summarized in the following chart. Bear in mind that these ranges are approximate, and may not encompass all compounds of a given class. Note also that the over 200 ppm range of chemical shifts shown here is much greater than that observed for proton chemical shifts.

14.6.7.2 ^{13}C chemical shift ranges

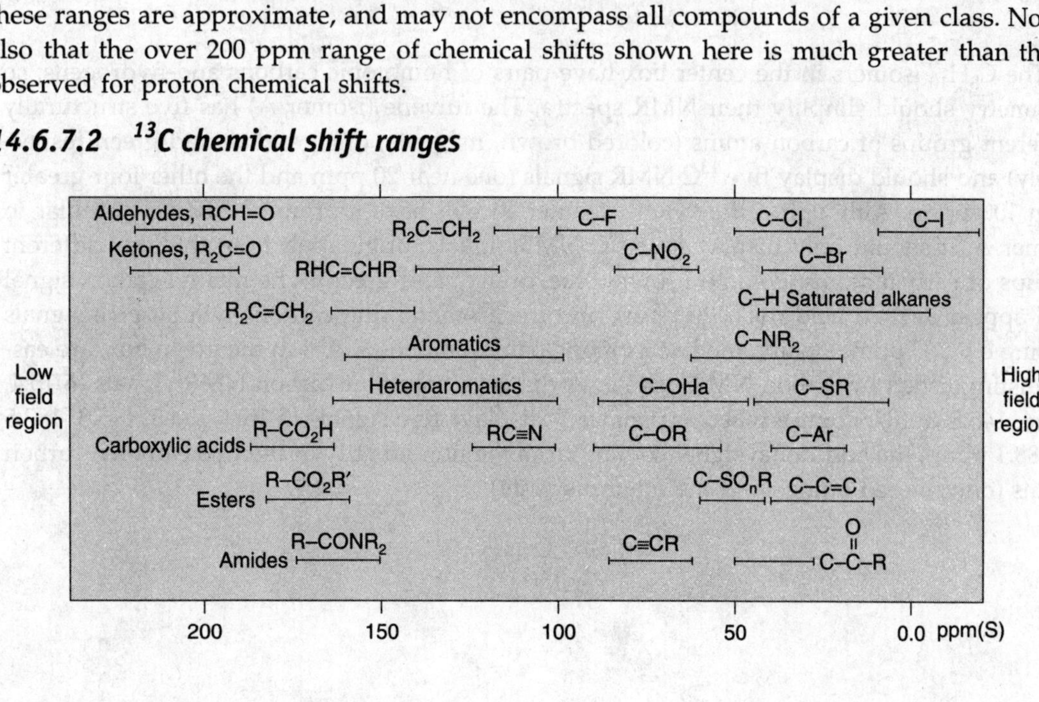

For samples in $CDCl_3$ solution. The δ scale is relative to TMS at δ = 0.

The isomeric pairs previously cited as giving very similar proton NMR spectra are now seen to be distinguished by carbon NMR. In the example on the left below, cyclohexane and 2,3-dimethyl-2-butane both give a single sharp resonance signal in the proton nmr spectrum (the former at δ 1.43 ppm and the latter at 1.64 ppm). However, in its carbon nmr spectrum cyclohexane displays a single signal at δ 27.1 ppm, generated by the equivalent ring carbon atoms; whereas the isomeric alkene shows two signals, one at δ 20.4 ppm from the methyl carbons, and the other at 123.5 ppm (typical of the green colored sp^2 hybrid carbon atoms).

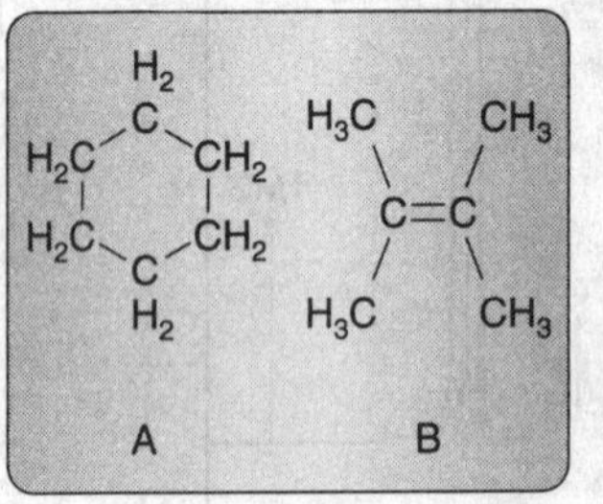

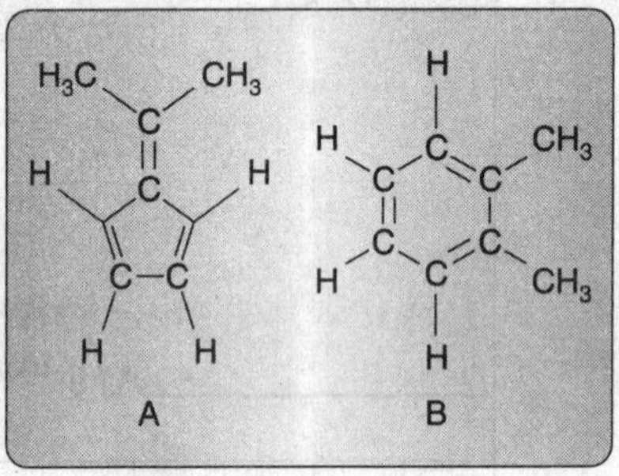

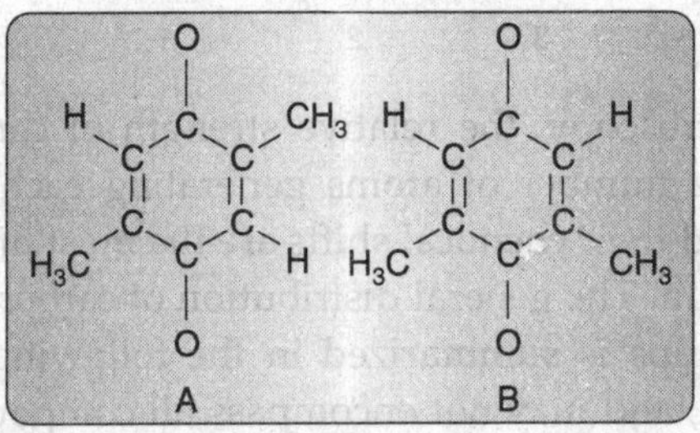

The C_8H_{10} isomers in the center box have pairs of homotopic carbons and hydrogens, so symmetry should simplify their NMR spectra. The fulvene (isomer A) has five structurally different groups of carbon atoms (colored brown, magenta, orange, blue and green respectively) and should display five ^{13}C NMR signals (one near 20 ppm and the other four greater than 100 ppm). Although ortho-xylene (isomer *B*) will have a proton NMR very similar to isomer A, it should only display four ^{13}C NMR signals, originating from the four different groups of carbon atoms (colored brown, blue, orange and green). The methyl carbon signal will appear at high field (near 20 ppm), and the aromatic ring carbons will all give signals having δ > 100 ppm. Finally, the last isomeric pair, quinones A & B in the green box, are easily distinguished by carbon NMR. Isomer A displays only four carbon NMR signals (δ 15.4, 133.4, 145.8 & 187.9 ppm); whereas, isomer B displays five signals (δ 15.9, 133.3, 145.8, 187.5 & 188.1 ppm), the additional signal coming from the non-identity of the two carbonyl carbon atoms (one colored orange and the other magenta).

14.7 APPLICATIONS OF NMR SPECTROSCOPY

NMR applications primarily are in the field of structure determination and delineation. There is literature of NMR spectra comparable to those for optical absorption spectra and X-ray diffraction spectra. Comparison of the NMR spectrum of an unknown compound with these in the literature may lead directly to identification of the unknown. Even if this is not the case, the NMR spectrum can help to distinguish between two possible isomeric structures and in many cases can furnish the critical information that the analyst needs to delineate the exact structure of the complicated compound.

Integration of the NMR spectrum furnished information that can be used for quantitative determination of one or more components present in a mixture.

14.7.1 Structural Diagnosis of NMR

A large number of principles are known which will decide the structure of an unknown compound from its NMR spectrum. Some of these are outlined as follows:

(i) The number of main NMR signals should be equal to the number of equivalent protons in the unknown compound.

(ii) The chemical shift indicates that what type of hydrogen atoms are present, e.g., methylene, methyl groups, olefins, ethers, etc.

(iii) The spin-spin splitting or multiplicity reveals the possible arrangement of groups in the molecule.

(iv) From the area of peaks, one can draw the conclusion about the number of hydrogen nuclei present in each group. For example, the relative areas of methyl (CH_3) and methylene (CH_2) peaks in $CH_3 - CH_2 - CH_3$ would be 6:2. In butane, it would be 6:4.

14.7.2 Quantitative Analysis

NMR spectroscopy has been used to determine the molar ratio of the components in a mixture.

If the NMR spectrum of a mixture of compounds is recorded, it may exhibit a resolved band due to one of the components and if the number of protons in the molecule giving rise to that band is known, then the integrated area of the band will give a direct measure of the relative concentration of that component in the mixture. This can be illustrated by taking the NMR spectrum of a mixture of benzene and cyclohexane which exhibits bands at 2.73 and 8.56 from benzene and cyclohexane respectively, and the integrated area of the former band will be proportional to six protons whereas that of the latter band will be proportional to twelve protons. Obviously if the integrated areas of the two bands in the spectrum of the mixture are equal, then the molecular ratio of the benzene to cyclohexane in the mixture must be 2:1.

NMR spectroscopy is very successful in analyzing a mixture of isomeric substituted components like ortho- and para-xylenes.

14.7.3 Hydrogen Bonding

The NMR can be used to study the hydrogen bonding in the metal chelates as well as in organic compounds. Protons signal is shifted towards low field in the case of hydrogen bonding. This reveals that hydrogen bond formation results in the decrease in the electron shielding of the proton.

If there occurs an upfield shift of the signal, this may be ascribed to the breaking of intermolecular hydrogen bonds. This can be seen in the case of ethanol on increasing temperature, or on diluting the ethanol with carbon tetrachloride.

Schneider, Bernstein and Pople consider that the shifts associated with hydrogen bonding arise from three terms:

(i) The electrostatic effect of the donor atom or group.

(ii) The diamagnetic anisotropy of the donor atom or group.

(iii) Changes in long-range shielding of the acceptor group.

In the case of strong hydrogen bonds, factor (i) is dominant. However, factor (ii) may become dominant in the case of weak hydrogen bonds. Factor (iii) possesses little experimental foundation. It is only operating in the case of intermolecular hydrogen bonding. Factor (iii) is important in explaining the gas-to-solution shifts for acetylene and hydrogen bonds.

14.7.4 Major Applications

Major application of NMR spectroscopy lies in the area of synthetic organic chemistry, inorganic chemistry, bioorganic chemistry, bioinorganic chemistry, polymer chemistry and organometallic chemistry. One of the main advantages of FT-NMR spectroscopy is that a number of FIDs can be accumulated and Fourier transformed, enhancing S/N ratio of the spectrum. This in turn facilitates the analysis of samples with low abundant nuclei and smaller gyromagnetic ratios, e.g., 13C, 29Si, 15N etc. Another major advantage of pulse NMR is that by suitably programming, a variety of pulse sequences can be generated and applied to obtain diverse information like relaxation times, connectivity through bond and through space, etc. Thus, by using FT-NMR it is not only possible to get the information about chemical shift and spin-spin coupling but also possible to deduce the conformational structural information of a particular molecule.

14.7.5 NMR Methodology to Elucidate Chemical Structure of Natural Compounds

Chemists typically employ NMR spectroscopy to acquire information about the structure of chemical and biochemical compounds. However, the correct structural configuration of some compounds, like small organic molecules, is very difficult — and sometimes impossible — to reveal using conventional NMR techniques. Gil, associate research professor of chemistry and director of the Department of Chemistry's NMR Facility, has developed an NMR methodology to measure a relativity new NMR parameter known as residual dipolar couplings (RDCs) in small molecules. Using RDCs, he was able to unambiguously determine the structure of the small molecule jaborosalactol 24, a naturally occurring steroid found in the Jaborosa parviflora plant. Researchers had unsuccessfully tried to elucidate the molecule's structure using conventional NMR techniques, but now, with the structure in hand, they are investigating the small molecule's function, and have found that it may have potential as an anti-cancer agent.

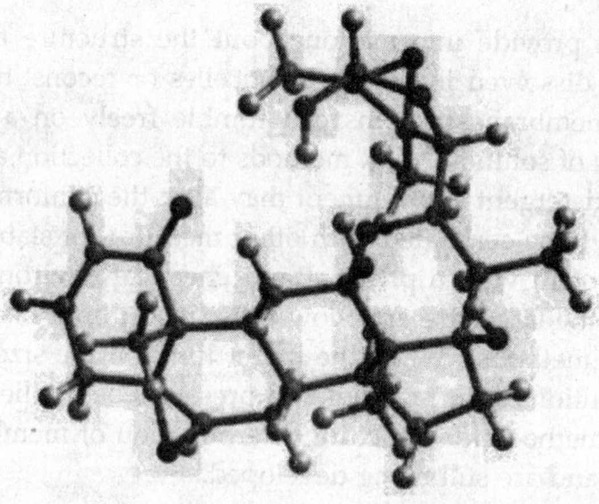

"It was impossible to figure out the compound's structure without RDCs," "conventional NMR techniques yielded three possible structural configurations for the compound. But RDCs revealed only one structure."

RDCs are readouts of the interaction between atomic nuclei induced to behave as tiny magnets in an NMR experiment. RDCs provide structural information about the molecule as well as the global orientation of all of the atoms in the molecule, even if the atoms are far away from each other. Traditional NMR methods cannot do this, according to Gil.

Over the past decade, RDCs have become an invaluable tool for determining the structure of large biological molecules, such as proteins and DNA, but the technique wasn't applicable to small molecules until recently. Small molecules need to be dissolved in an organic solvent before being analyzed in a typical NMR experiment. But to get a read out on RDCs, scientists first need to partially align the molecules that are tumbling around in the solution, which proved difficult. To overcome this problem, scientists — including Gil — began developing polymer gels that act like sponges that absorb the solution of small molecules. Gil, together with his Carnegie Mellon collaborators Nicolay Tsarevsky and Krzysztof Matyjaszewski, developed a gel called poly-methyl methacrylate (PMMA) that can be compressed or stretched, which causes the small molecules within the gel to partially align. This makes it possible to detect RDCs.

Although there are similar gels available, PMMA has a few advantages. During NMR experiments, the other polymer gels give off a background signal that is difficult to remove. With PMMA, the background signal from the polymer can be easily removed. Additionally, PMMA is easy to prepare and affordable.

"Our technique helped a group of natural product chemists to unambiguously determine the correct configuration of a natural compound that has since shown anti-proliferative properties in a breast cancer cell line. Our technique can help anyone that has small molecules with difficult configurations." – Gill said.

14.7.6 NMR pf Membrane Proteins

NMR spectroscopy can provide information about the structure and dynamics of membrane proteins that are dissolved in detergent micelles or reconstituted into lipid bilayers. Detergent-solubilized membrane proteins can tumble freely on a nanosecond timescale, allowing the application of solution NMR methods to the collection and assignment of NMR spectra. However, the detergent environment may alter the conformation of the protein, so sample conditions need to be correlated with other measures of stability or activity. A lipid bilayer provides the protein with a physical and chemical environment most similar to a natural membrane, but the large size and slow tumbling of lipid bilayer vesicles require the use of solid-state NMR methods. While the upper limit on the size of the macromolecular complexes to which solution NMR methods can presently be applied has been pushed well past 50 kD, solid state methods for structure determination of membrane proteins are more technically challenging and are still being developed.

Solution of NMR of α-helical membrane proteins

Standard triple- or quadruple-resonance solution NMR methods enable the assignment of resonances, the determination of backbone and side chain geometries using J couplings, and the measurement of inter-atomic distances using the nuclear Overhauser effect (NOE). These methods have been applied to several membrane protein systems, including the TM domain of glycophorin A, which forms a detergent-resistant parallel dimer of helices. While chemical shift dispersion of both backbone and side chain resonances is poor and line widths are generally broad, resulting in peak overlap, complexes containing at least four transmembrane helices should be analyzable by solution NMR if favorable detergent conditions can be identified.

The process of determining the structure of a complex of transmembrane helices might involve the following steps:

- expression of the sequence of interest in *E. coli*
- purification of ^{15}N labeled protein in milligram quantities
- reconstitution into detergent micelles
- collection of 2D ^{1}H-^{15}N correlation spectrum
- optimization of detergent environment
- collection of ^{15}N separated 3D NOESY-HSQC spectra
- purification of ^{15}N,^{13}C labeled protein in milligram quantities
- collection of 3D triple resonance HNCA, HNCACB (etc.) spectra
- assignment of backbone resonances
- collection of 3D ^{13}C TOCSY spectra

- assignment of side chain resonances
- collection of ^{13}C separated 3D NOESY-HSQC spectra
- collection of J coupling data
- structure calculations using NOEs and J couplings

14.8 SAMPLE PREPARATION

Biophysical investigations of macromolecules usually rely on the availability of a highly purified sample. The strongly hydrophobic nature of transmembrane domains of helical proteins necessitates the use of detergents or organic solvents; reverse-phase HPLC in organic solvents is a particularly useful purification method. The final samples will preferably be studied in detergent micelles. The choice of detergent and the path by which the protein is transferred from organic solvents into detergent micelles can greatly affect the structure and aggregation state of the TM domains, and hence their suitability for NMR structure determination.

Labeling the protein with stable isotopes ^{15}N, ^{13}C and/or ^{2}H can greatly improve the quality of the NMR data that can be obtained. This is usually accomplished by expression in bacteria grown in labeled media, since chemical synthesis of uniformly isotopically labeled peptides is extremely expensive. However, chemical synthesis can provide a convenient route to the production of peptides with labels incorporated at discrete sites.

14.9 REFERENCES

1. G.E Martin, A.S. Zekter (1988). *Two-Dimensional NMR Methods for Establishing Molecular Connectivity*. New York: VCH Publishers. p. 59.
2. J.W. Akitt, B.E. Mann (2000). *NMR and Chemistry*. Cheltenham, UK: Stanley Thornes. pp. 273, 287. ISBN 0748743448.
3. J.P. Hornak. "The Basics of NMR". http://www.cis.rit.edu/htbooks/nmr/. Retrieved 2009-02-23.
4. J. Keeler (2005). *Understanding NMR Spectroscopy*. John Wiley & Sons. ISBN 0470017864.
5. K. Wuthrich (1986). *NMR of Proteins and Nucleic Acids*. New York (NY), USA: Wiley-Interscience. ISBN 0471119172.
6. J.M Tyszka, S.E Fraser, R.E Jacobs (2005). "Magnetic resonance microscopy: recent advances and applications". *Current Opinion in Biotechnology* **16** (1): 93–99. doi:10.1016/j.copbio.2004.11.004. PMID 15722021.
7. J.C. Edwards. "Principles of NMR". Process NMR Associates. http://www.process-nmr.com/pdfs/NMR%20Overview.pdf. Retrieved 2009-02-23.
8. R.L Haner, P.A. Keifer (2009). *Encyclopedia of Magnetic Resonance*. John Wiley. doi:10.1002/9780470034590.emrstm1085.

CHAPTER 15

Bioreactor

15.1 INTRODUCTION

A Bioreactor is a vessel in which chemical process or biological process is carried out which involves organisms or biochemically active substances derived from such organisms. Bioreactors are commonly cylindrical, ranging in size from some liters to cube meters, and are often made of stainless steel. Bioreactor design is quite a complex engineering task. Under optimum conditions the microorganisms or cells will reproduce at an astounding rate. The vessel's environmental conditions like gas (i.e., air, oxygen, nitrogen, carbon dioxide) flow rates, temperature, pH and dissolved oxygen levels and agitation speed need to be closely monitored and controlled. A fixed bioreactor volume is maintained and ideally, the effluent stream should have the same compositions as the bioreactor contents. The culture is fed with fresh medium containing and sometimes two growth limiting nutrients such as glucose.

The concentration of the cells in the bioreactor is controlled by the concentration of the growth-limiting nutrient. A steady state cell concentration is reached where the cell density and substrate concentration are constant. The cell growth rate (μ) is controlled by the dilution rate (*D*) of growth limiting nutrient. Heat exchange is needed to maintain the bioprocess at a constant temperature. Biological fermentation is a major source of heat; therefore in many cases bioreactors need water refrigeration. This can be refrigerated with an external jacket, or for very large vessels, with internal coils. Optimal oxygen transfer is perhaps the most difficult task to accomplish. Oxygen is poorly soluble in water and even less in fermentation broths and is relatively scarce in air (20.8%). Oxygen transfer is usually helped by agitation that is also needed to mix nutrients and to keep the fermentation homogenous. There are however, limits to the speed of agitation, due both to high power consumption (that's proportional to the cube of the speed) and the damage to organisms due to excessive tip speed.

Bioreactor treatment may be performed using microorganisms growing in suspension in the fluid or attached on a solid growth support medium. In suspended growth systems, such as fluidized beds or sequencing batch reactors, contaminated water is circulated in an aeration basin where a microbial population aerobically degrades organic matter and produces carbon dioxide, water and biomass. The biomass is settled out in a clarifier, and then either recycles back to the aeration basin or disposed of as sludge. The increase in waste degradation and stabilization is accomplished through the addition of liquid and air to enhance microbial processes in Bioreactor.

15.2 HISTORICAL BACKGROUND

Bioreactor is a "fermenter". In general terms, a fermenter is something that, as its name would suggest ferments; this is, however, not a simple process. The process of fermentation has been known for thousands of years, but has been mainly used over this time to the glucose found in various fluids, seeds and tubers into alcohol, later used for human consumption. In recent times, however, with increased knowledge of bacteria and fungi, fermenters have been put to a more productive use.

The general idea behind the fermenter is to provide a stable and optimal environment for bacteria in which they can reproduce. More scientific uses of fementers would include to grow large quantities of genetically engineered bacteria. These bacteria, having had the genes that code for various proteins (human insulin, for example, spliced into them, will grow and reproduce, and will express the inserted gene. This will result in the desired protein being released into the growth medium from where it can be harvested, purified and hence sold and/or used. The growth medium for these fermenters has to be tailored specifically to the requirements of the bacterium that is being engineered. When a suitable subject has been chosen, the pH, temperature, light, pressure and nutrient concentration that would give that bacterium optimal growth rates are selected and prepared. A fermenter, then, is simply put on optimal environment for bacteria and/or fungi to grow in, and the cultivation of said organisms will yield a desirable substance.

The term "bioreactor" in the context of soil and water bioremediation refers to any vessel or container where biological degradation of contaminants is isolated and controlled. Bioreactors can range from crude devices such as lined depressions in the growth to advanced metal containers where environmental conditions can be monitored and controlled. The essential treatment mechanism in a bioreactor is natural degradation by existing and/or added populations of microorganisms. Bioreactors have proven to be effective in remediating soil and/or polluted water containing organic compounds under suitable controlled environmental conditions.

15.3 BIOREACTOR: BASIC CONCEPT

A bioreactor is a vessel in which is carried out a chemical process which involves organisms or biochemically active substances derived from such organisms. Bioreactors are commonly cylindrical, ranging in size from some liters to cube meters, and are often made of stainless steel.

Bioreactor design is quite a complex engineering task. Under optimum conditions the microorganisms or cells will reproduce at an astounding rate. The vessel's environmental conditions like gas (i.e., air, oxygen, nitrogen, carbon dioxide) flow rates, temperature, pH and dissolved oxygen levels, and agitation speed need to be closely monitored and controlled. Bioreactor manufacturer uses vessels, sensors, controllers, and a control system, digitally networked together for their bioreactor system.

Fouling can harm the overall sterility and efficiency of the bioreactor, especially the heat exchangers. To avoid it the bioreactor must be easily cleanable and must be as smooth as possible (therefore the round shape).

Heat exchange is needed to maintain the bioprocess at a constant temperature. Biological fermentation is a major source of heat; therefore in most cases bioreactors need water refrigeration. They can be refrigerated with an external jacket or, for very large vessels, with internal coils.

Optimal oxygen transfer is perhaps the most difficult task to accomplish. Oxygen is poorly soluble in water —and even less in fermentation broths— and is relatively scarce in air (20.8%). Oxygen transfer is usually helped by agitation that is also needed to mix nutrients and to keep the fermentation homogeneous. There are however limits to the speed of agitation, due both to high power consumption (that's proportional to the cube of the speed) and the damage to organisms due to excessive tip speed.

15.4 CONSTRUCTION OF THE LABORATORY BIOREACTOR

In terms of the construction, the following variants of the laboratory bioreactor can be made:

1. Glass bioreactor (without the jacket) with an upper stainless steel lid.
2. Glass bioreactor (with the jacket) with an upper stainless steel lid.
3. Glass bioreactor (without the jacket) with the upper and lower stainless steel lids.
4. Two-part bioreactor – glass/stainless steel. The stainless steel part has a jacket and ports for electrodes installation.
5. Stainless steel bioreactor with peepholes.

15.4.1 Upper Cover

The bioreactor's upper part has:

1. Ports for electrodes (pH, pO_2, T, p).
2. Ports for the supply of the titration and feeding medium.
3. Pipes for sampling and chemostate.
4. Ports for the connection of the outlet air condenser and filter.
5. Drive sealing (if the mixer's upper drive is used).
6. Air sparger mounting (if there is no lower lid).

The lid's ports, connections and mountings should ensure the airtightness and leak-proofness, sterility, convenient removal and installation of the sensors and other elements. The lid is usually connected with the reactor's rest part employing special screws with a circumference diameter (4-8 pieces) and a flat or "0" ring silicone rubber sealing.

15.4.2 Lid (lower)

If the bioreactor has a lower cover, then the following ports and elements should be placed and fastened there:

1. Discharge valve;
2. Sampling device;
3. Sparger;
4. Mixer's lower drive;
5. Heaters.

All these connections should secure airtightness and leak-proofness and should prevent the conditions for infection agents' reproduction ("pockets", unevennesses, etc.).

15.4.3 Mixers

The mixer is mechanically "put on" the mixer's axis. The mixer's diameter is normally 1/3 – 1/2 from the diameter of the reactor vessel. The location of the mixer's axis depends on the fact whether the bioreactor has the upper or lower drive. The upper drive is realizable more easily from the constructive viewpoint. In this case, the construction of the sealing is also easier than in the case of the lower drive. In this case, the servicing of the bioreactor is also simple: the motor is disconnected from the sealing connection places, and then the lid, together with the mixer axis and the mixers, is taken off. In its turn, the lower drive provides a whole range of attractions connected with the operation regimes. In this case, mixing can be ensured in the bioreactor at the removed upper lid. This is important when modelling mixing as well as when cleaning and washing the bioreactor. In this case, it is easier to optimize dispersion, as the mixer can be close to the bioreactor's bottom.

The standard turbine is most popular

The standard turbine is the most widespread mixer's construction. Its diameter *d* is 1/2 – 1/3 (most typically, 1/3) from the vessel's diameter. The turbine is a typical radial flow generating mixer. For the mixer axis, there are commonly 2-3 standard turbine mixers. The given mixer ensures the highest input power at fixed rotational speed values of the mixer (power index K_N = 5-7). This mixing system secures a sufficient course of mixing intensity, bubble dispersion and other mixing operations. This mixing system ensures the highest introduced power by constant rotations speed, comparing with another mixer. Therefore, the most part of bioreactors is equipped with such a mixing system.

The mixing of mechanical sensitive microorganisms

However, in the microorganism cultivation practice, there are fermentations in which the application of the standard turbine mixer does not provide the optimal cultivation result, as irreversible mechanical damages of the microorganisms' cells are caused (this applies mainly to mycelial organisms). For mixing of mechanically sensitive mycelial microorganisms, mixing systems are recommended, which generate dominating axial flows, thereby ensuring a more even mixing throughout the reactor's volume. Ekato Intermig mixing systems are among the most widespread ones. The so-called "counter-flow mixing system" is known. The approach to the mixing of more sensitive cells (tissue culture, animal cell, etc.) should be different, since, in this case, the mixing regime should have a laminar character.

The mixing by great airflows

Another case when the standard turbine is not effective enough is the dispersion of a great airflow at relatively low rotational speed values of the mixer. With increasing the inlet air amount, it is not dispersed any more, as the "flooding" regime sets in. In this case, to increase the dispersion efficiency is recommended. The efficiency of these mixers, in comparison with the standard turbine mixer, becomes even more pronounced in the cases of three-phase flows (intensive aeration and the presence of dispersive particles in the medium).

15.4.4 Baffles

Baffles are vertical radially located plates (their width is about 10% from the bioreactor vessel diameter). There are normally 3-4 such plates in the reactor. Baffles are necessary to prevent the formation of a funnel. As a result of funnel formation, the maximally possible rotational speed of the mixer can be essentially limited. Employing baffles, the consumed electric power of the mixer increases by about 20% at the same rotational speed.

15.4.5 Sparger

The supply of compressed air is realized through a sparger. The most widespread constructive solution is a loop pipe with small holes in the lower part (d = 0.05 - 0.15 mm). For mycelial culture fermentations, also loop spargers with a conical air outlet channel are employed. This construction prevents the possible overgrowing of cracks, since, in this case, the outlet diameter is greater, as well as pressure, because it is not distributed among the several pipes.

15.4.6 Bioreactor Vessel

Depending on the construction type, the bioreactor vessel is made of steel, metal or their combination. The relation between the height H and diameter D of the bioreactor is within 1.5-2.5. The bioreactors produced in Europe are commonly more "stately" than those produced in the U.S.A. The reactor filling is about 70%. Thus, it is also the part of the working volume in the reactor vessel. There are high requirements for the reactor vessel materials to prevent the inhibition of the microorganism growth. The same applies also to any other part (sensors, pipes, etc.), which are installed inside the bioreactor vessel. The glass should be 100% borosilicate, e.g., Pyrex® and Kimax®. All the metal parts should be made from stainless steel. The most widespread brand of the stainless steel applied in bioreactors is 316L. The letter L indicates that this steel is with a low composition of carbon. The inner surface of the stainless steel bioreactor should be polished to about a mirror surface quality to facilitate the washing and sterilization process. Welding should be carried out in a fully inert gas medium.

The inert gas should be argon, which fully replaces the air. With time, the application of the welding technology not corresponding to the requirements can cause the corrosion of the welds.

15.4.7 Condenser, Outlet Valve and Filter

The task of the condenser is to cool the liquid in the air outlet, resulting in the separation of the foam and other liquid particles.

For this purpose, a pipe is wound inside the condenser (or another solution that ensures, as possible, a longer air pathway); outside, there are two ports for cold water supply. The condenser is usually connected to the port located in the cover's center. The condenser outlet is connected with the outlet valve. With the outlet valve, it is possible to regulate the necessary overpressure in the bioreactor's vessel. The valve outlet can be connected with a porous microbiological filter.

15.4.8 Sampling

On the one hand, sampling may seem to be a simple procedure — just open the manual valve in the inlet of the bioreactor vessel, supply as much fermentation broth as required for the sample, and close the tap. In this sampling, one can quite easily give guarantee that infection will not be avoided.

The sampling construction should be such so that measures for preventing non-sterility before and after the sampling be avoided. In the sites of the infection origin, sterilization should be performed promptly with alcohol or steam.

15.5 KEY ISSUES IN BIOREACTOR DESIGN AND OPERATION

The goal of an effective bioreactor is to control, contain and positively influence the biological reaction. To accomplish this, the chemical engineer must take into consideration two areas. One is the suitable reactor parameters for the desired biological, chemical and physical (macrokinetic) system. The macrokinetic system includes microbial growth and metabolite production. Microbes can include bacteria, yeast, fungi, and animal, plant, fish and insect cells, as well as other biological materials.

The other area of major importance in bioreactor design involves the bioreaction parameters, including:

- controlled temperature
- optimum pH
- sufficient substrate (usually a carbon source), such as sugars, proteins and fats
- water availability
- salts for nutrition
- vitamins
- oxygen (for aerobic processes)

- gas evolution
- product and byproduct removal.

In addition to controlling these, the bioreactor must be designed to both promote formation of the optimal morphology of the organism and to eliminate or reduce contamination by unwanted organisms or mutation of the organism.

15.6 DESCRIPTION OF MODERN BIOREACTOR (FERMENTOR MINIFOR)

The bioreactor is a compact laboratory fermentor. The modern standard bioreactor is one which is easy to use and with the capacity to measure and control all the important parameters of the biological culture. The bioreactor has to take up minimum space on the bench but with a good access to all parts. Bioreactor should be able to work independently or be connected to a PC for advanced regulation and extensive data treatment.

- To keep the cost of the bioreactor low without compromising quality, several new ideas and innovations have been introduced in the modern standard bioreactors which include:
- The whole glass vessels with threaded fittings can be used in bioreactors, instead of a vessel with a stainless steel cover, which is expensive. They have been used for many years in cell culture and are proved to maintain perfect sterility.
- A new vibration mixer can be used in bioreactor, instead of a traditional propeller agitator, which requires an expensive motor and magnetic coupling. An electromagnet is an inexpensive membrane which perfectly assumes sterility and produces an efficient mixing without formation of a vortex. At the same time this type of mixing is gentler on cells and produces less foam.
- The culture can be heated by heat radiation produced in a parabolic radiator with a gold reflector placed under the fermentation vessel. The heat is absorbed gently in the culture on a similar way to the sun heating water. There is no overheating of the culture, as it is usually the case when a heater is placed in the medium. The expensive double wall vessels with thermostatic baths can be eliminated. At the same time pipes and cables disappear making the bioreactor less complex.
- New high performance plastics, as far as possible can be placed with the expensive pieces of equipment.
- All the electronics in the front part of the instrument can be put by using modern microprocessors. This makes the bioreactor unbelievably compact and eliminates the casing tower usual in other products.

15.7 TECHNICAL DESCRIPTION

15.7.1 Basic Unit

The main feature of the modern standard bioreactor is that all the electronics, power supply, IR heater, air valve, mass flow meter, cables and tubing are positioned in one base unit, which is used as a support for the bioreactor vessel and all other necessary equipment. The platform arrangement of the base unit makes all parts of the bioreactor clearly visible and easily accessible from all sides despite base dimensions of only 22 × 40 cm (approx. an A4 sheet of paper). Up to fine 750 mL reagent bottles in magnetic holders can be placed behind the bioreactor vessel and up to 4 pumps can be placed on adjustable holders mounted on bars at the rear part of the base unit. The sockets are located on the rear side at the base unit thus removing cables from the work area.

15.7.2 Fermentation Vessel

The standard fermentation vessel has a total volume of 1 liter (Fig. 15.1). Three other vessels are variable with a total volume of 0.5, 1, 3 liters and 5 liters. The last one is more for cell culture. The vessels are made of high quality pyrex glass with one large threaded central opening for the fixation of the membrane and vibromixer, 6 to 8 threaded side necks for the fixation of different sensors, air outlet, sample withdrawal, inoculation, etc. Two adjustable holders maintain the vessel in position. The vessel is sterilized in an autoclave.

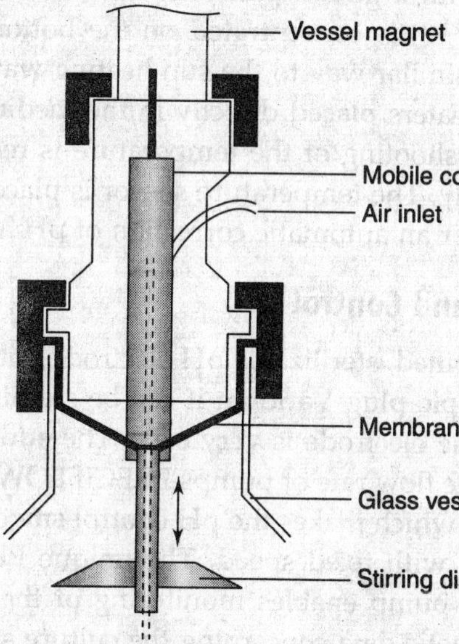

Fig. 15.1 *Fermenter vessel.*

15.7.3 Measurement and Regulation

The control panel consists of an LCD display and control buttons. All parameters (temperature, pH, PO_2, airflow rate, agitation and one free selectable parameters), for example, PCO_2, optical density, antifoam, etc.) are visible at a glance on a large LCD back light display (4 × 4 digits). The controls are simple and logical. On most parameters the limits of low or/and high alarm can be set. After alarm activation an acoustic signal is heard, an indication appears on the display and an electric signal appears on the rear side connector of the bioreactor. Each bioreactor can be operated in an autonomous way or can be coupled to a PC using a special fermentation program. This allows almost unlimited possibilities for control and data treatment — upto 6 bioreactors in parallel or more.

15.7.4 Agitation

Instead of a traditional propeller agitator a new vibromixer is used in the Minifor. A strong electromagnet moves one or several discs up and down. The major advantage is an efficient mixing and aeration of the culture medium together with very efficient separation of the inside of the vessel from the outside by a low cost silicone membrane. No vortex is built up and baffles are eliminated. This type of agitation is also gentler on the cells and foaming is reduced. The frequency of vibration is controlled by a microprocessor and can be varied through a broad range. The input of air is made through several capillaries fixed under the lowest disc.

15.7.5 Temperature Control

A new-patented IR radiator with a gilded parabolic reflector is used to warm the culture broth. The heat radiation (100 W) is concentrated on the bottom of the vessel where it is adsorbed by the medium in a similar way to the sun heating water. There is no overheating of the culture common with heaters placed directly in the medium. Thanks to the low heat capacity of the IR source, overshooting of the temperature is reduced and the temperature can be controlled more precisely. The temperature sensor is placed directly in the pH sensor and is used at the same time for an automatic correction of pH and pO_2 electrodes.

15.7.6 pH Measurement and Control

The pH is measured by a combined, sterilizable pH electrode with incorporated temperature sensor. Thanks to a new multiple plug Variopin it can be sterilised without any protection. The two-point calibration of the electrode is very easy. The addition of acid or base is controlled by a microprocessor. The flow rate of pumps PRECIFLOW, MULTIFLOW or HIFLOW is varied between 0 and 100%, which makes the pH control smoother than with the common ON/OFF switching of pumps with fixed speed. The unique PUMP FLOW INTEGRATOR (option) when coupled to the pump enables monitoring of the pumps activity during the process. This yields precise kinetic data concerning the culture state and activity.

The pH can also be controlled with CO_2, specially for cells culture. A special gas flow controller has been developped for such applications.

15.7.7 PO_2 Measurement and Regulation

A sterilizable Clark-type electrode measures the concentration of dissolved oxygen with a special reinforced membrane giving fast response times. The microprocessor performs two-point calibration and automatic temperature correction. The regulation of dissolved oxygen is obtained by a variation of the flow rate of air.

15.7.8 Air Input

The flow rate can be set from 0 to 5 l/min. in 0.1 steps. A precise mass-flow meter is used. This measurement is independent of the pressure and temperature variation of the air. Commonly used floating ball capillaries give inaccurate readings in this case. A proportional air valve using memory metals adjusts the flow rate.

15.7.9 Air Output

Used air is filtered by a filter. If necessary an optional condenser can be used to remove water vapor from the stream. The condenser based on a Peltier cell uses no cooling water.

15.7.10 Inoculation and Sampling Ports

Inoculation, addition of acid or base and sample removal is made through four stainless steel capillaries equipped with Luer-Lock adapters.

15.7.11 Sterilization

The fermentation vessel is sterilized as usual in an autoclave after disconnection of the cables from the electrodes and electromagnet.

15.7.12 Antifoam

A foam detection device is available to add an antifoam agent into the fermenter with a peristaltic pump or a syringe pump.

15.7.13 Continuous Culture

The weight of the fermenter can be kept constant by using a special weighing module placed under the front edge. The harvesting pump will keep automatically the volume of the culture constant. Continuous cultures allow a considerable increase of productivity.

15.7.14 Dimensions

A lot of effort was made to make the Minifor as small and as compact as possible.The platform arrangement of the fermenter makes control and access to all parts easy from all sides. The distance between the reagent bottles, pumps placed above them and the fermenter vessel is the shortest possible.

15.7.15 Pumps

Up to four pumps PRECIFLOW, MULTIFLOW or HIFLOW can be placed on holders fixed on two bars at the rear of the fermenter. They are connected by a single cable to the sockets on the rear side of the fermenter. Since the pumps are not bound to the fermenter they can be used for other applications elsewhere in the laboratory (e.g., for chromatography etc.). This can present a considerable economy for the user.

15.8 BIOREACTORS/FERMENTATION: TECHNOLOGIES

The various types of bioreaction systems covered here include batch, continuous, semi-continuous, surface/tray, submerged, air-lift loop and trickle-bed setups. As stated before, there are overlapping characteristics in several of the technologies discussed.

15.8.1 Batch Bioreactions

The majority of bioreactions are batch-wise. The first phase of batch bioreaction is commonly sterilization, after which the sterile culture medium is inoculated with microorganisms that have been cultivated to achieve a specific result. During this dynamic reaction period, cells, substrates (including the nutrient salts and vitamins) and concentrations of the products vary with time. Proper mixing keeps the differences in composition and temperature at acceptable levels. To promote aerobic cultivation, the medium is aerated to provide a continuous flow of oxygen. Gaseous by-products formed, such as CO_2, are removed, and aeration and gas-removal processes take place semicontinuously.

Next, an acid or alkali is added if the pH needs to be controlled. To keep foaming to acceptable levels, antifoaming agents may be added when indicated by a foam sensor. One of the first types of batch systems is the tray fermenter (Fig. 15.2), used in the early days of commercial aerobic bioreactions for products such as citric acid and penicillin. In this system, the trays are loaded with the culture medium and the organisms, and the airflow produces the bioreaction, during which exhaust gas is discharged. When the bioreaction is complete, the end product is removed from the trays. Because this method is inefficient for producing large commercial quantities, it fell quickly to the wayside with the emergence of submerged tank systems, which are designed to handle significantly higher volumes.

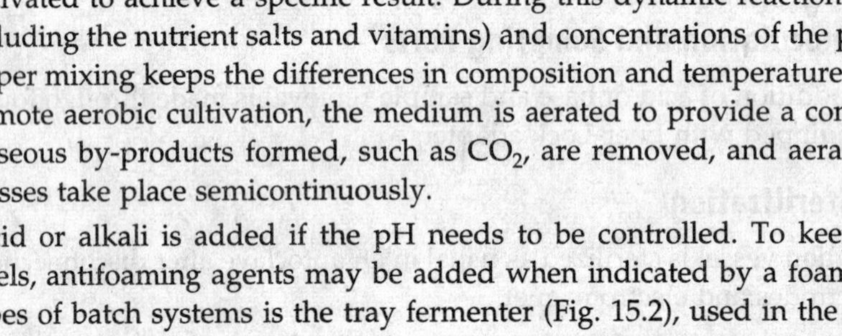

Fig. 15.2 *A tray bioreactor is loaded with the culture medium and organisms, then airflow is started to initiate the reaction.*

Overall, batch bioreaction systems provide a number of advantages, including:

- Reduced risk of contamination or cell mutation, due to a relatively brief growth period.
- Lower capital investment when compared to continuous processes for the same bioreactor volume.

- More flexibility with varying product/biological systems.
- Higher raw material conversion levels, resulting from a controlled growth period.

The disadvantages include:

- Lower productivity levels due to time for filling, heating, sterilizing, cooling, emptying and cleaning the reactor.
- Increased focus on instrumentation due to frequent sterilization.
- Greater expense incurred in preparing several subcultures for inoculation.
- Higher costs for labor and/or process control for this non-stationary process.
- Larger industrial hygiene risks due to potential contact with pathogenic microorganisms or toxins.

Common applications for batch bioreactors include:

- Products that must be produced with minimal risk of contamination or organism mutation.
- Operations in which only small amounts of product are produced.
- Processes using one reactor to make various products.
- Processes in which batch or semicontinuous product separation is adequate.

15.8.2 Continuous Bioreactions

The defining characteristic of continuous bioreaction is a perpetual feeding process. A culture medium that is either sterile or comprised of microorganisms is continuously fed into the bioreactor to maintain the steady state. Of course, the product is also drawn continuously from the reactor. The reaction variables and control parameters remain consistent, establishing a time-constant state within the reactor. The result is continuous productivity and output. These systems provide a number of advantages, including:

- Increased potential for automating the process.
- Reduced labor expense, due to automation.
- Less non-productive time expended in emptying, filling and sterilizing the reactor.
- Consistent product quality due to invariable operating parameters.
- Decreased toxicity risks to staff, due to automation.
- Reduced stress on instruments due to sterilization.

The disadvantages of continuous bioreactors include:

- Minimal flexibility, since only slight variations in the process are possible (throughput, medium composition, oxygen concentration and temperature).
- Mandatory uniformity of raw material quality is necessary to ensure that the process remains continuous.
- Higher investment costs in control and automation equipment, and increased expenses for continuous sterilization of the medium.

- Greater processing costs with continuous replenishment of non-soluble, solid substrates such as straw.
- Higher risk of contamination and cell mutation, due to the relatively brief cultivation period.
- Continuous bioreaction is frequently used for processes with high-volume production; for processes using gas, liquid or soluble solid substrates; and for processes involving microorganisms with high mutation-stability. Typical end products include vinegar, baker's yeast and treated wastewater.

15.8.3 Semicontinuous Bioreactions

This hybrid of batch and continuous operations is found in many types of processes. One of the more frequently used is initiating the bioreaction in the batch mode, until the growth-limiting substrate has been consumed. Then, the substrate is fed to the reactor as specified (batch) or is maintained by an extended culture period (continuous). For secondary metabolite production, in which cell growth and product formation often occur in separate phases, the substrate is typically added at a specified rate. Like batch reactors, semicontinuous reactors are non-stationary. These systems provide a number of advantages, including:

- Higher yield, resulting from a well-defined cultivation period during which no cells are added or removed.
- Increased opportunity for optimizing environmental conditions of the microorganisms in regard to the phase of growth or production and age of the culture.
- Nearly stationary operation, important with slightly mutating microorganisms and those at risk for contamination.

The disadvantages include:

- Lower productivity levels due to time-consuming procedures for filling, heating, sterilizing, cooling, emptying and cleaning the reactor.
- Greater expenses in labor and/or dynamic process control for the process.

Semicontinuous bioreactors are typically used when continuous methods are not feasible, for example, those in which slight mutation or contamination of the microorganism occurs. Such bioreactors are also used when batch methods do not offer the desired productivity levels.

15.8.4 Submerged Bioreactors — Stirred Tank

The most common type of aerobic bioreactor in use today is the stirred-tank reactor, which may feature a specific internal configuration designed to provide a specific circulation pattern. Ideal for industrial applications, this unit offers manufacturers both low capital and operating costs. For laboratory experiments with smaller volumes, the mixing vessel is typically made of glass. Stainless steel tank construction is the standard for industrial applications involving larger volumes. The height-to-diameter ratio of the vessel can vary, depending on heat removal requirements.

The operating principles of the stirred-tank bioreactor are relatively simple. As shown in Fig. 15.3, the sterile medium and inoculum are introduced into a sterilized tank, and the air supply typically enters at the bottom. For optimal mixing, the tank features not only an agitator system but also baffles, which help prevent a whirlpool effect that could impede proper mixing. In the early stages of the process, warm water may be circulated through the baffles to heat up the system; later, cool water may be circulated inside of them to keep the process from overheating. The number of baffles typically ranges from four to eight (Fig. 15.3).

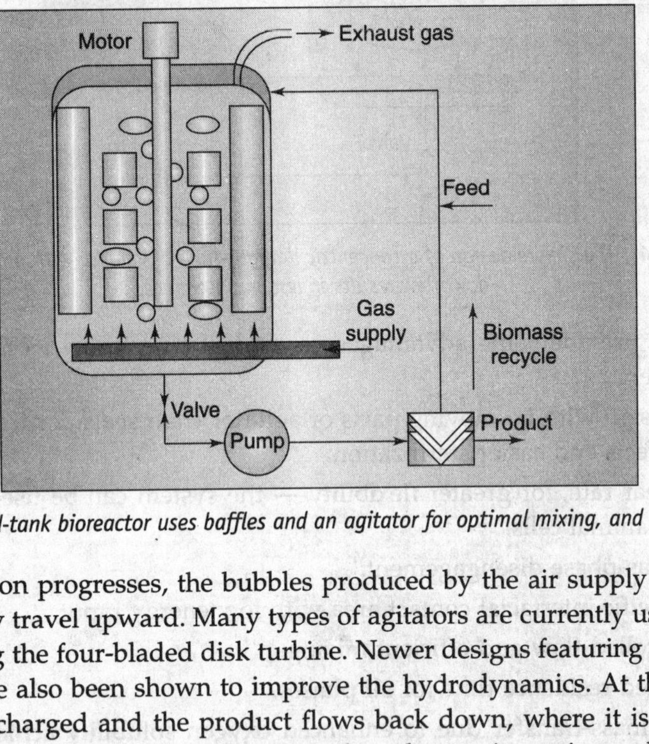

Fig. 15.3 *Stirred-tank bioreactor uses baffles and an agitator for optimal mixing, and recycles biomass.*

As the bioreaction progresses, the bubbles produced by the air supply are broken up by the agitator as they travel upward. Many types of agitators are currently used, with the most common one being the four-bladed disk turbine. Newer designs featuring 12 or 18 blades, or concave ones, have also been shown to improve the hydrodynamics. At the top of the tank, exhaust gas is discharged and the product flows back down, where it is drained from the tank. In a continuous flow stirred-tank reactor, the substrate is continuously fed into the system and the product is continually drawn out and separated, with the producing organism recycled back into the tank for reuse. As with conventional chemical reactors, bioreactors can be placed in series or parallel with controlled recycle streams.

15.8.5 Airlift Reactor Systems

Also known as a tower reactor, an airlift bioreactor can be described as a bubble column containing a draught tube. Many types of airlift bioreactors are currently in use today. Air is typically fed through a sparger ring into the bottom of a central draught tube that controls the circulation of air and the medium (Fig. 15.4). Air flows up the tube, forming bubbles, and exhaust gas disengages at the top of the column. The degassed liquid then flows downward and the product is drained from the tank. The tube can be designed to serve as an internal heat exchanger, or a heat exchanger can be added to an internal circulation loop.

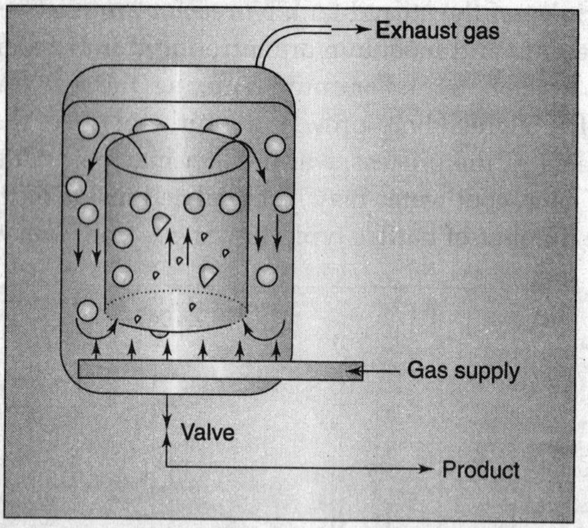

Fig. 15.4 *The simple design of a concentric draught-tube bioreactor with annular liquid downflow results in less maintenance.*

Airlift systems provide some advantages vs. more conventional bioreactors, such as the standard fermenter:

- Simple design with no moving parts or agitator shaft seals, for less maintenance, less risk of defects and easier sterilization.
- Lower shear rate, for greater flexibility — the system can be used for growing both plant and animal cells.
- Efficient gas-phase disengagement.
- Large, specific interfacial contact area with low energy input.
- Well-controlled flow and efficient mixing.
- Well-defined residence time for all phases.
- Increased mass-transfer due to enhanced oxygen solubility achieved in large tanks with greater pressures.
- Large-volume tanks possible, increasing the output.
- Greater heat-removal vs. conventional stirred tanks.

The main disadvantages are:

- Higher initial capital investments due to large-scale processes.
- Greater air throughput and higher pressures needed, particularly for large-scale operation.
- Low friction with an optimal hydraulic diameter for the riser and downcomer.
- Lower efficiency of gas compression.
- Inherently impossible to maintain consistent levels of substrate, nutrients and oxygen with the organisms circulating through the bioreactor and conditions changing.
- Inefficient gas/liquid separation when foaming occurs.

However, these disadvantages can and must be minimized in designing airlift systems. For example, if only one location serves as the feed source, the organism would experience continuous cycles of high growth, followed by starvation, resulting in the production of undesirable by-products, low yields and high mortality rates. A design with multiple feed points eliminates this risk, especially for large-scale operations. The same risks are inherent in a single entry point for oxygen, which must be delivered at various places within the vessel, with the majority of the air entering the bottom to circulate the fluid through the reactor.

15.8.6 Airlift External-Loop Reactors

Another type of airlift system is the airlift external-loop reactor system (Fig. 15.5), used primarily for batch operation. Figure 15.5 shows a cut-away view, that is, the vessel and downcomer are actually taller than shown for the particular diameter drawn. A variation of the airlift system uses induced circulation to direct air and liquid throughout the vessel. This system consists of a riser and an external downcomer, which are connected at the bottom and the top, respectively. As the injected air at the bottom of the riser creates gas bubbles that begin to rise through the main tank, exhaust gas disengages at the top and the resulting heavier solution descends through the downcomer.

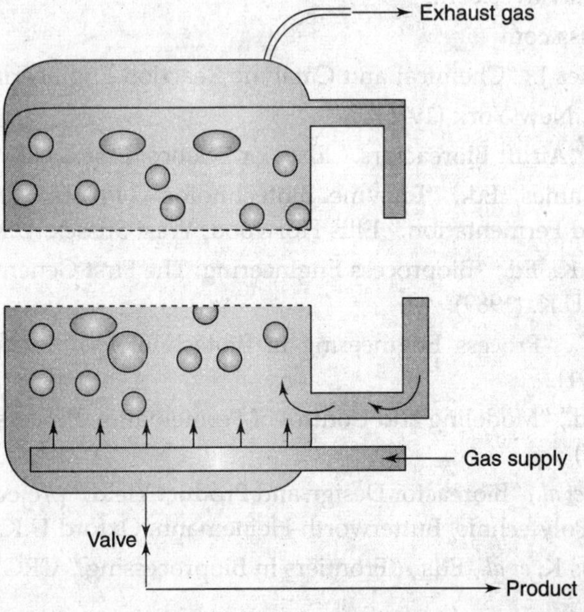

Fig. 15.5 *An airlift external-loop reactor has induced circulation that directs air/liquid throughout vessel.*

This has some advantages over standard airlifts:

- Effective heat-transfer and efficient temperature control.
- Low friction with an optimal hydraulic diameter for both the riser and downcomer.
- Well-defined residence time in the individual section.

- Increased opportunity for measurement and control in the riser and the downcomer.
- Independent control of the gas input-rate and liquid velocity by a throttling device between riser and downcomer.

15.8.7 Anaerobic Bioreactions

Anaerobic bioreactions are used in applications such as ethanol production, winemaking, beer brewing and wastewater treatment. Technologies are well established in wine, ethanol and beer production, with improvements resulting from product improvements and manufacturing cost reductions. Continuous bioreactors for beer have been commercialized; however, batch fermenters continue to receive capital investments. Waste treatment is a field largely considered to have matured with fully functional technology in place, therefore, not as much attention has been paid to developing newer wastewater treatment processes.

15.9 REFERENCES

1. www.keofitt.dk
2. www.alnab.se
3. www.strahmanvalves.com
4. www.schufausa.com
5. Carberry, James J., "Chemical and Catalytic Reaction Engineering,"
6. McGraw-Hill, New York (1976).
7. Chisti, M. Y., "Airlift Bioreactors," Elsevier Science, Essex, U.K. (1989).
8. Crabbe, M. James, Ed., "Enzyme Biotechnology: Protein Engineering, Structure Prediction and Fermentation," Ellis Horwood, West Sussex, U.K. (1990).
9. Ghose, Tarun K., Ed., "Bioprocess Engineering: The First Generation," Ellis Harwood, West Sussex, U.K. (1989).
10. Jackson, A. T., "Process Engineering in Biotechnology," Prentice Hall, Englewood Cliffs, NJ (1991).
11. Leigh, J. R., Ed., "Modeling and Control of Fermentation Processes," Peter Peregrinus, London (1987).
12. Mijnbeek, G., *et al.*, "Bioreactor Design and Product Yield," project of Open Universiteit and Thames Polytechnic, Butterworth-Heinemann, Oxford U.K. (1992).
13. Sikdar, Subhas K, *et al.*, Eds., "Frontiers in Bioprocessing," CRC Press, Boca Raton, FL (1990).
14. Sinclair, C. G., and B. Kristiansen, "Fermentation Kinetics and Modeling," Taylor and Francis, New York (1987).
15. Underkofler, Leland A., and Richard J. Hickey, Eds., "Industrial Fermentations," Chemical Publishing, New York (1954).

16. van't Reit, Klaas, and Johannes Tramper, "Basic Bioreactor Design," Marcel Dekker, New York (1991).
17. Wang, Daniel I. C., *et al.*, "Fermentation and Enzyme Technology," John Wiley, New York (1979).
18. www.capmazagine.org

CHAPTER 16

Gel Electrophoresis

16.1 INTRODUCTION

Gel electrophoresis is an important and widely used tool in the analysis of biological molecules. Along with the discovery of restriction enzymes and the polymerase chain reaction, it has been an integral part of the explosion of bioscience research and engineering. Gel electrophoresis is a widely used technique for separating electrically charged molecules. It is a central technique by molecular biology and genetics laboratories, it lets researchers to separate and purify the nucleic acids DNA and RNA and proteins, and they can be studied individually. Gel electrophoresis is often followed by staining and blotting procedures used to identify the separated molecules.

In electrophoresis, an electric field is generated to separate charged molecules that suspended in a matrix or gel support. Negatively charged molecules move towards anode are one side of the gel, and positively charged molecules move toward the cathode on the other side. The gel itself is a porous matrix, or meshwork, often made of carbohydrate chains. Molecules are pulled through the open spaces in the gel, but slowed down by the meshwork based on their differing properties. The parameters that determine the migration rate of these molecules through the meshwork are the strength of the electric field, the composition of the gel support matrix, the composition of the liquid buffer solution the gel sits in, and the size, charge, and chemical composition of the molecules being separated. Smaller molecules move faster than larger molecules, because they encounter less frictional drag. The size of the pores in the gel can be changed so this frictional drug is increased or decreased, allowing faster separation, or finer resolution.

The electrophoretic technique can analyze and purify a variety of biomolecules, but used to separate nucleic acids and proteins. A basic consideration for choosing technique is the composition of the sample to be separated, for example, does it contain nucleic acid (DNA or RNA), or is it composed of proteins, or carbohydrates? What sizes of the molecules to be separated? Another important consideration is the separation for qualitative, where the technique is being used to evaluate the composition of the sample, or is it quantitative, in that the separated materials are collected for further analysis? Cellulose or starch is used as a support medium for molecular-weight biomolecules such as amino acids and carbohydrates, whereas separation of proteins and nucleic acids are almost always done in gels made of a insoluble material such as agarose, an acrylamide. The introduction of the technique of electrophoresis in acrylamide or agarose gels was a major advance in nucleic acid technology. Previously, large and expensive ultracentrifuge equipment was required to separate macromolecules on the basis of sedimentation coefficients or buoyant densities and scintillation controls were required to measure the radioactivity of the physically separated fractions. The advent of gel electrophoresis, ethidium bromide staining and autoradiography essentially brought an end to the era of intensive centrifugal analysis of nucleic acids. The development of these techniques was combined with the use of restriction enzymes and cloning in bacterial plasmids or phages to obtain large quantities of pure fragments of DNA molecules.

Gel electrophoresis is simple, rapid, non-destructive and inexpensive and when combined with autoradiography, blot hybridization and elution of specific fragments from the gel, is an extremely powerful tool in the arsenal of recombinant DNA technology both for preparative and for analytical purposes.

Two-dimensional electrophoresis (2DE) is an integral tool in the study of proteomics and is often described as the proteomics "work horse". 2DE enables the separation of complex protein mixtures by combining isoelectric focusing (IEF) in the first dimension and SDS-PAGE in the second dimension. Isoelectric focusing is a sophisticated and reproducible technique where proteins are immobilized at their isoelectric point. It can be used to separate large number of proteins that are isolated all at once after being expressed in response to a hormone, drug, or other stimulus. 2D-PAGE can be used in combination with mass spectrometry to systematically identify all the major proteins present in the sample.

16.2 BACKGROUND

The term "electrophoresis" was originally meant to refer to the migration of charged particles in an electrical field. The alternative term "ionophoresis" had been reserved for the migration of lower molecular weight substances in stabilized media such as gels and powders (Tiselius, 1959). Today, the general term electrophoresis covers all applications regardless of the material being studied and the medium being used.

Studies by W.B. Hardy around the turn of the twentieth century established that many biologically important molecules, such as enzymes and other proteins, displayed characteristic electrophoretic mobilities. The demonstration that biomolecules migrated in predictable ways in an electrical field attracted a great deal of interest among biochemists. For example, Michaelis (1909) was able to use electrophoresis to determine the isoelectric points of various enzymes. The isoelectric point, pI, is loosely defined as the pH at which a protein will no longer migrate in an electrical field. It is important to note here that electrophoresis made such determination possible before researchers were able to actually purify the proteins themselves. In fact, as electrophoresis methods improved, it was found that some substances that many had believed were well-defined entities via other means such as crystallization were yielding multiple components in an electrical field. This meant that some "pure" substances were actually composed of two or more previously unsuspected subunits. While the majority of the early electrophoretic studies of biomolecules were carried out in liquid phase, von Klobusitzky and Konig (1939) successfully applied an electrophoretic field to paper strips saturated with an electrolytic solution to separate components of snake venom. This "paper electrophoresis" pre-dates by several years the technique of "paper chromatography" with which it is usually associated. During the late 1940s and early 1950s, this solid phase electrophoresis was used to characterize a vast array of biologically important substances including the amino acids (Wunderly, 1959). The advantage of the solid support electrophoresis over free solution electrophoresis lies in the ability of the solid support to stabilize migration. Such stabilization led to the development of a precise mathematical theory of molecular migration. It was clear from the outset that time and the intensity of the electrical field were major determinants of the mobility of a substance on paper strips. If, however, those two factors were to be held constant, the mobility of a molecule, *u*, can be written as,

$$U = Qd \;/ 4\pi r^2 n$$

where Q is the charge on the molecule, d is distance, r is the radius of the molecule, and η is the viscosity of the solution used to wet the strips. The radius of the molecule could be experimentally determined from the observed diffusion constant. Note here that the radius of the molecule is in the denominator of the mobility expression and is squared. This means that the relationship between molecular size and mobility is non-linear. That is, all other factors being equal, mobility increases as the inverse square of the size of the molecule. This will become important later in discussions of electrophoretic separations of nucleic acids.

The types of apparatus designed for paper electrophoresis are nearly identical to those in use today for gel electrophoresis.

While paper and other solid support materials proved to be an advantage over free solutions for the electrophoretic analysis of biomolecules, gels were adopted later because gels not only minimized diffusion better than paper supports they actually participated in the separation process by interacting with the migrating particles. Gels can be thought of as semisolid matrices whose pore sizes aid in separation. The semisolid nature of the gel participates through a process known as molecular sieving. The three common media for gel electrophoresis are starch, polyacrylamide, and agarose. Of these, the starch gel medium is the least versatile whereas a wide range of separation effects can be achieved using the other two media.

There are limitations to the use of both polyacrylamide and agarose but these are effectively minimized when the material to be analyzed is a nucleic acid. Even very small nucleic acids (i.e., oligonucleotides) are easily separated in an electrical field by one or the other medium. The primary criteria for choosing polyacrylamide or agarose gel electrophoresis are length and whether or not the nucleic acid is single stranded or double stranded. Short, single-stranded DNAs like oligonucleotides require polyacrylamide where long (> 100 bp), double-stranded DNA are best resolved on agarose.

16.3 PRINCIPLES OF GEL ELECTROPHORESIS

Electrophoresis is a technique used to separate and sometimes purify macromolecules — especially proteins and nucleic acids — that differ in size, charge or conformation. As such, it is one of the most widely used techniques in biochemistry and molecular biology. When charged molecules are placed in an electric field, they migrate toward either the positive or negative pole according to their charge. In contrast to proteins, which can have either a net positive or net negative charge, nucleic acids have a consistent negative charge imparted by their phosphate backbone, and migrate toward the anode. Proteins and nucleic acids are electrophoresed within a matrix or "gel". Most commonly, the gel is cast in the shape of a thin slab, with wells for loading the sample. The gel is immersed within an electrophoresis buffer that provides ions to carry a current and some type of buffer to maintain the pH at a relatively constant value.

The gel itself is composed of either agarose or polyacrylamide, each of which have attributes suitable to particular tasks.

- Agarose is a polysaccharide extracted from seaweed. It is typically used at concentrations of 0.5 to 2%. The higher the agarose concentration, the "stiffer" the gel. Agarose gels are extremely easy to prepare: you simply mix agarose powder with buffer solution, melt it by heating, and pour the gel. It is also non-toxic.
- Agarose gels have a large range of separation, but relatively low resolving power. By varying the concentration of agarose, fragments of DNA from about 200 to 50,000 bp can be separated using standard electrophoretic techniques.
- Polyacrylamide is a cross-linked polymer of acrylamide. The length of the polymer chains is dictated by the concentration of acrylamide used, which is typically between 3.5 and 20%. Polyacrylamide gels are significantly more annoying to prepare than agarose gels. Because oxygen inhibits the polymerization process, they must be poured between glass plates (or cylinders).

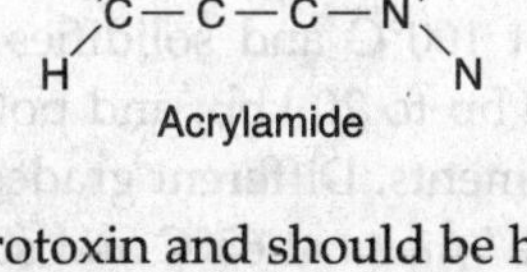

Acrylamide

- Acrylamide is a potent neurotoxin and should be handled with care. Wear disposable gloves when handling solutions of acrylamide, and a mask when weighing out powder. Polyacrylamide is considered to be non-toxic, but polyacrylamide gels should also be handled with gloves due to the possible presence of free acrylamide.
- Polyacrylamide gels have a rather small range of separation, but very high resolving power. In the case of DNA, polyacrylamide is used for separating fragments of less than about 500 bp. However, under appropriate conditions, fragments of DNA differing is length by a single base pair are easily resolved. In contrast to agarose, polyacrylamide gels are used extensively for separating and characterizing mixtures of proteins.

16.4 THEORY OF GEL ELECTROPHORESIS

If a molecule of net charge, q, is placed in an electric field, a force, F, is exerted upon it:

E = potential difference

q = net charge

d = distance between the positive and negative electrodes

E/d = field strength

This force is opposed by friction:

n = viscosity

v = velocity $F = \frac{E}{d} q \frac{E}{d} q = 6\pi$ mv v = Eq/do π

However, this equation is not adequate to explain electrophoresis in a gel, since it does not take into account the existence of gel pores. In reality, electrophoresis in a gel matrix can be thought of as a type of "gel filtration" where the driving force is the electric field. Penetration of the gel matrix by a mixture of nucleic acid molecules of different sizes results in the retardation of larger molecules. The electrophoretic mobility of nucleic acid molecules through a gel matrix is a function not only of molecular configuration (duplex or single stranded nucleic acid molecules), gel porosity, and ionic strength of the electrophoresis buffer.

Exclusion of nucleic acid macromolecules from the gel occurs if the average pore size is smaller than the size of the macromolecule. However, in the case of duplex DNA, extremely large molecules can penetrate a gel by "snaking" through the pores in an extended fashion. This behavior gives rise to the non-linear migration and pileup of large DNA molecules which is normally observed at the top of an agarose gel; this phenomenon has been used in the development of an elegant technique for the separation of extremely large duplex linear DNA molecules by field inversion and orthogonal field electrophoresis (FIGE and OFAGE). (Carle & Olson, 1984; Carle et al., 1986).

Standard gel electrophoresis is used to separate nucleic acid molecules ranging from oligonucleoticles to molecules approximately 20 kbp in size. Agarose, a natural product derived from kelp, which melts at 100°C and solidifies at 42°C, is used for separation of molecules from approximately 100 bp to 20 kbp, and polyacrylamicle, an artificial polymer, is used for separation of small fragments. Different grades of agarose are available for different purposes: "low melting agarose" melts at 65°C and solidifies at 37°C; "Nusieve" agarose (Marine Colloids Division of FMC Corp.) can be prepared at concentrations of 2-10% and can separate molecules down to 10 bp.

"Ultrapure" agarose is used for preparative gel etectrophoresis and recovery of bands with fewer contaminants. Denaturing acrylamide gels can separate oligonucleotides differing by a single nucleoticle. Both double stranded and single stranded nucleic acids can be separated by electrophoresis. Conditions for denaturation usually involve high concentrations of urea and elevated running temperatures, formamide, methyl mercury, formaldehyde, or glyoxyol pretreatment.

The mobility of nucleic acid fragments in a gel matrix exhibits a complex but characteristic profile, as shown in Fig. 16.1.

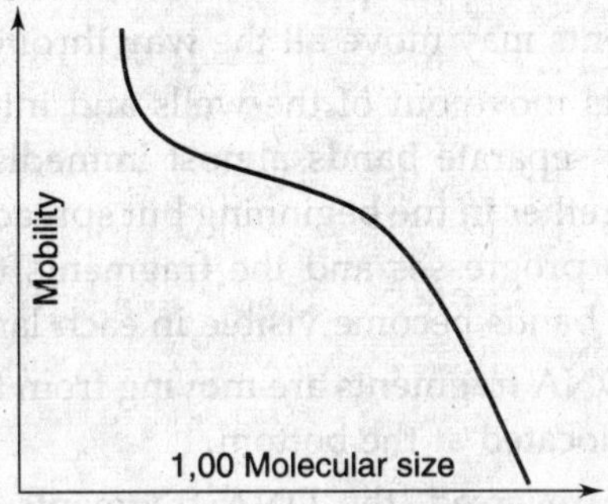

Fig. 16.1 *Electrophoretic Mobility of DNA or RNA in an Agarose or Acrylamide Gel as a Function of the Size of the Molecule. Note the Regions of Nonlinearity at the Top and Bottom of the Gel and the Central Region of Approximate Linearity.*

16.4.1 Concept

1. Researchers should record their answers as a drawing with labels and description, as indicated below (Fig. 16.2).

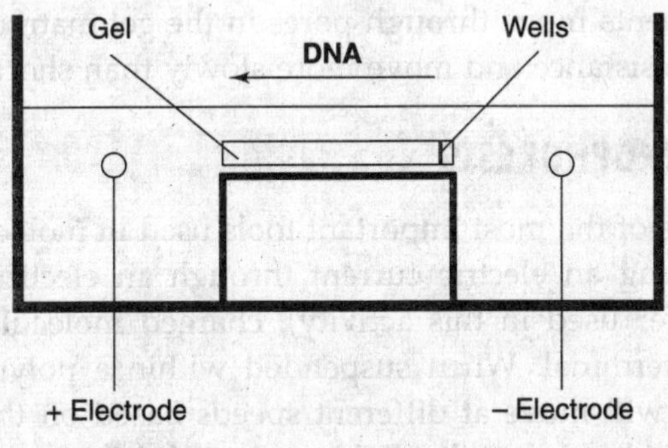

Fig. 16.2 *Demonstrate gel elecrophoresis.*

The blue dye allows the scientist to see the DNA solution as it is loaded into the wells of the gel. The blue dye molecules are also visible as they move through the gel and, because they are very small, move faster than most of the DNA molecules. The band of blue dye molecules (the dye front) indicates when the molecules are near the end of the gel.

2. The smaller the molecule the faster it travels through the agarose gel matrix.
3. The agarose molecules form a mesh-like matrix with holes (pores) between the molecules. A large DNA molecule fragment will encounter more resistance moving through the pores in the matrix than a small molecule will. This resistance will cause the large molecule to move more slowly.

4. Because DNA molecule fragments are invisible to the naked eye as they travel through the gel, the researcher cannot see how far they have traveled and will not know when to turn off the electric power. If the electrophoresis is allowed to run too long, the DNA fragments may move all the way through the gel and off the end.
5. As the DNA fragments move out of the wells and into the gel matrix, they start to separate and move as separate bands almost immediately. The separated bands of fragments are close together in the beginning but spread farther apart toward the end. As the electrophoresis progresses and the fragments that are close in size separate from each other, more bands become visible in each lane.
6. Because the bands of DNA fragments are moving from the top toward the bottom, the positive pole must be located at the bottom.
7. If the electrodes were reversed, the DNA fragments would move in the opposite direction.
8. There are two bands in Lane 3 and three bands in Lane 5.
9. The bands farthest from the wells contain the shortest DNA fragments, and the bands closest to the wells contain the longest fragments.
10. DNA molecule fragments are introduced into a gel matrix, and an electric field is applied across the gel. Because DNA molecules are negatively charged, they will move through the gel toward the positive pole of the electric field. As different-sized DNA fragments move through pores in the gel matrix, longer DNA fragments encounter more resistance and move more slowly than shorter fragments.

16.5 GEL ELECTROPHORESIS

Gel electrophoresis is one of the most important tools used in molecular biology and genetic engineering. By conducting an electric current through an electrolyte buffer (such as the sodium bicarbonate buffer used in this activity), charged molecules will migrate towards the oppositely charged terminal. When suspended within a polymer matrix (such as the agar gel), the molecules will move at different speeds based on their size. In this activity, negatively charged dye molecules are loaded into the gel. When a current is passed through the gel, the molecules migrate towards the positive terminal, with smaller molecules moving faster than larger ones. Look at the ingredients list of the food coloring you used to see what molecules are inside. We used Smart and Final brand dyes that have Blue #1, Red #40, and Yellow #5. Using gel electrophoresis, they discovered that McCormick brand "blue" food coloring is made of a blue dye plus a small amount of red. McCormick "red" dye is actually made of two different red colored molecules. Test different brands to discover the true composition of each color.

In molecular biology, this method is used to separate biological compounds such as DNA or proteins based on size. Like the colored dyes, DNA and proteins are negatively charged, so will migrate towards the positive electrode at different speeds depending on their size. Scientists can use special enzymes to cut a large strand of DNA into many smaller pieces.

The size of the pieces will depend on the specific base sequence of the large DNA strand. This idea is used in DNA fingerprinting to identify people based on the genetic code of their DNA.

Example The process of DNA fingerprinting by using electricity to separate colored dyes.

Materials

For the electrophoresis chamber:

small plastic box approximately 8 × 12 cm

(empty micropipette tip boxes are perfect – available at RAFT)

2 regular popsicle sticks

2 narrow popsicle sticks (coffee stirrer kind)

scissors

masking tape

two 5" pieces of stainless steel wire

2 electrical leads with alligator clips

five 9V batteries ("super heavy duty" kind work fine)

For the gel and buffer:

water

baking soda

agar-agar powder (available from Asian grocery stores) or agarose (available from chemical supply companies)

mat knife or razor blade

For the samples:

water

food coloring

glycerin (available from pharmacies)

needle-tip disposable pipet (Flinn #FB1260; $29.95 for 400)

(optional) beaker of water for rinsing tips between samples

Assembly

1. Make a comb to create wells in the gel that will eventually hold the samples. Cut the narrow popsicle sticks so that they sit just above the bottom of the base when hung from a regular popsicle stick (~3-4 cm depending on the depth of your box). Cut 5 teeth and tape them to a regular popsicle stick so that they are evenly spaced and hang down to the same level. Tape the other regular popsicle stick on the other side to secure the teeth, and check to see that they hang evenly when placed on the box without touching the bottom. Place the comb vertically near the top of your box.
2. Make a 0.2% sodium bicarbonate buffer by dissolving 2 g of baking soda in 1 L of water. You will need ~100 mL per set up.

3. Make a 1% gel solution by adding 1 g of agar-agar powder to 100 mL of sodium bicarbonate buffer. You will need 40–50 mL of gel solution per set up. To dissolve the powder, heat the solution in the microwave, stopping every so often to vigorously stir the solution. Watch the solution carefully, as it will quickly boil over when hot enough. When you see bubbles, stop the microwave, and swirl the solution until the agar-agar particles completely dissolve. The solution should be translucent and will solidify at room temperature.
4. Once the solution is cool enough to pour, add just enough into the box so that ~0.5 cm of the comb teeth are submerged. Adjust the comb by sliding it so that it is ~1.5 cm from the top of the box. Thinner gels will yield better separations.
5. The actual gel only needs to be half the size of the box. In addition, you need to make space to place the electrodes. Once the gel sets (~5-10 min), use a knife to cut off the bottom half of the gel. Also, without disturbing the comb, cut out a thin strip from the top of the gel to make room for a wire electrode. Your gel should now be around 6 cm long and 8 cm wide (still the full width of your box). You can recycle the pieces of gel that you cut out by reheating them in the microwave.
6. Bend each piece of stainless steel wire to run along the width of the box and hook over the side. Place one on either side of the gel. Use tape to secure them to the box if you need to. These will be the positive and negative electrodes.
7. Make a high voltage power supply by connecting the five 9 V batteries. Clip two batteries together by inserting the positive terminal of one into the negative terminal of another. Attach the remaining batteries one by one in this way until you have a five-battery pack. Clip an electrical lead to each of the exposed terminals of the pack. You should now be able to use the battery pack to power something by attaching the other ends of the electrical leads.
8. Prepare 5 different samples by mixing 1-2 drops of food coloring with 1 mL glycerin and 1 mL water in a small tube. We used blue, red, green, yellow, and purple (made by mixing blue and red food coloring).

Operation

1. When your gel set-up is ready, pour just enough buffer to cover the solidified gel. Make sure you fill up the space left from the cut gels and that the gel is completely submerged.
2. Gently remove the comb by pulling straight up without tearing the gel. The wells should fill with buffer.
3. Use the needle tip pipet to transfer ~10 μL of each sample to an empty well. The volume of the thin tip of the pipette is about 10 μL. Submerge the tip in the buffer directly above the well and gently squeeze the sample into the well. It should fall into the well since it is denser than the surrounding buffer. You should use a new pipette for each sample to prevent contamination between samples. If you only have a few pipettes, rinse out the tip well in a large beaker of water before reusing.

4. Once all the samples are loaded, connect the leads from the power supply to the stainless steel wire electrodes attached to the box. Connect the negative terminal to the electrode at the top of the gel (near the combs) and the positive terminal to the electrode at the bottom of the gel. Observe bubbles forming along the electrodes when a complete circuit is made.
5. Allow the samples to run for 15-20 minutes and observe what happens to each sample.

16.6 SDS-PAGE : GEL ELECTROPHORESIS OF PROTEINS

Technique is to set up gel plates before you mix the gel mixes. Use the THIN spacers and choose a comb–number of wells varies. Make both resolving gel and stack (NO APS or TEMED)–then add APS and TEMED to resolving gel, mix and pour about 3-3.5 mL per gel. Before it polymerises, now add APS and TEMED to your stack mix and pour it gently on top of the resolv. gel (using a Pasteur pipette works well). Put in the comb and let it solidify. It takes 15-20 minutes. You can keep gels overnight at 4°C if well wrapped to prevent drying out and if you keep the comb in.

Note: Just before you want to load gels, wash out wells with distilled water to remove unpolymerised acrylamide.

GEL RECEPIES (enough for 2 thin mini-gels)

RESOLVING GEL (NOTE pH 8.8)

Percentage of Gel	*8%*	*10%*	*12.5%*
30: 0.8% w/v acrylamide:bisacrylamide	2 mL	2.5 mL	3.1 mL
1.0 M Tris-Cl pH 8.8	3 mL	3 mL	3 mL
20% SDS	38 μL	38 μL	38 μL
dH_2O	2.43 mL	1.9 mL	1.3 mL
Mix together. Add APS and TEMED just before pouring			

10% APS	36 μL	36 μL	36 μL
TEMED	5 μL	5 μL	5 μL
	7.5 mL	7.5 mL	7.5 mL

STACKING GEL (NOTE pH 6.8)--use 4% stack for <10% res. gel and 6% stack (easier to handle) for >10% resolving gels.

Percentage of Stack	*4%*	*6%*
30:0.8% w/v acryl:bisacryl	660 μL	1 mL
1M Tris-Cl pH6.8	630 μL	630 μL
20% SDS	25 μL	25 μL
dH_2O	3.6 mL	3.6 mL
Mix together. Add APS and TEMED just before pouring.		

10% APS	25 μL	25 μL
TEMED	5 μL	5 μL
	5 mL	5 mL

Preparations of sample: Protein samples are in 1X sample loading buffer (also called Laemmli dye), i.e. to 6 μL protein sample, add 3 μL 3X Laemmli dye stock. Boil 3 minutes before loading gel.

Sample loading buffer (Laemmli loading dye) 3X stock:

1 M Tris-Cl pH 6.8 2.4 mL

20% SDS 3 mL

Glycerol (100%) 3 mL

B-mercaptoethanol 1.6 mL

Bromophenol blue 0.006 g

10 ml (store 4°C)

10X Running buffer (also called Laemmli buffer):

Tris base 30.3 g

Glycine 144 g

SDS 10 g

make to 1 L with dH_2O

For BioRad apparatus: need 500 ml of 1X buffer so dilute 50 mL 10X stock + 450 mL dH_2O.

Running gels: Using the BioRad apparatus, run gels at 200 V (constant voltage) until the bromophenol blue dye is just off. Takes 50-60 minutes. Gels are run at room temperature.

16.7 AGAROSE GEL ELECTROPHORESIS OF DNA

Preparing and running standard agarose DNA gels

The equipment and supplies necessary for conducting agarose gel electrophoresis are relatively simple and include:

- **An electrophoresis chamber and power supply**
- **Gel casting trays**, which are available in a variety of sizes and composed of UV-transparent plastic. The open ends of the trays are closed with tape while the gel is being cast, then removed prior to electrophoresis.
- **Sample combs**, around which molten agarose is poured to form sample wells in the gel.
- **Electrophoresis buffer**, usually Tris-acetate-EDTA (TAE) or Tris-borate-EDTA (TBE).
- **Loading buffer**, which contains something dense (e.g., glycerol) to allow the sample to "fall" into the sample wells, and one or two tracking dyes, which migrate in the gel and allow visual monitoring or how far the electrophoresis has proceeded.

- **Ethidium bromide**, a fluorescent dye used for staining nucleic acids.

  Note: Ethidium bromide is a known mutagen and should be handled as a hazardous chemical — wear gloves while handling.

- **Transilluminator** (an ultraviolet lightbox), which is used to visualize ethidium bromide-stained DNA in gels.

  Note: Always wear protective eyewear when observing DNA on a transilluminator to prevent damage to the eyes from UV light.

To pour a gel, agarose powder is mixed with electrophoresis buffer to the desired concentration, then heated in a microwave oven until completely melted. Most commonly, ethidium bromide is added to the gel (final concentration 0.5 ug/mL) at this point to facilitate visualization of DNA after electrophoresis. After cooling the solution to about 60°C, it is poured into a casting tray containing a sample comb and allowed to solidify at room temperature or, if you are in hurry, in a refrigerator.

After the gel has solidified, the comb is removed, using care not to rip the bottom of the wells. The gel, still in its plastic tray, is inserted horizontally into the electrophoresis chamber and just covered with buffer. Samples containing DNA mixed with loading buffer are then pipetted into the sample wells, the lid and power leads are placed on the apparatus, and a current is applied. You can confirm that current is flowing by observing bubbles coming off the electrodes. DNA will migrate towards the positive electrode, which is usually colored red.

The distance DNA has migrated in the gel can be judged by visually monitoring migration of the tracking dyes. Bromophenol blue and xylene cyanol dyes migrate through agarose gels at roughly the same rate as double-stranded DNA fragments of 300 and 4000 bp, respectively.

When adequate migration has occurred, DNA fragments are visualized by staining with ethidium bromide. This fluorescent dye intercalates between bases of DNA and RNA. It is often incorporated into the gel so that staining occurs during electrophoresis, but the gel can also be stained after electrophoresis by soaking in a dilute solution of ethidium bromide. To visualize DNA or RNA, the gel is placed on a ultraviolet transilluminator. Be aware that DNA will diffuse within the gel over time, and examination or photography should take place shortly after cessation of electrophoresis.

Migration of DNA fragments in agarose

Fragments of linear DNA migrate through agarose gels with a mobility that is inversely proportional to the $\log_{10}$ of their molecular weight. In other words, if you plot the distance from the well that DNA fragments have migrated against the $\log_{10}$ of either their molecular weights or number of base pairs, a roughly straight line will appear.

Circular forms of DNA migrate in agarose distinctly differently from linear DNAs of the same mass. Typically, uncut plasmids will appear to migrate more rapidly than the same plasmid when linearized. Additionally, most preparations of uncut plasmid contain at least two topologically different forms of DNA, corresponding to supercoiled forms and nicked circles. The image to the right shows an ethidium-stained gel with uncut plasmid in the left lane and the same plasmid linearized at a single site in the right lane.

Several additional factors have important effects on the mobility of DNA fragments in agarose gels, and can be used to your advantage in optimizing separation of DNA fragments. Chief among these factors are:

Agarose Concentration: By using gels with different concentrations of agarose, one can resolve different sizes of DNA fragments. Higher concentrations of agarose facilites separation of small DNAs, while low agarose concentrations allow resolution of larger DNAs.

The image to the right shows migration of a set of DNA fragments in three concentrations of agarose, all of which were in the same gel tray and electrophoresed at the same voltage and for identical times. Notice how the larger fragments are much better resolved in the 0.7% gel, while the small fragments separated best in 1.5% agarose. The 1000 bp fragment is indicated in each lane.

Voltage: As the voltage applied to a gel is increased, larger fragments migrate proportionally faster than small fragment. For that reason, the best resolution of fragments larger than about 2 kb is attained by applying no more than 5 volts per cm to the gel (the cm value is the distance between the two electrodes, not the length of the gel).

Electrophoresis Buffer: Several different buffers have been recommended for electrophoresis of DNA. The most commonly used for duplex DNA are TAE (Tris-acetate-EDTA) and TBE (Tris-borate-EDTA). DNA fragments will migrate at somewhat different rates in these two buffers due to differences in ionic strength. Buffers not only establish a pH, but provide ions to support conductivity. If you mistakenly use water instead of buffer, there will be essentially no migration of DNA in the gel. Conversely, if you use concentrated buffer (e.g., a 10X stock solution), enough heat may be generated in the gel to melt it.

Effects of Ethidium Bromide: Ethidium bromide is a fluorescent dye that intercalates between bases of nucleic acids and allows very convenient detection of DNA fragments in gels, as shown by all the images on this page. As described above, it can be incorporated into agarose gels, or added to samples of DNA before loading to enable visualization of the fragments within the gel. As might be expected, binding of ethidium bromide to DNA alters its mass and rigidity, and therefore its mobility.

Agarose gels, as discussed above provide the most commonly used means of isolating and purifying fragments of DNA, which is a prerequisite for building any type of recombinant DNA molecule.

16.8 DENATURING AGAROSE GEL ELECTROPHORESIS OF RNA

The overall quality of an RNA preparation may be assessed by electrophoresis on a denaturing agarose gel; this will also give some information about RNA yield. A denaturing gel system is suggested because most RNA forms extensive secondary structure via intramolecular base pairing, and this prevents it from migrating strictly according to its size. Be sure to include a positive control RNA on the gel so that unusual results can be attributed to a problem with the gel or a problem with the RNA under analysis. RNA molecular weight markers, an RNA sample known to be intact, or both, can be used for this purpose.

Ambion's NorthernMax® reagents for Northern Blotting include everything needed for denaturing agarose gel electrophoresis. These products are optimized for ease of use, safety, and low background, and they include detailed instructions for use.

An alternative to using the NorthernMax reagents is to use the procedure described below. This denaturing agarose gel method for RNA electrophoresis is modified from "Current Protocols in Molecular Biology", Section 4.9 (Ausubel et al., Eds.). It is more time-consuming than the NorthernMax method, but it gives similar results.

1. **Prepare the gel.**

(a) Heat 1 g agarose in 72 mL water until dissolved, then cool to 60°C.

(b) Add 10 mL 10X MOPS running buffer, and 18 mL 37% formaldehyde (12.3 M).

WARNING: Formaldehyde is toxic through skin contact and inhalation of vapors. Manipulations involving formaldehyde should be done in a chemical fume hood.

10X MOPS running buffer:

0.4 M MOPS, pH 7.0

0.1 M sodium acetate

0.01 M EDTA

(c) Pour the gel using a comb that will form wells large enough to accommodate at least 25 μL.

(d) Assemble the gel in the tank, and add enough 1X MOPS running buffer to cover the gel by a few millimeters. Then remove the comb.

2. **Prepare the RNA sample.**

(a) To 1-3 μg RNA, add 0.5-3X volumes Formaldehyde load dye.

- To simply check the RNA on a denaturing gel, as little as 0.5X Formaldehyde Load Dye can be used, but to completely denaturate the RNA, e.g. for Northern blots, use 3X volumes of Formaldehyde Load Dye.
- Ethidium bromide can be added to the Formaldehyde Load Dye at a final concentration of 10 μg/mL. Some size markers may require significantly more than 10 μg/mL ethidium bromide for visualization. To accurately size your RNA, however, it is important to use the same amount of ethidium bromide in all the samples (including the size marker) because ethidium bromide concentration affects RNA migration in agarose gels.

(b) Heat denature samples at 65-70°C for 5-15 min.

Denaturation for 5 min is typically sufficient for simply assessing RNA on a gel, but a 15 min denaturation is recommended when running RNA for a Northern blot. The longer incubation may be necessary to completely denature the RNA.

3. **Electrophoresis**

Load the gel and electrophorese at 5-6 V/cm until the bromophenol blue (the faster-migrating dye) has migrated at least 2-3 cm into the gel, or as far as 2/3 the length of the gel.

Results

Visualize the gel on a UV transilluminator. (If ethidium bromide was not added to the Formaldehyde Load Dye, the gel will have to be post-stained and destained.)

Intact total RNA run on a denaturing gel will have sharp 28S and 18S rRNA bands (eukaryotic samples). The 28S rRNA band should be approximately twice as intense as the 18S rRNA band (Fig. 16.3, lane 3). This 2:1 ratio (28S:18S) is a good indication that the RNA is intact. Partially degraded RNA will have a smeared appearance, will lack the sharp rRNA bands, or will not exhibit a 2:1 ratio. Completely degraded RNA will appear as a very low molecular weight smear (Fig. 16.3, lane 2). Inclusion of RNA size markers on the gel will allow the size of any bands or smears to be determined and will also serve as a good control to ensure the gel was run properly (Fig. 16.3, lane 1).

Note: Poly(A) selected samples will not contain strong rRNA bands and will appear as a smear from approximately 6 kb to 0.5 kb (resulting from the population of mRNAs, and depending on exposure times and conditions), with the area between 1.5 and 2 kb being the most intense (this smear is sometimes apparent in total RNA samples as well).

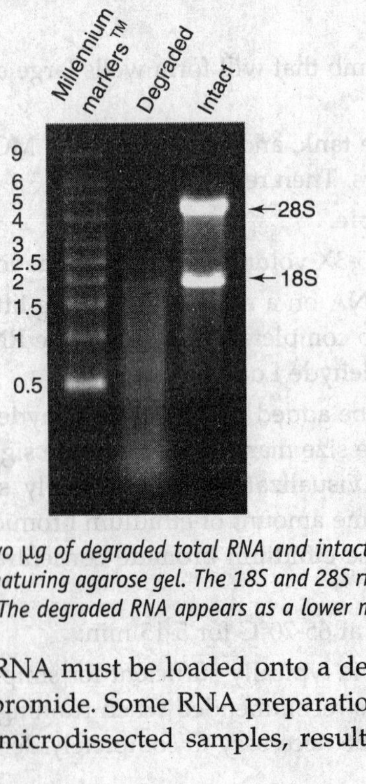

Fig. 16.3 *Intact vs. Degraded RNA. Two μg of degraded total RNA and intact total RNA were run beside Ambion's RNA millennium markers™ on a 1.5% denaturing agarose gel. The 18S and 28S ribosomal RNA bands are clearly visible in the intact RNA sample. The degraded RNA appears as a lower molecular weight smear.*

Generally, at least 200 ng of RNA must be loaded onto a denaturing agarose gel in order to be visualized with ethidium bromide. Some RNA preparations, such as those from needle biopsies or from laser capture microdissected samples, result in very low yields. In these cases, it may be impossible to spare 200 ng of RNA to assess integrity. Alternative nucleic acid stains, such as SYBR® Gold and SYBR® Green II RNA gel stain from Molecular Probes, offer a significant increase in sensitivity over ethidium bromide. Using a 300 nm transilluminator (6 × 15 watt bulbs) and a special filter, as little as 1 ng and 2 ng of RNA can be detected with SYBR Gold and SYBR Green II RNA gel stain, respectively.

16.9 PROTEIN FRACTIONATION WITH 1D GEL ELECTROPHORESIS

Protocol: 1d gel electrophoresis

SDS-PAGE is the most commonly used low-level protein separation and purification technique in biological research. The reason of using this technique in LC MS is to reduce sample complexity and to purify contaminants as detergents, salts, DNA and other non-protein compounds

The procedures given below were optimized to enable the sensitive characterization of proteins directly from one-dimensional polyacrylamide gels of varying thickness (0.5 – 1.5 mm) and acrylamide concentration (7 – 18%). For best results, gels should be used no earlier than one hour after casting to reduce acrylamidation of cysteine residues during electrophoresis.

Recipe for the separating gel

Amount			*Reagent*
10% gel	12.5% gel	15% gel	
4.43 mL	3.387 mL	2.23 mL	H_2O
3.15 mL	3.15 mL	3.15 mL	4x separating buffer pH 8.8
4.2 mL	5.243 mL	6.3 mL	30% acryl/bisacryl amide
0.6 mL	0.6 mL	0.6 mL	Glycerol
0.123 mL	0.123 mL	0.123 mL	10% ammonium persulfate
10 μL	6 μL	6 μL	TEMED

Recipe for the stacking gel

Amount	*Reagent*
2.2 mL	H_2O
1.0 mL	4x separating buffer pH 8.8
0.533 mL	30% acryl/bisacryl amide
0.224 mL	Glycerol
40 μL	10% ammonium persulfate
4 μL	TEMED

Protocol: Coloring of the gel

Coloring can be done with silver staining or Colloidal Coomassie. In case of silver staining, use the following protocol, for it is optimized to be compatible with the purification and mass spectrometry methods used. Colloidal Coomassie can be performed according to standard protocols.

Protocol: Optimized protocol for silver staining of polyacrylamide gels

Reagents

- Fixation solution (40 : 10 : 50 v/v/v ethanol : acetic acid : water)
- Washing solution (30% v/v ethanol in water)
- Sensitizing solution (0.02% sodium thiosulfate)
- 0.1% $AgNO_3$
- Developing solution (0.05% formaldehyde in 2% sodium carbonate)
- 1% acetic acid

Procedure

1. After the gel has been run, fix the protein by incubating the gel slab in fixation solution for 20 - 30 minutes.
2. Incubate the gel in washing solution on a shaking platform for two times 20 minutes. Extended washing time helps to eliminate yellowish background usually observed after long developing of the gel.
3. Rinse the gel slab with water (2 changes, 1 - 2 minutes per change) and then leave it further in water for 30 minutes on a shaking platform.
4. Sensitize the gel with sensitizing solution for 1 - 2 minutes. Discard solution and quickly rinse the gel slab with three changes of water (10 seconds each).
5. Incubate the gel in chilled 0.1% $AgNO_2$ for 30 minutes at 4°C (fridge).
6. Discard silver nitrate solution and quickly rinse the gel with two changes of water (30 seconds per each change).
7. Change the gel chamber.
8. Develop the gel with developing solution. Discard the developing solution as soon as it turns yellow and replace it with a fresh portion.
9. When a sufficient degree of staining has been obtained, quench staining by discarding the developing solution and replacement with 5% acetic acid. Wash the gel with 1% acetic acid several times and store in the same solution.
10. Wash the gel with milliq water several times before cutting it to remove acetic acid.

Protocol: Fractionation of 1D gel

Reagents

- Scanner
- Real-size print of the gel
- Glass plate
- Milliq

- Clean scalpel

Method

1. Wear gloves in order to avoid keratin contaminations. Never touch the gel with bare hands.
2. Clean the scanner before use to avoid keratin contaminations. Make a scan of the gel. Print the scan in real size.
3. Plan the slices by drawing them on the real-size print. Code the slices. Prepare tubes for each slilce.
4. Clean a glass plate to cut the gel on with soap and alcohol.
5. Place the gel on the glass plate and place the real-size print of the gel underneath.
6. Slice the lanes as planned. Put each slide in a separate tube. Keep the gel wet with milliq to avoid breaking of the gel.
7. Prepare the slices for in-gel reduction, alkylation and digestion by chopping them in approximately 1 × 1 mm particles.

16.10 TWO-DIMENSIONAL GEL ELECTROPHORESIS

Two-dimensional gel electrophoresis (2-D electrophoresis) is a powerful and widely used method for the analysis of complex protein mixtures extracted from cells, tissues, or other biological samples. This technique separates proteins in two steps, according to two independent properties: the first dimension is isoelectric focusing (IEF), which separates proteins according to their isoelectric points (pI); the second dimension is SDS-polyacrylamide gel electrophoresis (SDS-PAGE), which separates proteins according to their molecular weights (MW). In this way, complex mixtures consisted of thousands of different proteins can be resolved and the relative amount of each protein can be determined.

The procedure involves placing the sample in gel with a pH gradient, and applying a potential difference across it. In the electrical field, the protein migrates along the pH gradient, until it carries no overall charge. This location of the protein in the gel constitutes the apparent pI of the protein.

There are two alternative methods to create the pH gradient-carrier ampholites and immobilized pH gradient (IPG) gels. For Proteome Research the IEF can be performed with commercial IPGs for highly reproducible results.

The IEF is the most critical step of the 2-D electrophoresis process. The proteins must be solubilize without charged detergents, usually in high concentrated urea solution, reducing agents and chaotrophs. To obtain high quality data it is essential to achieve low ionic strength conditions before the IEF itself. Since different types of samples differ in their ion content, it is necessary to adjust the IEF buffer and the electrical profile to each type of sample.

The separation in the second dimension by molecular size is performed in slab SDS-PAGE. Twelve parallel gels can be separated in a fixed temperature to minimize the separation variations between individual gels.

16.10.1 Gel Electrophoresis

The two-dimensional electrophoresis capabilities include 7 cm mini gel, 13 cm 18 cm and 24 cm large format gel systems. Routinely one run a wide variety of complex protein mixtures on these various formats, with the mini-gel systems generally used to obtain an initial overview of a sample's protein distribution. The large gel systems are then used to analyse and investigate regions of interest with the aim of visualizing detailed sections of the proteome. Proteins profiles are visualized using either Coomassie or mass spectrometry compatible silver stains.

Theory

- 2D gels offer outstanding resolving power for complex samples and current in gel digestion techniques and MALDI based PMF makes it an essential proteomics tool.
- The use of 1D gels should not be overlooked as it is useful for the analysis of samples not suited to 2D gels such as membrane proteins.
- This method is used for the separation and identification of proteins in a complex mixture using two separate dimensions that are run perpendicular to one another.
- This allows a complex biological sample to be separated over a larger area, increasing the resolution of each component.

The 1st Dimension

- The 1st dimension uses isoelectric focusing (IEF) which separates proteins based on their overall charge or the pI which is a property of a protein due to the amino-acid composition.
- IEF uses pH gradients that can be established by adding ampholytes or immobilines to an acrylamide gel. Ampholytes are a mixture of different amphoteric molecules that have a range of isoelectric point values.
- Immobilized strips are dehydrated which allows rehydration of the strips directly with the sample to be separated.

The 2nd Dimension

- The 2nd dimension separates proteins based their molecular weight.
- A number of gel gradients (polyacrylamide SDS-PAGE) concentrations and different buffering systems can be used in the second dimension.
- This method is essentially the same as a 1D SDS-PAGE gel, proteins separated by mass based on a near equivalent charge to mass ratio due to the presence of the SDS detergent anion.

Example 2D gel

- The investigation of global protein expression of an organism or a specific tissue which allows for rapid protein identification using mass spectrometry to generate peptide fingerprint data.

- Differential expression is a technique that compares two or more samples to study differences in their protein expression in a differential system. An example could be that a diseased and healthy tissue are differentially examined to elucidate proteins responsible for the pathology of disease, or the effects of a drug on treated versus untreated tissue samples.

Sample Information and Preparation Required

- **Protein samples** should be provided as concentrated as possible and details of concentration and buffer type should be given. If these are unknown then for and extra charge the sample can be cleaned up and quantitated.
- **Amount of Material:** The following table provides a guideline to the amounts of protein required for various gel types:
- - mini gel 7 cm 10-50 µg total protein

  - large gel 24 cm 100-500 µg total protein
- **Sample information:** Any useful details relating to the potential behaviour of your proteins of interest in a PAGE system such as the nature of the proteins e.g., membrane or soluble.
- **Sample buffer:** Samples should be supplied free of the following components which interfere with the first dimension isoelectric focusing step: salt >50 mM, nucleic acids, polysaccharides, lipids, SDS and other ionic detergents. A desalting procedure should be carried out prior to sample submission or if necessary we can perform an appropriate clean up.
- **Approximate amount of protein:** An estimated concentration of protein in ug/mL is essential. The recommended lower limit is 350 ug/mL. If unknown we can include a quantitation in the procedure.
- **IPG Strip pH range:** If the pI of the target proteins are known then a narrow pI range may be chosen to maximize resolution otherwise, a broad range (3-10) strip will be used.
- **SDS-PAGE gel percentage:** An appropriate gel must be chosen for obtaining an optimal separation of your proteins of interest depending on the estimated molecular weight range.
- **Protein pre-fractionation:** In some cases such as human serum a single to a few proteins may be present in a sample in such high abundance that they interfere with proper electrophoresis or inhibit resolution of other proteins.
- **Proteases:** Protease activity can result in poor quality gels. Their activity varies between different organisms and tissues. To limit their activity, the sample should be kept on ice at all times during sample preparation or a protease inhibitor cocktail can be added.
- **Sample Solubilisation:** The preparation of sample for 2D electrophoresis generally requires the use of a buffer that solubilizes and denatures proteins such as 8 M urea. This separates the proteins into individual species and improves the separation by IEF.

- **Centrifugation:** All samples should be centrifuged before IEF to sediment particulates present that can cause smearing and blockage of the polyacrylamide gel matrix.
- **Protein Estimation** is an important step to know the concentration of proteins in a sample before it can be loaded onto the IEF gel to avoid overloading or underloading the IPG strip.
- **Urea a chaotrope** used to solubilize proteins disrupts hydrogen and hydrophobic bonds, and has no effect on the native charge of a protein. In solution, at higher temperatures, urea forms an equilibrium with isocyanate which can cause carbamylation of primary amino groups such as the N-terminus or lysine or arginine sample proteins and thus modifies the protein charge resulting in streaks of spots across the gel caused by titration of the charge of a particular protein.
- **Staining**

  - Coomassie blue staining is a quick, easy and cheap and multiple protocols and pre-made reagents are available. This stain is not specific and will also detect non-protein components.

  - Silver staining can give excellent results but is a time-consuming and expensive technique that can interfere with MS analysis if not done properly. The advantage of silver staining is its sensitivity. MS friendly silver staining protocols are available.

  - An alternative to silver stain is Sypro Ruby has the same level of sensitivity and is MS compatible however is more expensive.

16.10.2 2-D Agarose Gel Electrophoresis of DNA Replication Intermediates

Two-dimensional (2-D) agarose gel electrophoresis is adapted from the procedure by Bell and Byers (Anal. Biochem. 130:527, 1983). The first dimension gel is intentionally run at low voltage in low percentage agarose to separate DNA molecules in proportion to their mass. The second dimension is run at high voltage in a gel of higher agarose concentration in the presence of Ethidium bromide so that the mobility of a non-linear molecule is drastically influenced by its shape. Demonstrate for interpretations of 2-D gel patterns obtained for different replication intermediates (Fig. 16.4).

First Dimension: 0.4% agarose in 1x TBE (0.089 M Tris-borate, 0.089 M Boric Acid, 0.002 M EDTA); electrophoresis buffer, 1x TBE.

Pour agarose in horizontal slab gel apparatus with a comb ~1.5 mm from the bottom support plate. The dimensions of the slab are somewhat arbitrary, convenient dimensions are 13 cm wide, 20 cm long, and ~0.6 cm thick. Use a comb with narrow, thin teeth (for example, 4 mm wide and 1.2 mm thick) to produce a narrow first dimension gel lane with tight, sharp bands. Load alternate lanes of the gel. The empty lanes prevent accidental cross-contamination when the lanes are excised for the second dimension. Carry out the electrophoresis, with the gel submerged by a few mm, at 0.7 to 1.0 V/cm for 15 to 30 hr. (If a voltmeter is not available, estimate the voltage required by measuring the distance between the electrodes of the gel box and set the voltage accordingly.) The exact conditions depend upon the size of

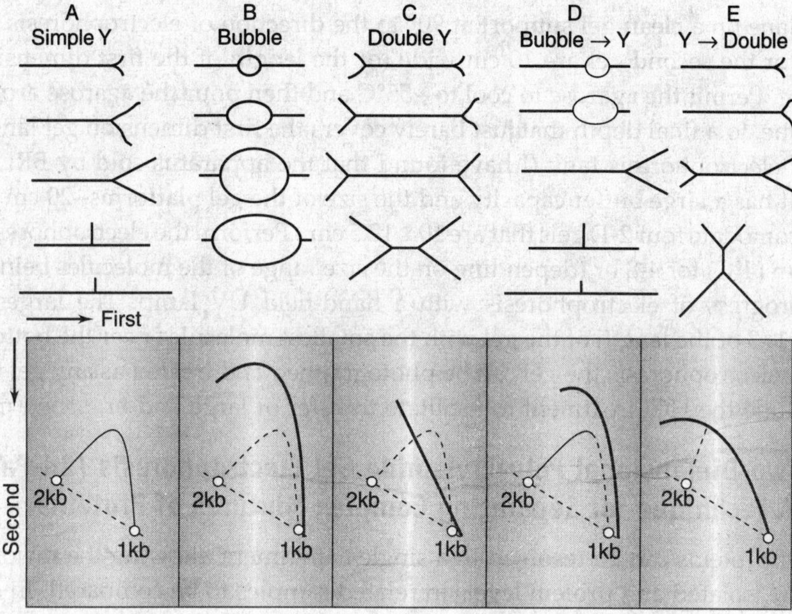

Fig. 16. 4 *2-D gel patterns for different DNA replication intermediates.*

the molecules to be analyzed. The band containing the fragment of interest and the band of twice that size should be separated by 3-5 cm in the first dimension gel. For example, a 3 kb fragment can be successfully separated from a 6 kb fragment by electrophoresis at 1V/cm for 15 to 18 hr. To achieve the same separation for a 6 kb fragment and a 12 kb fragment might require 24-30 hr at the same or lower voltage. It becomes appropriate to lower the voltage when large molecules are being examined since branched intermediates in this size range are more likely to retarded in their mobility. Thus they will not run at their true molecular weight have modified conditions for examining molecules larger than 6 kb.

Stain the first dimension slab gel in 1 × TBE with 0.3 μg/mL ethidium bromide for ~20 min. Examine/photograph the gel on a long wave UV light box and mark lanes to be run in the second dimension. (Do not expose the gel to excessive doses of UV light as the DNA will become nicked and replicating structures can be lost.) Excise the lanes of interest with a clean razor blade--making at least one edge very straight, vertical, and close to the lane of interest. It is through this cut edge that the DNA will run in the second dimension. The lane can be trimmed at either end to include just that portion that contains the size range of interest. When possible, include the entire lane from the well to at least 1 cm beyond the band of interest.

Second Dimension: 1.0 to 1.2 % agarose in 1 × TBE and 0.3 μg/mL ethidium bromide; electrophoresis buffer is 1 × TBE with 0.3 μg/mL ethidium bromide (2-3 L total) circulated from anode to cathode at 50 to 100 mL/min.

Place the lane on a clean gel support at 90° to the direction of electrophoresis. Convenient dimensions for the second gel are 10 cm wide (or the length of the first dimension gel lane) by 13 cm long. Permit the agarose to cool to ~55°C and then pour the agarose around the first dimension lane, to a final depth that just barely covers the first dimension gel lane. Submerge the gel in an electrophoresis tank (I have found that the apparatus sold by BRL, Model H1, works well. It has a large buffer capacity and the size of the gel platforms--20 cm wide, 25 cm long–will accomodate four 2-D gels that are 10 × 12.5 cm.) Perform the electrophoresis at 5 V/cm in a cold room (4°C) for 4-8 hr (depending on the size range of the molecules being analyzed). Check the progress of electrophoresis with a hand-held UV lamp. The largest molecules should run ~1/3 of the length of the gel with the smallest molecules near the bottom corner of the gel. After electrophoresis the gel can be photographed and treated as any gel for Southern analysis. Include the HCl treatment to facilitate transfer of large and branched molecules.

16.10.3 Two-Dimensional Polyacrylamide Gel Electrophoresis (2d-Page) — A Technique for Separating Complex Mixtures of Proteins

Thousands of proteins can be resolved in a single experiment allowing the major proteins in a sample to be isolated and protein levels in related samples to be compared. In combination with mass spectrometry, the proteins can also be identified:

Key principles

- Proteins differ from each other in terms of their mass and charge.
- Both these properties can be used to separate proteins by gel electrophoresis.
- The successive application of both techniques in perpendicular directions (two dimensions) provides maximum separation and allows thousands of proteins to be resolved.
- Staining the gel reveals the positions of individual proteins as spots or smudges. These can be picked and analyzed by mass spectrometry.
- There are tens of thousands of proteins in the cell, differing in abundance over six orders of magnitude. 2D-PAGE is not sensitive enough to detect the rare proteins and many proteins will not be resolved. Splitting the sample into different fractions is often necessary to reduce the complexity of protein mixtures prior to 2D-PAGE.

Working

The polyacrylamide gel provides a supporting matrix through which proteins can migrate. This gel has a pH gradient from top to bottom—the top is more acidic than the bottom.

A complex protein mixture is loaded in the middle of the left side, where the pH is neutral, and a voltage is applied across the gel. The proteins then migrate through the gel until they reach their isoelectric point (the point at which their charge is the same as the surrounding pH). This technique is called isoelectric focusing (Fig. 16.5).

The gel is then soaked in a denaturing solution (which causes the proteins to unfold) containing the detergent sodium dodecylsulfate (SDS). This molecule has a very strong negative charge and binds to all proteins, effectively making them all the same charge.

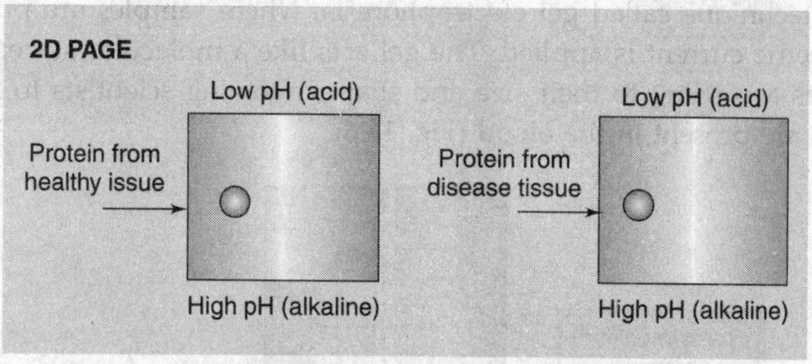

Fig. 16.5 *Working of 2D page.*

A voltage is then applied across the gel from side to side with the anode (positive terminal) at the right. The proteins all move towards the anode, but smaller proteins move faster through the gel than larger ones, thus separating the proteins according to their size.

Two-dimensional polyacrylamide gel electrophoresis (2D-PAGE) is used to separate mixtures of proteins, and is particularly useful for comparing related samples Đ such as healthy and diseased tissue. The first step is to load the proteins onto the gel, which has a pH gradient from top to bottom.

2D-PAGE WITH MASS SPECTROMETRY

2D-PAGE can be used in combination with mass spectrometry to systematically identify all the major proteins present in a sample.

However, the technique is particularly powerful when comparing related samples, such as healthy tissue vs disease tissue. For example, proteins that are more abundant in disease tissue may represent novel drug targets or diagnostic disease markers.

Comparative 2D-PAGE can also be used to look for proteins whose expression varies similarly under the same set of conditions (these may have related functions) and to identify proteins produced in response to drug therapy (these may be responsible for drug-related side effects).

Two-dimensional polyacrylamide gel electrophoresis, (2D-PAGE) is used to separate mixtures of proteins, and is particularly useful for comparing related samples – such as healthy and diseased tissue. The first step is to load the proteins onto the gel, which has a pH gradient from top to bottom.

16.11 NEW GEL ELECTROPHORESIS TECHNIQUE ALLOWS SEPARATION OF GLYCATED PROTEINS

University of Bath researchers have developed a methodology for monitoring protein glycation, a process where a protein binds to a sugar molecule. Alzheimer's, diabetes, and many cancers involve various glycated proteins, and so the new technology may help in diagnosis and development of treatments for these diseases.

They used a technique called gel electrophoresis, where samples are put into a thin gel layer and an electric current is applied. The gel acts like a molecular sieve, sorting proteins from the samples according to their size and shape, allowing scientists to identify whether specific proteins are present in the blood (Fig. 16.6).

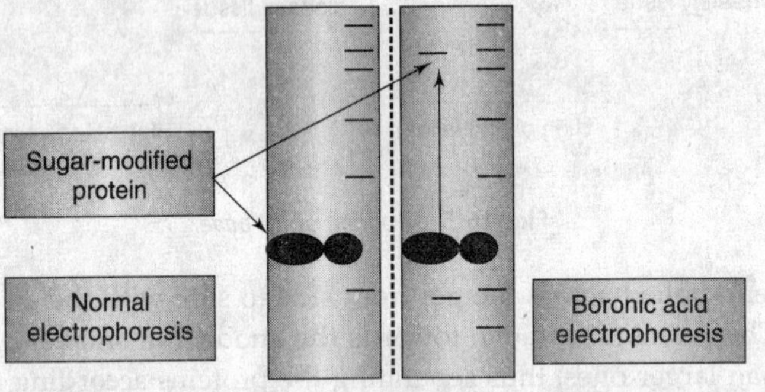

Fig. 16.6 *New gel electrophoresis technique allows separation of glycated proteins.*

For this study, the researchers have patented a new type of gel electrophoresis, which uses boronic acid to distinguish between the glycated and unmodified proteins.

Whilst the technique has only been assessed in the lab at present, the researchers say it has the potential to be developed into a test for these conditions in patients.

Dr James added: "Currently there is no blood test for Alzheimer's disease.

"If we can develop this technique into a test, doctors could potentially diagnose patients at an early stage before their symptoms show up in a brain scan."

The method could also be used to diagnose diabetes, which also leads to elevated levels of glycated proteins in the blood.

16.12 REFERENCES

1. Molecular Cell Biology by Lodish et. al., W. H. Freeman (2000)
   A classic molecular biology text available FREE online at:
   http://www.ncbi.nlm.nih.gov/books/bv.fcgi?rid=mcb
   http://en.wikipedia.org/wiki/Gel_electrophoresis
   http://en.wikipedia.org/wiki/Food_dye
2. Krysan and Calos (1991, MCB 11; 1464-1472) and Raghuraman et al. (2001, Science 294; 115–121)
3. Carle GF and Olson MV 1984 Separation of chromosomal DNA molecules from yeast by orthogonal-field-alternation gel electrophoresis. Nucleic Acids Res. 25:564–567.
4. Carle GF, Frank M and Olson MV 1986 Electorphoretic separations of large DNA molecules by periodic inversion of the electric field. Science 232: 65–68.
5. Consden R, Gordon AH and Matrin AJP 1944 Qualitative analysis of proteins: a partition chromatography method using paper. Biochem J 38: 224–232.

6. Grassmann W, Hannig K, and Knedel M 1955 Über ein verfahren zur elektrophoretischen bestimmung der serum proteine auf filtrierpapeir. Deutsche Medizinische Wochenschrift 76: 333–339.
7. Hardy WB 1899 On the coagulation of proteid by electricity. J Physiol 24: 288–304.
8. Hardy WB 1905 Colloidal solution. The globulins. J Physiol 33: 251–337.
9. Lai E, Birren BW, Clark SM, Simon MI, and Hood L 1989 Pulsed field gel electrophoresis. Biotechniques 7: 34–42.
10. Lerman LS and Frisch HL 1982 Why does the electrophoretic mobility of DNA in gels vary with the length of the molecule? Biopolymers 21: 995–997.
11. Lumpkin OJ, Dejardin P, and Zimm BH 1985 Theory of gel electrophoresis of DNA. Biopolymers 24: 1573–1593.
12. Maniatis T, EF Fritsch, and J Sambrook 1982 Molecular Cloning: A Laboratory Manual. Cold Spring Harbor: Cold Spring Harbor Laboratory Press.
13. Michaelis L 1909 Elektrische überführung von fermenten. Biochemische Zeitschrift 16: 81-86.
14. Smith SB, Aldridge PK and Callis JB 1989 Observation of individual DNA molecules undergoing gel electrophoresis. Science 243: 203–206
15. Tiselius A 1959 Introduction. In: M. Bier (ed.) Electrophoresis: Theory, Methods, and Applications. New York: Academic Press. Pp. xv-xx.
16. von Klobusitzky D and König P 1939 Biochemische studien über die gifte der schlangengattung Bothrops. VI. Arch Exp Pathol Pharmakol 192:271–275.
17. Williams FG, Pickles EG, and Durrum EL 1955 Improved hanging-strip paper electrophoresis technique. Science 121: 829–832.
18. Wunderly C 1959 Gel structure and the diffusion of proteins. Clin Chim Acta. 4: 754–759.
19. Berg JM, Tymoczko JL Stryer L (2002). Biochemistry (5th edn.). WH Freeman. ISBN 0-7167-4955-6. http://www.ncbi.nlm.nih.gov/books/bv.fcgi? &rid=stryer.section.438#455.
20. Robyt, John F.; White, Bernard J. (1990). Biochemical Techniques Theory and Practice. Illinois: Waveland Press. ISBN 0-88133-556-8.
21. Lodish H, Berk A, Matsudaira P, et al. (2004). Molecular Cell Biology (5th edn.). WH Freeman: New York, NY. ISBN 978-0716743668. http://www.ncbi.nlm.nih.gov/books/bv.fcgi?&rid=mcb.section.1637#1648.
22. Troubleshooting DNA agarose gel electrophoresis. Focus 19:3 p. 66 (1997). http://www.ncbi.nlm.nih.gov/pubmed/5806584
23. Milan Bier (ed.) (1959). Electrophoresis. Theory, Methods and Applications (3rd printing edn.). Academic Press. pp. 225. LCC 59-7676. OCLC 1175404.
24. Schaffer, H. and Sederoff, R. (1981). Improved estimation of DNA length from agarose gels. Anal. Biochem. 115: 113 122.

25. Prunell, A. (1980). A photographic method to quantitate DNA in gel electrophoresis. Methods Enzymol. 65: 353–358.

26. Carmichel, G. and McMaster, G. (1980). The analysis of nucleic acids in gels using glyoxal and acrinine orange. Methods Enzymol. 65: 380–391.

27. Parker, R. (1980). Conversion of circular DNA to linear strands for mapping. Methods Enzymol. 65. 415-428. Southern, E. (1979) Gel electrophoresis of restriction fragments. Methods Enzymol. 68: 152–182.

28. Brunk, C. and Simpson, L. Comparison of various ultraviolet sources for fluorescent detection of ethidium bromide-DNA complexes in polyacrylamide gels. Anal. Biochem. 82: 455–462.

29. Smith, M., Devine, C., Cohn, S. and Lieberman, M. (1984). Quantitative transfer of DNA from polyacrylamide or agarose gels to nitrocellulose. Anal. Biolchem. 137:120-124.

30. Howley et a] (1979). A rapid method for detecting and mapping homologies between heterologous DNAs. J. Biol. Chem. 254:4876-83.

31. Carle, G., Frank, M. and Olson, M. (1986). Electrophoretic separations of large DNA molecules by periodic inversion of the electric field. Science 232:65-68.

32. Carle, G. and Olson, M. (1984). Separation of chromosome DNA molecules of yeast by orthogonal field alternating gel electrophoresis. Nucl. Acids Res. 12:5647-64.

33. Smith, G. and Summers, M. (1980). The bidirectional transfer of DNA and RNA to nitrocellulose or diazobenzyloxymethyl paper. Anal. Biochem. 109:123-129.

34. Hassur, S. and Whitlock, H. (1974). UV shadowing-a new and convenient method for the localization of uv-absorbing species in polyacrylamide gels. Anal. Biochem. 59:162-167.

35. Vogelstein, B. and Gillespie, D. (1979). Preparative and analytical purification of DNA from agarose. Proc. Antl. Acad. Sci. USA 76:615-619.

36. Maniatis, T., Fritsch, E. and Sambrook, J. (1982). Molecular cloning: a laboratory manual. Cold Spring Harbor Laboratory, Cold Spring Harbor, NY.

37. Marini, J., Levene, S., Crothers, D. and Englund, P. (1982). Bent ehlical structure in kinetoplast DNA. Proc. Natl. Acad. Sci. USA 79:7664-7668.

38. Kidane, G., Hughes, D. and Simpson, L (1984). Sequence heterogeneity and anomalous electrophoretic mobility of kinetoplast minicircle DNA in Leishmania tarentolae. Gene 27:265-277.

CHAPTER

17

Polymerase Chain Reaction

17.1 INTRODUCTION

The Polymerase Chain Reaction (PCR) created a revolution in molecular biology research and its applications. PCR is powerful technique in which from a single copy of DNA molecule, millions of copies can be obtained with high accuracy, specificity in very short time. So, in few years since its invention, the polymerase chain reaction has become a convenient and trust worthy alternative to traditional methods of cloning of DNA. PCR is an *in vitro* method that enzymatically amplifies specific DNA sequences using oligonucleotide primers that flank the region of interest in the target DNA. It involves a repetitive series of cycles each of which consists of template denaturation, primer annealing and extension of the annealed primers by a DNA polymerase to create the expontial accumulation of a specific fragment whose ends are determined by the 5′ ends of the primers. PCR is so named because it involves a polymerase and the products synthesized in each cycle can serve as template in the next so the number of DNA copies approximately doubles at every cycle to create a chain reaction.

Polymerase enzyme used in PCR reaction is isolated from Thermus aquaticus, therefore it is named Taq polymerase. Taq polymerase is a thermostable enzyme that enabled the amplification reaction to be carried out by cycling the temperature within the reaction tube after mixing all the reaction components. The Taq polymerase enzyme works with high efficiency in relatively simple buffer systems. Thus, the essential components of the polymerase chain reaction include: the DNA template, two defined oligos to act as primers, the four deoxynucleutide triphosphates (dNTPS) for incorporation into the new DNA strands. Taq polymerase and magnesium, which is the cationic cofactor of the enzyme, in a Tris buffer of appropriate pH and salt concentration. The salt most frequently used is potassium chloride. A layer of light mineral oil is always placed on top of each reaction to prevent evaporation. These basic components remain largely unchanged for most application of PCR, but their concentrations and the buffering pH of the reaction vary according to the type of experiment; the native of the primers and/or template, and the experience of the experimenter.

Polymerase chain reaction is a technique widely used in molecular biology, microbiology, genetics, diagnostics, clinical laboratories, forensic science, environmental science, hereditary studies, paternity testing and many other applications.

17.2 HISTORICAL BACKGROUND

Development of the Polymerase Chain Reaction (PCR) as a basic component of the molecular biology has occurred very rapidly from its inception in 1985. Upto 1985, cloning of gene of interest into prokaryotic or eukaryotic cell with the help of vector was considered to be essential tool for getting large number of identical copies of the genes or molecular geology laboratories. Yet another powerful technique for amplifying DNA was discovered by Kary Mullis of Catus Corporation, San Fransisco, USA in 1985. Though Catus published and developed PCR according to Hoffmann La Roche, they did not invent it. PCR was described in 1973. "Science Published PCR as Major Scientific Development and Taq DNA Polymerase". Enzyme used in the technique was described as molecule of the year 1989. Since then, more than 15,500 articles have been published in which this technique was used. As PCR became more widely used, scientists rapidly learned more about it and, as a result, learned that the PCR had its strong points and its deficiencies. Very quickly, PCR demonstrated its power to amplify very small amounts (e.g., a single copy) of template nucleic acid and to amplify different nucleic acids (e.g., DNA and RNA). At the same time, laboratory personnel learned that this biochemical reaction had a unique deficiency; namely, a strong susceptibility to contamination from its own product. Early experience with the PCR soon showed that additional precautions were needed.

The name, polymerase chain reaction, comes from the DNA polymerase used to amplify (replicate many times), a piece of DNA by *in vitro* enzymatic replication. The original molecule or molecules of DNA are replicated by the DNA polymerase enzyme, thus each doubling the number of DNA molecules. Then each of these molecules is replicated in a second "cycle" of replication, resulting in four times the number of the original molecules. Again, each of the molecule is replicated in a third cycle of replication. This process is known as a "chain reaction" in which the original DNA template is exponentially amplified. With PCR it is possible to amplify a single piece of DNA or a very small number of piece of DNA over many cycle generating millions of copies of the original DNA molecule. PCR has been extensively modified to perform a wide array of genetic manipulations, diagnostic tests, and or many other uses. The polymerase chain reaction is used by a wide spectrum of scientists in an ever-increasing range of scientific disciplines.

PCR is a Nobel Prize winning technology used to amplify, or copy specific sequences of genetic material (DNA). Using this technology, scientists can produce large enough quantities of DNA to be able to examine it in detail for a variety of different purposes.

17.3 PRINCIPLES

PCR – Polymerase Chain Reaction is an *in vitro* method of enzymatic synthesis of specific DNA sequences, developed by Kary Mullis in 1983. It is a very simple and inexpensive technique for characterizing, analyzing and synthesizing any specific piece of DNA or RNA from virtually any living organism plant/animal/virus/bacteria. It exploits the natural function of the polymerases present in all living things to copy genetic material or perform "molecular photocopying".

PCR consists of three basic steps:

Denaturation: During this step, the two strands melt open to form single-stranded DNA and all enzymatic reactions stop. This is generally carried out at 90°C - 98°C.

Annealing: Annealing of primers to each original strand for new strand synthesis is carried out between 45°C - 55°C.

Extension of 72°C: The polymerase adds dNTPs complementary to the template at the 3′ and of the primers, since both strands are copied during PCR, there is an exponential increase in the number of copies of the gene.

These 3 steps are repeated 20 - 30 times in an automated thermocycler that can heat and cool the reaction mixture in tubes within a very short time. This results in exponential accumulation of specific DNA fragments, ends of which are defined by 5′ ends of the primers. The doubling of number of DNA strands corresponding to the target sequences allows us to estimate the amplification associated with each cycle using the formula.

Amplification = z', where z = number of cycles.

PCR can amplify a desired DNA sequence of any origin, hundreds of millions of times in matter of hours, a task that would require days with recombinant technology. It is especially valuable because the reaction is highly specific, easily automated and very sensitive (Fig. 17.1).

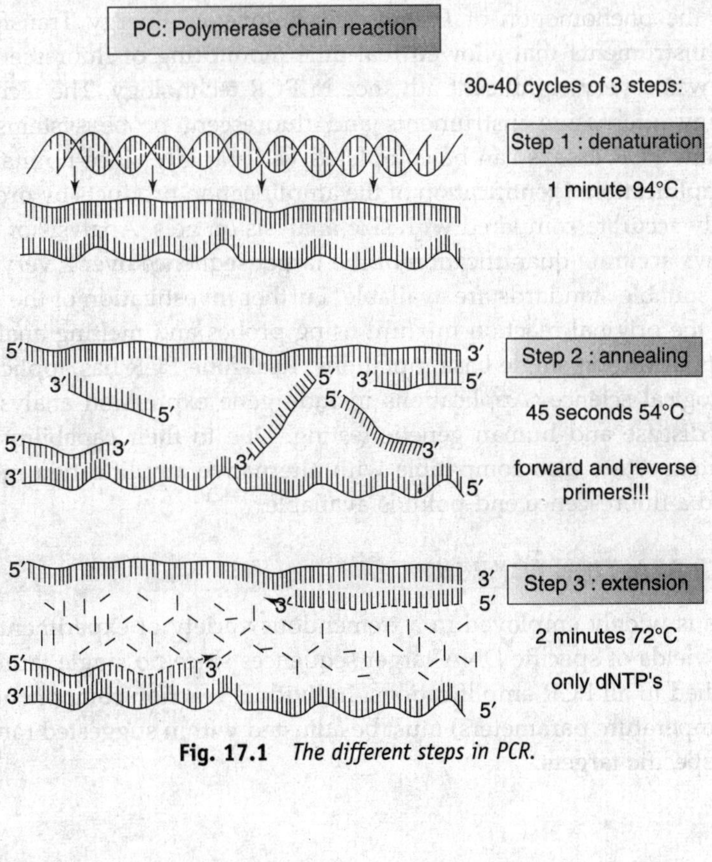

Fig. 17.1 *The different steps in PCR.*

17.4 REAL-TIME PCR

Traditionally, PCR is performed in a tube and when the reaction is complete the products of the reaction (the amplified DNA fragments) are analyzed and visualized by gel electrophoresis. However, Real-Time PCR permits the analysis of the products while the reaction is actually in progress. This is achieved by using various fluorescent dyes which react with the amplified product and can be measured by an instrument. This also facilitates the quantitation of the DNA. Not only can one tell instantly "what" DNA is present in the sample but also "how much". Quantitative PCR (Q-PCR), as this technique is known, is used to measure the quantity of a PCR product (usually in a Real-Time PCR procedure). It is the method of choice to quantitatively measure starting amounts of DNA, cDNA or RNA. PCR is therefore often used to determine whether a DNA sequence is present in a sample and the number of its copies in the sample. Another advantage of Real-Time PCR is the rapidity of the assay, since it is not necessary to perform electrophoresis or other procedure after the DNA amplification reaction.

The development of fluorescent methods for a closed tube polymerase chain reaction has greatly simplified the process of quantification. Current approaches use fluorescent probes that interact with the amplification products during the PCR to allow kinetic measurements of product accumulation. These probe methods include generic approaches to DNA quantification such as fluorescent DNA binding dyes. There are also a number of strand-specific probes that use the phenomenon of Fluorescent Resonance Energy Transfer (FRET). The development of instruments that allowed real-time monitoring of fluorescence within PCR reaction vessels was a very significant advance in PCR technology. The technology is very flexible and many alternative instruments and fluorescent probe systems are currently available. Real-time PCR assays can be completed very rapidly since no manipulations are required post-amplification. Identification of the amplification products by probe detection in real-time is highly accurate compared with size analysis on gels. Analysis of the progress of the reaction allows accurate quantification of the target sequence over a very wide dynamic range, provided suitable standards are available. Further investigation of the Real-Time PCR products within the original reaction mixture using probes and melting analysis can detect sequence variants including single base mutations. Real-time PCR has applications in many branches of biological science. Applications include gene expression analysis, the diagnosis of infectious disease and human genetic testing. Due to their capability in fluorimetry the real-time machines are also compatible with alternative amplification methods such as NASBA provided a fluorescence end-point is available.

17.5 PCR REACTION COMPONENTS

The PCR process is widely employed in a tremendous variety of experimental applications to produce high yields of specific DNA target sequences since no single target set of conditions can be applied to all PCR amplifications. Individual reaction component concentration (and time and temperature parameters) must be adjusted within suggested ranges of efficient amplification of specific targets.

17.5.1 Sample Volume and Reaction Tubes

Most PCR protocols are performed at the 25 μL – 50 μL scale in 0.2 mL or 0.5 mL microcentrifuge tubes; however, 5 μL reactions have also been successful. Larger (> 100 μL) volumes are usually not recommended, but if they are used, longer incubation times may be needed to assure adequate thermal equilibration of the reaction mixture. Reaction tubes are designed for optimal fit in each thermal cycler. Reaction tubes are used in conjunction with the heated covers on oil-free PCR. The optimized PCR vessels are available with attached caps, multicaps, strips or full plate covers. Reaction volumes (5 μL - 50 μL) are selected on the cycler at run-time so that sample temperature control is precise.

17.5.2 Vapor Barrier and Thermal Transfer Fluid

In some PCR system, the need for mineral oil as a vapor barrier and mineral oil or glycerol as a thermal transfer fluid in the wells, is eliminated since there is a tight fit between the thin walls of the reaction tubes in the block, and the heated cover of each instrument.

17.5.3 Template DNA or RNA

PCR sample may be single or double-stranded DNA or RNA. If the starting sample is RNA, total RNA, Poly (A^+) RNA, viral RNA or rRNA can be used. A reverse transcriptase (for example, MULV or rTth DNA polymerase) is used to prepare first-strand cDNA prior to conventional PCR amplification. For RNA templates with high G + C content or complete secondary structure, the high-temperature reverse transcriptase activity of thermostable rTth DNA polymerase is effective. When using DNA as the starting template, nanogram amounts of cloned templates, up to microgram amounts of genomic DNA, or up to 20,000 target copies are chosen to start optimization trials. However, even very low levels of sample (i.e. mRNA from tons of cells, DNA from single cells or individual viral genomes) may be sufficient for PCR amplification.

The quality of the template influences the outcome of the PCR. For instance, large amounts of RNA in a DNA template can chelate Mg^{2+} and reduce the yield of the PCR. Also, impure templates may contain polymerase inhibitors that decrease the efficiency of the reaction. The integrity of the template is also important. Template DNA should be of high molecular weight. To check the size and quality of the DNA run at an aliquot on an agarose gel. When testing a new template, always include a positive control with primers that amplify a product of known size and produce a good yield. The amount of template in a reaction strongly influences performance in PCR. The recommended amount of template for standard PCR is:

- a maximum of 500 ng of human genomic DNA
- 1 – 10 ng bacterial DNA
- 0.1 – 1 ng plasmid DNA

Low amounts of template, e.g., < 10 mg human genomic DNA will require specific reaction modification, such as changes include number, redesign of primers, use of "Hot start", etc.

17.5.4 Primers

Primer Design: Generally, accepted criteria for PCR primer design should be observed, as guidelines:

- Primers ideally should be no closer than 30 – 50 bp from the end of the sequence to be analyzed for mutation (e.g. intron/exon boundary)
- Primers should be 18 - 30 bp in length
- Primers should contain 45 - 60% GC
- The Tm difference between primers in a pair ideally should be less than 2°C
- Avoid primer self-complementary (e.g., hairpin dimmers)
- Avoid primer sequences that contain runs of three or more nucleotides of the same base (e.g., GGGG)
- Avoid the use of primers with non-templated toils such as universal sequencing primers, M13-or-T7-primers, etc.

In most PCR applications, it is the sequence and the concentration of the primers that determine the overall assay success. For convenience, several primer design software programs are available.

PCR Primers: PCR primers are oligonucleotides, typically 15-30 bases long hybridizing to opposite strands and flanking the region of interest in the target DNA. When choosing the PCR primers it is important that they do not contain bases complementary to themselves or with each other. Complementarity at the 3′ ends should especially be avoided to minimize the formation of an artifactual product called "primer-dimer" or "primer-oligomer". A 40% – 60% GEC content is recommended for each primer, avoiding internal secondary structure and long stretches of any one base. Also, primers should not fit on regions of secondary structure (within the target) having a higher melting point than the primer. Non-template, complementary 5′ extensions may be added to primers to allow a variety of useful post-amplification manipulations of the PCR product without significant effect on the amplification itself. These 5′ extensions can be restriction sites, promoter sequences, etc.

Optimal annealing temperatures are often higher than the Tm of the primer (approx. 5° – 10°C) and have to be determined empirically. For both primers, the Tm should be similar. Bases that do not hybridize to the template may be added at the 5′ end of the primer (e.g., for introducing restriction sites into the amplification products. Primer concentrations between 0.1 and 0.6 μM are generally optimal. Higher primer concentrations may promote this priming and accumulation of non-specific product. Lower primer concentrations may be exhausted before the reaction is completed, resulting in lower yields of desired product. For some systems, a higher primer concentration (upto 1μM may improve results. When testing new primers, always include positive control reaction with a template that has been tested for function in PCR.

17.5.5 DNA Polymerase

The choice of a DNA polymerase can profoundly affect the outcome of PCR. For most routine PCR, Taq DNA polymerase has long been the standard PCR enzyme. However, Taq DNA polymerase has its limitations. Product selection guide will provide more information on choosing the appropriate DNA polymerase. Some general guidelines are:

- To increase the fidelity of PCR, use DNA polymerase or enzyme blends (e.g., the Expand PCR systems which contain blends of Taq DNA polymerase and a thermostable polymerase with proofreading activity.
- To improve PCR yield, use the Expand High Fidelity PCR system.
- To amplify GC-rich PCR system, or the Expand Long Template PCR system with supplied buffer.
- To amplify targets upto 5 kb, use the Expand High Fidelity PCR systems, or the High Fidelity PCR Master.
- To amplify targets > 5 kb, use Expand Long Template PCR Systems or Expand 20 kb PLUS PCR System.

For most assays, the optimum amount of thermostable DNA polymerase (or a blend of polymerase) will be between 0.5 and 2.5 units/50 µL reaction volume; increased enzyme concentration sometimes lead to decreased specificity.

17.5.6 Deoxy Nucleoside Triphosphates

In the standard PCR protocol, each deoxy-nucleoside triphosphate (dNTP) concentration is 200 µM. It is important to keep the four dMTS concentrations above the estimated km of each dMTP (1 µM – 10 µM) and balanced for best base incorporation. However, deviations (from these standard recommendations) may be beneficial in certain applications. For example, when random mutagenesis of a specific target is desired, unbalanced dNTP concentrations promote a higher degree of misincorporation by Taq DNA polymerase. Lowering the dNTP and magnesium ion by an equal molar concentration can improve fidelity. If you increase the concentration of dNTPs you must also increase Mg^{2+} concentration. Increase in dNTP concentration reduces free Mg^{2+}, thus interfering with polymerase activity and decreasing primer annealing. The final dNTP concentration should be 50 – 500 µM (each dNTP).

17.5.7 pH

Generally, the pH of the reaction buffer supplied with the corresponding thermostable DNA polymerase (pH 8.3 – 9.0) will give optimal results. However, for some systems, raising the pH may stabilize the template and enhance results. Determining the optimal reaction pH can be simplified with the PCR optimization kit.

17.5.8 PCR Buffers

PCR buffers is composed of 500 mM potassium chloride and 100 mM Tris HCl (pH 8.3 at room temperature). It is supplied with a separate solution of 20 25 mM magnesium chloride for optimization titrations. Gene Amp 10 × PCR Gold Buffer is composed of 500 mM potassium chloride and 150 mM Tris-HCl (pH 8.05 at room temperature). It is supplied with a separate solution of 25 mM magnesium chloride for optimization titrations. The addition of co-solvents to these standard buffers such as DMSO, betaine, formamide or glucol may be useful when trying to qualify *G* + *C* rich target or through regions of strong secondary structure.

The recommended dilution buffer for Taq DNA polymerase is 0.15% NP – 40, 0.15% tween 20, 0.1 mM EDTA one 20 mM Tris-HCl (pH 8.3 at room temperature).

Special buffer formulations are recommended for use with stoffer fragment. rTth DNA polymerase and rTth DNA polymerase, XL. The addition of co-solvents to differing salt concentrations of buffers will have an effect on Tm of primers, thus optimization of annealing temperatures will be necessary.

17.5.9 $MgCl_2$ Concentration

$MgCl_2$ forms soluble complexes with drops to produce the actual sustrate that the polymerase recognizes. The concentration of free Mg^{2+} depends on the concentration of compounds that bind the ion, including dNTP, free pyrophosphate (ppi) and EDTA. For best results, always determine the optimal Mg^{2+} concentration empirically. The optimal Mg^{2+} concentration may vary from approximately 1mM to 5 mM. The most commonly used Mg^{2+} concentration is 1.5 mM (with dNTPs at a concentration of 200 μM each). Mg^{2+} influences enzymes activity and increases the Tm of double stranded DNA. Excess Mg^{2+} in the reaction can increase the non-specific primer binding and increase the non-specific background of the reaction.

17.5.10 Reaction Additives

In some cases, adding the following compounds can enhance the efficiency or specificity of PCR:

- Betaine (0.5 – 0 2 m)
- Bovine serum albumin (BSA 100 ng/50uL)
- Detergents
- Dimethylsulfoxide (DMSO; 2-10%) (v/v)
- Gelatine
- Glycol (1-5%) (v/v)
- Pyrophosphatase (0.001 – 0.1 units/reactions)
- Spermidine
- T4 gene 32 protein

The effect of these additives must be determined empirically.

17.6 CONTAMINATION CONTROL

A variety of approaches can be used to control PCR amplicon contamination. They can be grouped into two broad methods: (1) methods that use physical means to prevent dispersion of PCR amplicons and (2) methods that exploit some type of chemical reaction to render the amplicons incapable of serving as templates in a new round of PCR. Each of these has a place in the PCR laboratory, and most successful PCR laboratories use a spectrum of these methods to effectively control contamination.

17.6.1 Physical Methods

This category includes the physical barrier approaches. The most popular is the use of either positive displacement or barrier pipette tips to prevent aerosols. These barrier methods prevent the reintroduction of small amounts of a contaminating aerosolized sample into the next sample that is pipetted. Use of these tips is generally recommended in the pre-PCR areas of the laboratory where samples are being processed and template nucleic acids (DNA and RNA) are being isolated and purified. Use of these tips is necessary and cost-effective in the post-PCR laboratory because there is already a large amount of amplicon present.

An adjunct to these tips is the use of a laminar flow hood or biological safety cabinet to facilitate preparation of PCR samples and reagents. When they are prepared in such an enclosure, there is much less chance of an external source of PCR amplicon contaminating the samples and reagents being manipulated for the subsequent PCRs.

17.6.2 Chemical Methods

A number of chemical approaches have been developed during the last 20 years. However, only a few have seen any real success in becoming routine in their use and application for controlling PCR contamination.

- **UV photolinking:** This approach can be used in both a pre-PCR and post-PCR setting. The basis for this reaction is that adjacent pyrimidines on a DNA strand can be cross-linked when exposed to UV light of 254 nm. The reaction is very fast and can be effective for bigger amplicons; i.e., those greater than 700 bp. Smaller amplicons are harder to inactivate because there are fewer adjacent pyrimidines. There is also a role for interstrand cross-linking in UV inactivation. Once cross-linked, the pyrimidine dimers cannot be excised and so the DNA polymerase is sterically blocked, or the DNA cannot completely denature, and the synthesis reaction is effectively halted.

UV photolinking is most often used in a pre-PCR setting in which the equipment is installed in a small tabletop cabinet that is used for sample preparation. All of the items to create a PCR are placed inside the cabinet and then illuminated before the PCRs are assembled. There are some caveats with this approach, however, that are worth mentioning. First, there is a safety concern about exposure to UV light, and this must be addressed. Second, the photoreaction favors thymidine over cytidine by about 10:1.

Therefore, amplicons that are AT-rich are more efficiently disabled than AT-poor sequences. Third, decreasing length of the amplicon usually gives a lower rate of protection; therefore, short amplicons are not well controlled. The approach is nevertheless effective and should be used when possible.

• **Uracil-DNA-glycosylase:** This enzyme (also known as UDG) is very effective at destroying PCR amplicons when vigorously used for sample preparation. Briefly, at the pre-PCR step, dTTP is substituted with dUTP, and UDG is included in the reaction mix. All other reaction components remain the same. During the PCR, DNA polymerase substitutes dU for dT in the growing DNA strand. In the final product, there is now dU instead of dT in the DNA sequence. Before any new sample is processed, it is first exposed to UDG enzyme. If UDG comes across any U-containing DNA strands, the U's are cleaved, leaving the strand with gaps. Following heating in the next PCR, the basic strands fall apart and cannot be amplified. The use of UDG provides the added advantage of a hot start by degrading all PCR products made prior to the first full cycle (Fig. 17.2).

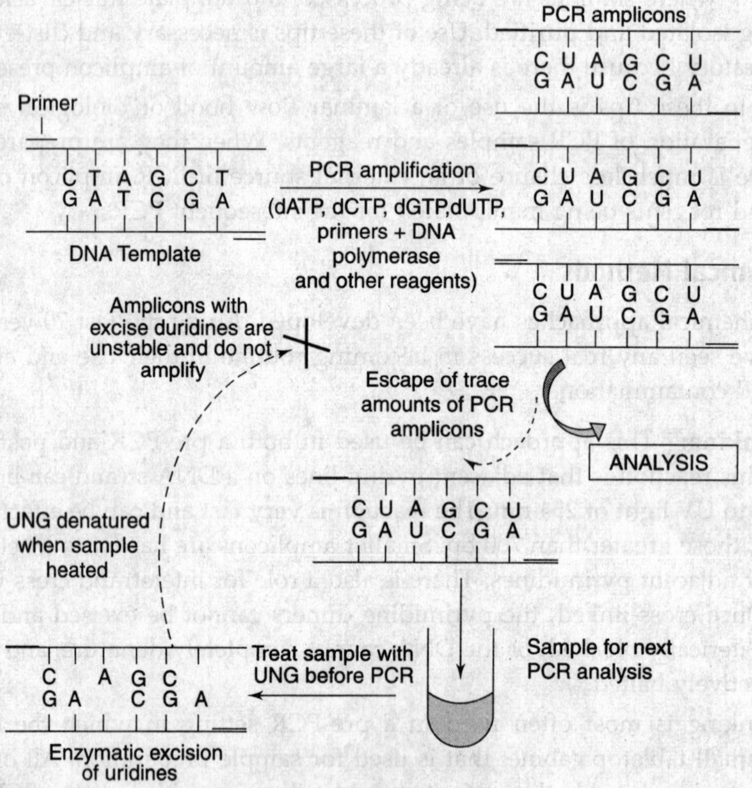

Fig. 17.2 *Use of the UDG reaction to prevent PCR contamination.*

Most PCR laboratories use one or more of these reactions in conjunction with laboratory operation and design to control PCR amplicon contamination effectively. Selection of the best combination of these techniques is driven by factors such as types of templates, their prevalence, and others.

17.7 CHARACTERIZATION OF THE PRODUCT OF PCR

After PCR reaction has been carried out it is important to assess the purity of amplified product. Contamination of the reagents by foreign DNA or annealing of primers to alternative sites in the template DNA can produce variety of unwanted DNA fragments. It is also important to know the degree of amplication achieved. Before the PCR product is used in further applications, it has to be checked if:

1. **There is a product formed**
   Though biochemistry is an exact science, not every PCR is successful. There is, for example, a possibility that the quality of the DNA is poor, that one of the primers doesn't fit, or that there is too much starting template.
2. **The product is of the right size**
   It is possible that there is a product, for example, a band of 500 bases, but the expected gene should be 1800 bases long. In that case, one of the primers probably fits on a part of the gene closer to the other primer. It is also possible that both primers fit on a totally different gene.
3. **Only one band is formed**
   As in the description above, it is possible that the primers fit on the desired locations, and also on other locations. In that case, you can have different bands in one lane on a gel.

17.7.1 Gel Electrophoresis

Assess homogeneity (purity) of the product. Multiple bands suggested primers annealing to multiple sites. Smear of DNA suggests presence of excess template DNA (which does not allow reaction to proceed in exponential phase and only long strands produced) (Fig. 17.3).

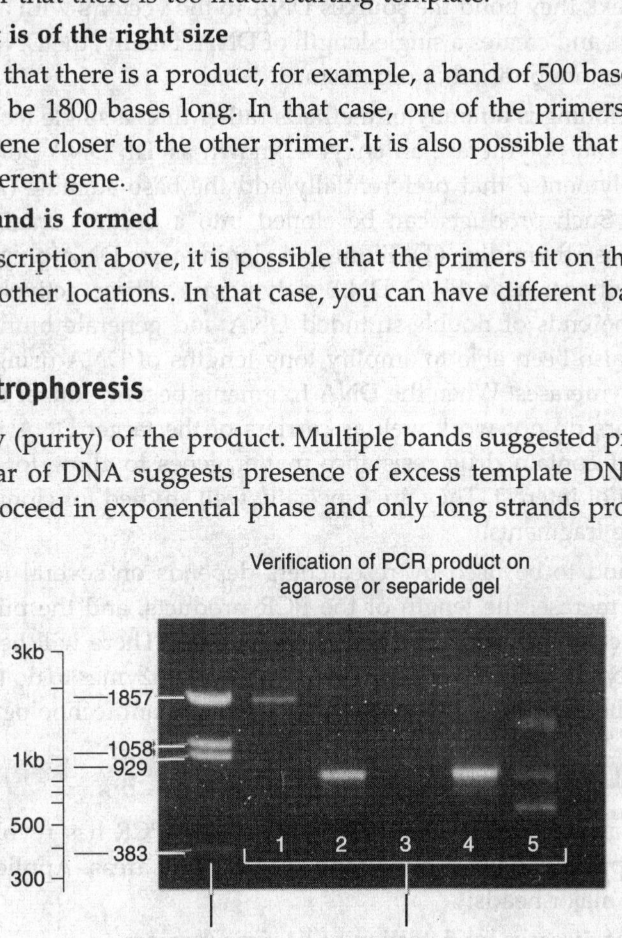

Fig. 17.3 *Verification of the PCR product on gel.*

The ladder is a mixture of fragments with known size to compare with the PCR fragments. Notice that the distance between the different fragments of the ladder is logarithmic. Lane 1:

PCR fragment is approximately 1850 bases long. Lane 2 and 4 : the fragments are approximately 800 bases long. Lane 3: no product is formed, so the PCR failed. Lane 5: multiple bands are formed because one of the primers fits on different places.

17.8 PCR – CLONING TECHNOLOGY

PCR had an immediate impact on cloning technology. It could produce large quantities of DNA that could be rapidly cloned and subsequently used to study the functions and behaviour of genes in living systems. DNA cloning involves four basic steps. Scientists first isolate the source and vector DNA and free them from contaminants. Then they use restriction enzymes to cut these two DNAs, creating ends that can connect the source DNA with the vector. Next, they bond the sources DNA to the vector's with a DNA ligase enzyme that repairs the cuts and creates a single length of DNA. Finally, the DNA is transformed into a host cell – a bacterium or another organism.

PCR-mediated cloning is a family of methods rather than a single technique – TA cloning, for example, uses Taq polymerase an enzyme known as Tth DNA polymerase, or one of a group of other polymerase that preferentially add the base adenine (A) to particular ends of PCR products. Such products can be cloned into a vector containing complementary overhangs of the base thymidine (T). Blunt-end cloning uses DNA polymerases that possess proofreading activity such as PWO DNA polymerase. These actively remove mispaired nucleotide from the ends of double-stranded DNA and generate blunt-end PCR products. Researchers have also been able to amplify long lengths of DNA using mixtures of several different DNA polymerases. When the DNA fragments become longer than 10,000 bases, the conventional vectors do not work well as carriers of the target DNA instead scientists use hybrid vectors that contain drug resistance matter genes to allow for positive selection of the DNA fragment of interest. They are especially well studied for cloning large mammalian genes or multigene fragments.

A cloning method to be used by researchers depends on several factors, including the type of DNA polymerase, the length of the PCR products, and the purpose of the cloning experiment. Now cloning method continues to emerge. "There will be a push to get away from traditional cloning methods towards other types of enzymes to do the cloning". A major development will be able to use the site-specific recombinant technology.

17.9 APPLICATIONS OF PCR

PCR technique is a vital tool in many areas of science. PCR has revolutionized molecular biology and has applications in a huge range of scientific areas. Application of PCR can be put into following major heads:

(i) Diagnosis of diseases by detection of bacteria/viruses
(ii) Diagnosis of genetic diseases
(iii) DNA fingerprinting for forensic medicine
(iv) For research purpose

(v) For molecular mapping

(vi) For preparation of molecular markets

(vii) To monitor genetic engineering and gene therapy experiments

(viii) Study of polymorphism

With PCR, detection of mycobacteria can be done in few hours but differentiation between live and dead is not possible.

Detection of pathogens in food, water and other samples can be done with PCR and specific detection is possible in spite of very low number of organisms. Sensitive and specific detection of genes for pathogenicity factors, e.g., gene for Staphyloccal enterotoxm. Listeria hemolysin gene, campylobacter, *E* coli in soil Chlamydia in throat, specimen, etc., is reported. Indian researchers under Dr. V. Sritharam at Anna University, Tamil Nadu have developed near instant diagnostic test for tuberculosis. PCR is used to identify DNA strands if any in the sample. Detection of AIDS virus itself (direct) instead of detection of antibody of AIDS (indirect) can be done with PCR.

Prenatal diagnosis of sickle cell anemia (earlier Southern blot technique with probe for hybridization was used) diagnosis of phenlyketonuria (PKU), β thalassemia, hemophilia, Tay Sach's disease. Dychennes muscular dystrophy etc. can be done with the help of PCR. Labeled probes or RFLP patterns of healthy and detective PCR products are means of diagnosis. Ensuring the absence of genetic disorder with the help of PCR can be useful while going for *in vitro* fertilization too. Choosing of right egg can be done.

PCR can be used for determination of sex of human and livestock embryos, fertilized *in vitro* before implantation. For this purpose, PCR primers and probes specific for sex hormones are used. Sex linked disorders can also be detected by this method.

Identification of criminals, disputed parentage are some of the problems which can be tackled effectively by DNA fingerprinting through PCR. From a piece of hair, blood stain, seminal fluid having partially degraded DNA, etc. characterization of individual is possible through DNA amplification via PCR.

Study of genetic profile of animals, human being which lived few thousands years earlier (mummies) is possible through use of PCR. Evolutionary and cultural lineage can be traced.

Genetic and physical chromosome maps can be prepared both in plants and animals using PCR. PCR has also been used for developing molecular markers closely linked to specific genes of economic importance. PCR may be used for plant breeding by selection of desirable plants in segregating populations and also for isolation of genes. PCR may also be used for amplification of gene that has been cloned in cells or organisms (via transfer by vectors). PCR will prove extremely useful to monitor a gene in genetic engineering or gene therapy experiments.

One of the important applications of PCR is to study DNA polymorphism. Instead of amplifying a specific DNA fragment it is possible to use PCR to generate arbitrary products, which can then be used to obtain DNA fingerprints of given isolate. Because of the polymorphism in the DNA of bacterial isolates they can be distinguished on the basis of fingerprints.

This method is known as Random Amplification of Polymorphic DNA (RAPD). RAPD is more simple and efficient than RFLP. It has been used for study of polymorphism and chromosome mapping in bacteria, maize, soybean, mouse humans, etc.

In future automation, better reproducibility, less contamination, more accuracy, addition of mixture of primers, sorting out the product for different genetic disorders, simultaneous identification of pathogens from same sample, etc. will be some of the goals to be achieved from PCR.

The human genome project and the commercial Sequencing effort led by Celera genomics have made significant progress in determining DNA sequences of humans. PCR technology promises advances in human genetics. Fields of science beyond traditional molecular biology laboratories have benefited from PCR. PCR has become a useful tool on many scientific disciplines.

Cloning technology has also enabled the routine study of gene function. IT is now a relatively simple process to isolate large DNA fragments that contain genes and then to express the genes in transgenic hosts. In addition, the RT-PCR technique permits scientists to determine the genes responsible for producing very low-level of messenger RNA that may play an important role in cellular metabolism and the disease process. The tools of PCR and cloning are geared toward understanding, treating and preventing the diseases that affect the quality of life.

17.10 REFERENCES

1. Gordon L.K. and Haseltine W.A. 1982. Quantitation of cyclobutane pyrimidine dimer formation in double-and single-stranded DNA fragments of defined sequence. Radiat. Res. 89: 99–112.
2. Kwok S. and Higuchi R. 1989. Avoiding false positives with PCR (erratum Nature [1989] 339: 490). Nature 339: 237–238.
3. Lo Y.M., Mehal W.Z., and Fleming K.A. 1988. False-positive results and the polymerase chain reaction. Lancet 2: 679.
4. Longo M.C., Berninger M.S., and Hartley J.L. 1990. Use of uracil DNA glycosylase to control carryover contamination in polymerase chain reactions. Gene 93: 125–128.
5. Mullis K.B. 1987. Process for amplifying nucleic acid sequences. U.S. Patent #4,683,202.
6. Mullis K.B., Erlich H.A., Arnheim N., Horn G.T., Saiki R.K., and Scharf S.J. 1987. Process for amplifying, detecting, and/or cloning nucleic acid sequences. U.S. Patent #4,683,195.
7. Newton C.R. 1995. Setting up a PCR laboratory. In PCR: Essential data. (Ed. C.R. Newton), p. 216. Wiley, New York.

8. Pellett P.E., Spira T.J., Bagasra O., Boshoff C., Corey L., de Lellis L., Huang M.L., Lin J.C., Matthews S., Monini P., Rimessi P., Sosa C., Wood C., and Stewart J.A. 1999. Multicenter comparison of PCR assays for detection of human herpesvirus 8 DNA in semen. J. Clin. Microbiol. 37: 1298–1301.
9. Scherczinger C.A., Ladd C., Bourke M.T., Adamowicz M.S., Johannes P.M., Scherczinger R., Beesley T., and Lee H.C. 1999. A systematic analysis of PCR contamination. J. Forensic Sci. 44: 1042–1045.
10. Thornton C.G., Hartley J.L., and Rashtchian A. 1992. Utilizing uracil DNA glycosylase to control carryover contamination in PCR: Characterization of residual UDG activity following thermal cycling. Biotechniques 13: 180–184.
11. Allen, R., et al. 1991. Analysis of the VNTR locus AmpliFLP D1S80 by the PCR followed by high resolution PAGE. American Journal of Human Genetics 48:137.
12. Chamberlain, J.S., et al. 1990. Multiplex PCR for the diagnosis of Duchenne muscular dystrophy. In PCR Protocols: A Guide to Methods and Applications (M.A. Innis, et al., eds.). Academic Press, New York, pp. 272–281.
13. Micklos, D.A. and G.A. Freyer. 1990. DNA Science: A First Course in Recombinant DNA Technology. Carolina Biological Supply Company, Burlington, NC and Cold Spring Harbor Laboratory Press, Cold Spring Harbor, NY.
14. Mullis, K.B. 1990. The unusual origin of the polymerase chain reaction. Scientific American 262:56-65.
15. Paabo, S., R.G. Higuchi, and A.C. Wilson. 1989. Ancient DNA and the polymerase chain reaction. Journal of Biological Chemistry 264:9709-9712.
16. Saiki, R.K., et al. 1985. Enzymatic amplification of beta-globin sequences and restriction site analysis for diagnosis of sickle cell anemia. Science 230:1350-1354.

CHAPTER

18 Biosensors

18.1 INTRODUCTION

A biosensor is an analytical device which converts a biological response into an electrical signal. The term "biosensor" is often used to cover sensor devices used in order to determine the concentration of substances and other parameters of biological interest even where they do not utilize a biological system directly. Biosensors represent a rapidly expanding field, at the present time, with an estimated 60% annual growth rate; the major impetus coming from the health-care industry (e.g., 6% of the western world are diabetic and would benefit from the availability of a rapid, accurate and simple biosensor for glucose) such as food quality appraisal and environmental monitoring.

Biosensors use a combination of biological receptor compounds (antibody, enzyme, nucleic acid, etc.) and the physical or physio-chemical transducer directing, in most cases, "real-time" observation of a specific biological event (e.g., antibody antigen interaction). They allow the detection of a broad spectrum analytes in complex sample matrices, and have shown great promise in areas such as clinical diagnostic, food analysis, bioprocess and environmental monitoring. Biosensors may be divided into six basic groups, depending on the method of signal transduction; optical mass, electrochemical, magnetic micromechanical and thermal sensors.

Research and development in this field is wide and multidisciplinary, spanning biochemistry, bioreactor science, physical chemistry, electrochemistry, electronics and software engineering.

18.2 PRINCIPLE OF BIOSENSORS

A biosensor is a sensor that is based on the use of biological material for its sensing function. The biocomponent specially reacts or interacts with the analyte of interest resulting in a detectable chemical or physical change. A biosensor is an analytical device which converts a biological response into an electrical signal. The term "biosensor" is often used to cover sensor devices used in order to determine the concentration of substance and other parameters of biological interest even where they do not utilize a biological system directly.

18.3 HISTORY

Many biosensors (Cass, 1990; Suzuki, 1990; Buerk, 1993) have been constructed since the first report describing development of the first enzyme sensor for glucose measurement by Clark in 1962 (Buerk, 1993). This biosensor measured the product of an enzyme reaction using an electrode, which was a remarkable achievement even though the enzyme was not immobilized on the electrode.

Today's basic biosensors originated from the investigation of Updike and Hicks in 1967 (Buerk, 1993). Their sensor combined membrane-immobilized glucose oxidase with an oxygen electrode and oxygen measurements were carried out before and after the enzyme reaction. Many other biological elements and transducers have been examined since this first glucose biosensor was developed. For example, in 1977 Karube reported the first microbial sensor which used the whole cells as a biological sensing element. Many immunosensors using antibodies were fabricated in 1980s and recently DNA, RNA and even artificial recognition elements have been employed. Transducers have also been improved, and novel transducer elements such as optical detectors, including surface plasma resonance (SPR) detectors are widely used in modern biosensor.

18.4 WORKING OF BIOSENSORS

A biosensor is a detection system composed of biological sensing elements, such as enzymes, antidotes, microorganisms, and DNA, and an electronic signal transducing element (Fig. 18.1). A transducer can convert a change in the concentration of a product of a biological reaction into an electronic signal. For example, an oxygen electrode converts a change in the concentration of a sample caused by enzymes or biodegradation into a change in electric current. The transducers used in biosensors include electrodes, piezoelectric quartz, crystals, and optics. Many biosensors can be applied to important fields such as diagnosis and environmental monitoring.

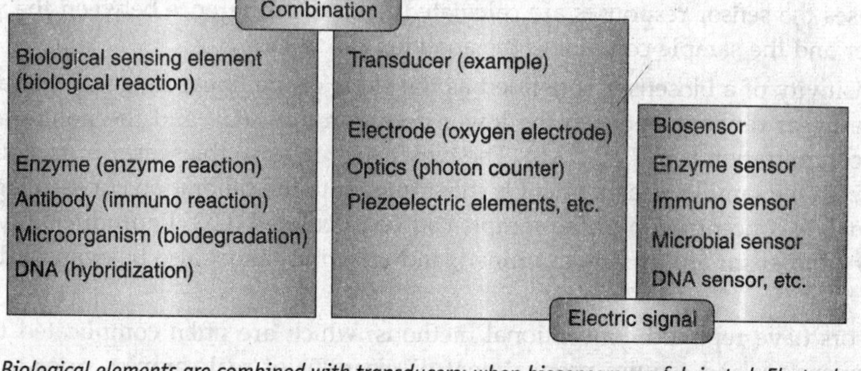

Fig. 18.1 *Biological elements are combined with transducers: when biosensors are fabricated. Electrodes and optical devices are widely used as transducers in biosensors. A piezoelectric device such as a quartz crystal detects a weight change before and after the biological reaction as applied to immuno and DNA sensors. Transducers convert a change in the concentration of the compound into an electronic signal.*

18.4.1 Working Principle of Glucose Sensor

The principle of a simple glucose sensor using Glucose Oxidase (GOD) is shown in Fig. 18.2. The target analyte, glucose is a substrate of GOD. Glucose diffuses into membrane and a biological reaction occurs within the membrane. When GOD catalyzes the glucose oxidation, dissolved oxygen in the sample is consumed and gluconic lactone and hydrogen peroxide are produced. The oxygen electrode detects a change in dissolved oxygen concentration of the sample as a change in electric current. Since the change in dissolved oxygen concentration is a result of the glucose oxidation, the glucose concentration of the sample can be measured. The biosensor does not directly detect the target analyte (glucose). Instead, it measures the change in the concentration of a co-reactant (oxygen) or a co-product (hydrogen peroxide) of the reaction catalyzed by the immobilized biological-sensing material (i.e., GOD).

Biosensor configurations are categorized into two types; (a) batch type and (b) flow type as depicted in Fig. 18.3. In flow-type sensors, a bioreactor, a column stuffed with biomaterial-immobilized beads, can be incorporated separately from a transducer. Flow-type biosensors are very useful for continuous monitoring of target analytes. Immobilization of biomaterials is usually required for biosensor fabrication. Typical immobilization carriers are glass, alginate and artificial resin beads, and membranes.

Disposable biosensors such as amperometric glucose sensors for medical diagnosis are examples of a batch-type sensor. Flow injection analysis (FIA) is a technique sometimes used in flow-type biosensors. Because biosensors using FIA yield very rapid and accurate measurements, these have been applied to various fields, such as environmental monitoring and food quality control.

In a typical biosensor study, parameters such as the amount of immobilized biological-sensing elements (e.g., enzymes), pH, and temperature of the buffer are optimized, and then a calibration curve is generated using standard solutions containing known concentrations of the analyte. Figure 18.4 shows a typical calibration curve of a biosensor. Calibration curves are normally obtained by one of the two methods: from reaction curves of the biological co-reaction measurement (Fig. 18.4a) or the biological co-product measurement (Fig. 18.4b). In both cases the sensor responses are calculated from the difference between the values of pure buffer and the sample containing the analyte.

The sensitivity of a biosensor is defined as the slope of the linear range of the calibration curve. The linear range is between the lower detection limit (Cl) and the nonlinear profile at higher concentration (C6, 7, and - 8). The biosensor response (biosensor output signal Rs) obtained from the sample measurement is substituted into the calibration curve in Fig. 18.4 (c) and the analyte concentration of the sample can be calculated (Cx). Reproducibility and the lifetime of a biosensor are usually examined, and selectivity must also be evaluated in some cases.

Biosensors have replaced conventional methods, which are often complicated time-consuming expensive, and require pretreatment or clean-up of real samples prior to analysis Biosensors generally have the following advantages compared to other analytical methods:

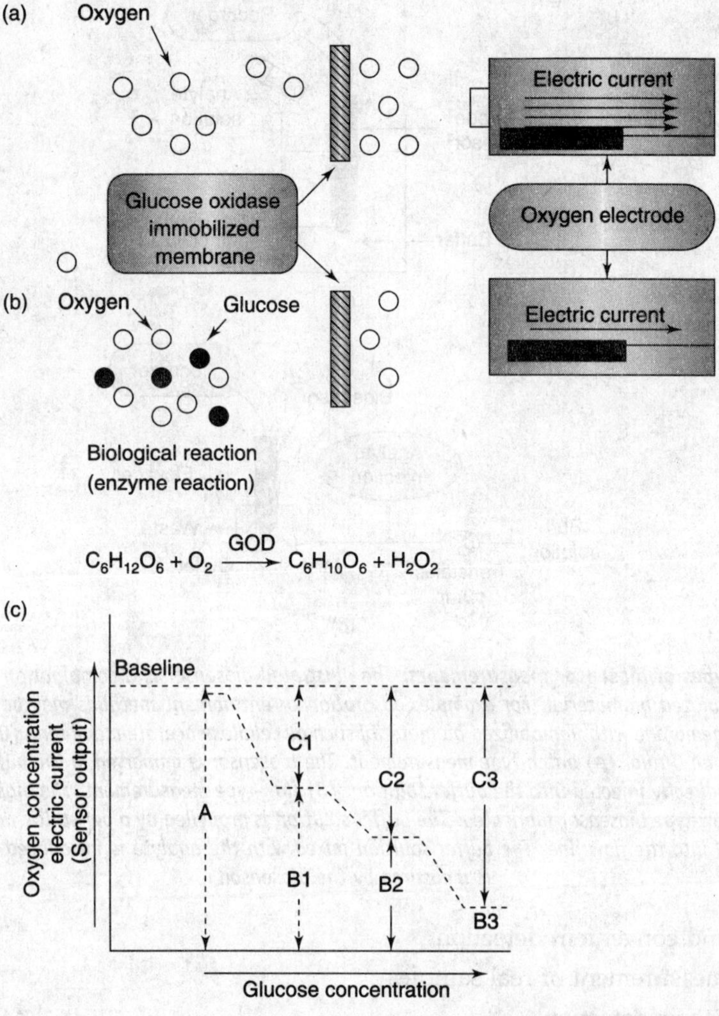

Fig. 18.2 *Glucose sensor-principle: the analyte is glucose, and the biological-sensing element of the biosensors is glucose oxidase (GOD). The membrane-immobilized GOD is combined with an oxygen electrode as a transducer. An oxygen electrode generates electric current depending on the dissolved oxygen concentration. Glucose and oxygen permeate into the membrane-immobilized GOD. (a) Dissolved oxygen is not consumed when a sample which does not contain glucose such as pure buffer solution is measured by the glucose sensor. The sample is air-saturated and the oxygen electrode produces stable high current. This value is defined as the baseline current of the sensor. (b) A sample containing glucose is measured by the glucose sensor. GOD catalyzes the glucose oxidation using oxygen and dissolved oxygen is consumed. The oxygen concentration is lower than that in (a) and the electric current decreases from the baseline current. (c) Typical reaction curve of the sensor is illustrated. A, the baseline current: b1-b3, the electric currents obtained from samples containing glucose. The differences between a and b1-b3 are the sensor responses (c1-c3). As glucose concentration increases, the sensor response will also increase.*

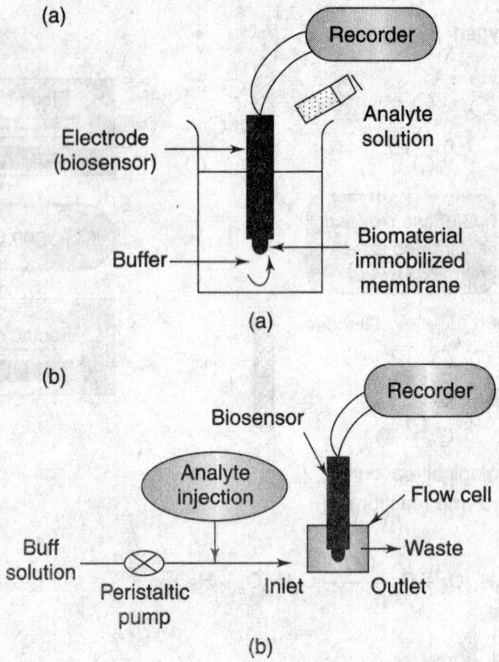

Fig. 18.3 ***Two types of biosensor measurements:*** *The illustrated biosensor is a combination of an electrode and a membrane-immobilized biomaterial. For example, adsorption or entrapment methods may be used to immobilize biomaterials. The membrane with immobilized biomaterial such as cellulose acetate membrane (0.45 μm) is attached to the electrode by an O-ring. (A) batch-type measurement. The biosensor is immersed in the buffer solution and the analyte solution is directly injected into the buffer solution. (B) flow-type measurement. The biosensor is attached to a flow cell and a flow-type biosensor fabricated. The buffer solution is propelled by a peristaltic pump and the analyte solution is injected into the flow line. The buffer solution mixed with the analyte is introduced to the flow cell and is measured by the biosensor.*

1. Rapid and convenient detection
2. Direct measurement of real samples
3. Very specific detection

On the other hand, stability and reproducibility have been problematic for biosensors due to the inherent instability of biomaterials used as sensing elements. Although biosensors have disadvantages numerous investigations on biosensors to date have helped overcome at least some of these difficulties, allowing practical application of many biosensors in the real world.

18.5 BENEFITS OF BIOSENSOR

A successful biosensor must possess at least some of the following beneficial features:

1. The biocatalyst must be highly specific for the purpose of the analyses, be stable under normal storage conditions and, except in the case of colorimetric enzyme strips

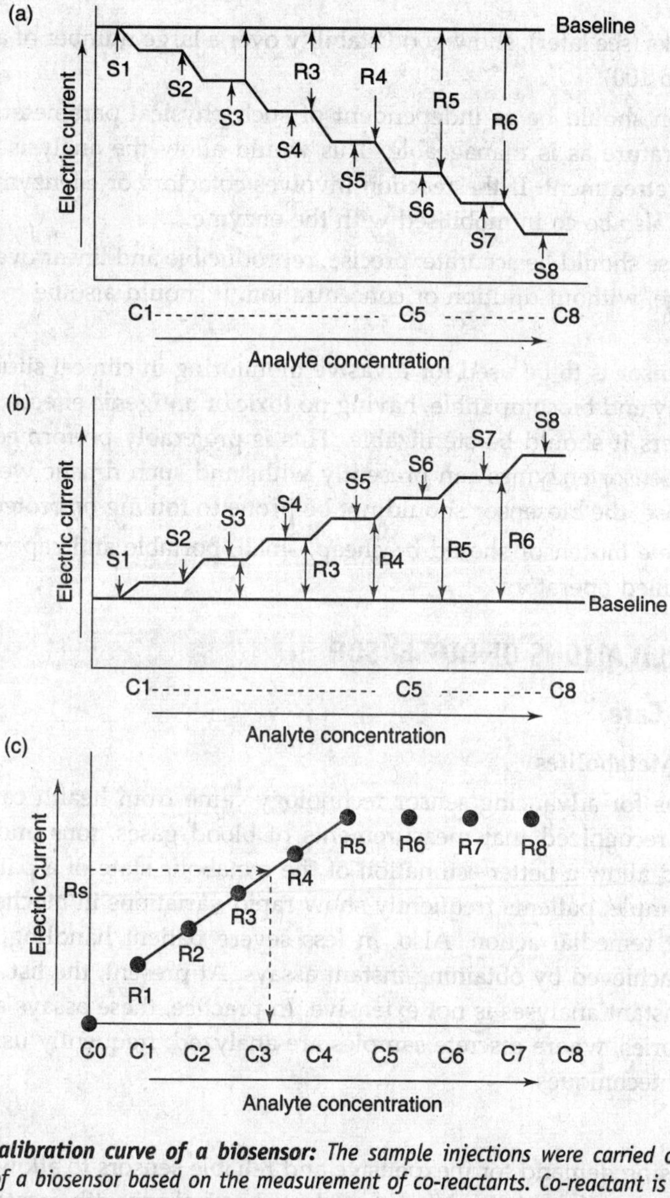

Fig. 18.4 ***Typical calibration curve of a biosensor:*** *The sample injections were carried out eight times (sn). (A) a reaction curve of a biosensor based on the measurement of co-reactants. Co-reactant is consumed during the measurement and the sensor output such as electric current in the figure decreases as the analyte concentration increases. The baseline current is higher than that of the sample containing the analyte. The sensor response (rn) is calculated by subtracting the measurement value (e.g., the electric current) of the sample containing the analyte from the baseline value. (b) A reaction curve of a biosensor based on the measurement of co-products. Co-product or sensor output increases as the analyte concentration increases. The baseline value is less than that of the sample containing the anlyte. The sensor response (rn) is obtained by subtracting the baseline value from the measurement value of the sample containing the anlyte. (c) A calibration curve. The dynamic range of a biosensor is the linear correlation from the detection limit c1 to c5. The sensitivity of the sensor, for example, is calculated as (r4 – r2)/ c4 – c2). Rs is obtained when an unknown concentration sample is measured by the biosensor and it is substituted into the linear correlation and the sample concentration cs is obtained.*

and dipsticks (see later), show good stability over a large number of assays (i.e., much greater than 100).

2. The reaction should be as independent of such physical parameters as stirring, pH and temperature as is manageable. This would allow the analysis of samples with minimal pretreatment. If the reaction involves cofactors or coenzymes these should, preferably, also be co-immobilised with the enzyme.
3. The response should be accurate, precise, reproducible and linear over the useful analytical range, without dilution or concentration. It should also be free from electrical noise.
4. If the biosensor is to be used for invasive monitoring in clinical situations, the probe must be tiny and biocompatible, having no toxic or antigenic effects. If it is to be used in fermenters it should be sterilizable. This is preferably performed by autoclaving but no biosensor enzymes can presently withstand such drastic wet-heat treatment. In either case, the biosensor should not be prone to fouling or proteolysis.
5. The complete biosensor should be cheap, small, portable and capable of being used by semi-skilled operators.

18.6 APPLICATIONS OF BIOSENSOR

18.6.1 Health Care

Measurement of Metabolites

The initial impetus for advancing sensor technology came from health care area, where it is now generally recognized that measurements of blood gases, ions and metabolites are often essential and allow a better estimation of the metabolic state of a patient. In intensive care units, for example, patients frequently show rapid variations in biochemical levels that require an urgent remedial action. Also, in less severe patient handling, more successful treatment can be achieved by obtaining instant assays. At present, the list of the most commonly required instant analyses is not extensive. In practice, these assays are performed by analytical laboratories, where discrete samples are analyzed, frequently using the more traditional analytical techniques.

Market Potential

There is an increasing demand for inexpensive and reliable sensors to allow not only routine monitoring in the central or satellite laboratory, but also analysis with greater patient contact, such as in the hospital ward, emergency rooms, and operating rooms. Ultimately, patients themselves should be able to use biosensors in the monitoring and control of some treatable condition, such as diabetes. It is probably true to say that the major biosensor market may be found where an immediate assay is required. If the costs of laboratory maintenance are counted with the direct analytical costs, then low-cost biosensor devices are desirable in the whole spectrum of analytical applications from hospital to home.

Diabetes

The 'classic' and most widely explored example of closed-loop drug control is probably to be found in the development of an artificial pancreas. Diabetic patients have a relative or absolute lack of insulin, a polypeptide hormone produced by the beta-cells of the pancreas, which is essential to the metabolism of a number of carbon sources. This deficiency causes various metabolic abnormalities, including higher than normal blood glucose levels. For such patients, insulin must be supplied externally. This has usually been achieved by subcutaneous injection, but fine control is difficult and hyperglycaemia cannot be totally avoided, or even hypoglycaemia is sometimes induced, causing impaired consciousness and the serious long-term complications to tissue associated with this intermittent low glucose condition.

Insulin Therapy

Better methods for the treatment of insulin-dependent diabetes have been sought and infusion systems for continuous insulin delivery have been developed. However, regardless of the method of insulin therapy, its induction must be made in response to information on the current blood glucose levels in the patient. Three schemes are possible, the first two dependent on discrete manual glucose measurement and the third a 'closed-loop' system, where insulin delivery is controlled by the output of a glucose sensor which is integrated with the insulin infuser. In the former case, glucose has been estimated on 'finger-prick' blood samples with a colorimetric test strip or more recently with an amperometric 'pen'-size biosensor device by the patient themselves. Obviously, these diagnostic kits must be easily portable, very simple to use and require the minimum of expert interpretation. However, even with the ability to monitor current glucose levels, intensive conventional insulin therapy requires multiple daily injections and is unable to anticipate future states between each application, where diet and exercise may require modification of the insulin dose. For example, it was shown that administration of glucose by subcutaneous injection, 60 min before a meal provides the best glucose/insulin management.

Artificial Pancreas

The introduction of a closed-loop system, where integrated glucose measurements provide feedback control on a pre-programmed insulin administration based on habitual requirement, would therefore relieve the patient of frequent assay requirements and perhaps more desirably frequent injections. Ultimately, the closed-loop system becomes an artificial pancreas, where the glycaemic control is achieved through an implantable glucose sensor.

18.6.2 Industrial Process Control

Bioreactor Control

Real-time monitoring of carbon sources, dissolved gases in fermentation processes could lead to optimization of the procedure giving increased yields at decreased materials cost. While as real-time monitoring with feedback control involving automated systems does exist, currently only a few common variables are measured on-line (e.g., pH, temperature, CO_2, O_2)) which are often only indirectly related with the process under control.

Three different methods of controlling a bioreactor are:

1. Off-line distant: central laboratory coarse control with significant time lapse
2. Off-line local: fine control with short time lapse
3. On-line: real-time monitoring and control

On-Line Control

Method 3 is most desirable, which allows the process to follow an ideal pre-programmed fermentation profile to give maximum output. However, many problems exist with on-line measurements including *in situ* sterilization, sensor lifetime, sensor fouling, etc. Some of the problems can be overcome if the sensor is situated so that the sample is run to waste, but this causes a volume loss, which can be particularly critical with small volume fermentations.

Off-Line Control

Although Method 3 may be the ultimate aim, considerable advantage can be gained in moving from Method 1 to Method 2 giving a rapid analysis and thus enabling finer control of the fermentation. The demands of the sensor are perhaps not as stringent in Method 2 as in Method 3.

Benefits of Control

The benefits which are achievable with process-control technology are considerable:

- Improved product quality; reduction in rejection rate following manufacture
- Increased product yield; process tuned in real time to maintain optimum conditions throughout and not just for limited periods
- Increased tolerance in quality variation of some raw materials. These variations can be compensated in the process-control management
- Reduced reliance on human 'seventh sense' to control process
- Improved plant performance-processing rate and line speed automated, so no unnecessary dead-time allocated to plant
- Optimized energy efficiency
- The use of biosensors in industrial process in general could facilitate plant automation, cut analysis costs and improve quality control of the product.

18.6.3 Environmental Monitoring

Air and water monitoring

Another assay situation which may involve a considerable degree of the unknown is that of environmental monitoring. The primary measurement media here will be water or air, but the variety of target analyst is vast. At sites of potential pollution, such as in factory effluent, it would be desirable to install on-line real-time monitoring and alarm, targeted at specific analytes, but in many cases random or discrete monitoring of both target species and general hazardous compounds would be difficult. The possible analytes include Biological Oxygen Demand (BOD) which provides a good indication of pollution, atmospheric acidity and river water pH, detergent, herbicides and fertilizers (organophosphates, nitrates, etc.). The survey of market potential has identified the increasing significance of this area and this is now sustained by a strong interest from industry.

18.7 REFERENCES

1. Compendium of Chemical Terminology, 2nd edn. (the "Gold Book") (1997). Online corrected version: (2006-) "biosensor". (http://goldbook.iupac.org/B00663.html)
2. Cavalcanti A, Shirinzadeh B, Zhang M, Kretly LC (2008). "Nanorobot Hardware Architecture for Medical Defense". (http://www.mdpi.org/sensors/papers/s8052932.pdf) Sensors **8** (5): 2932–2958. doi:10.3390/s8052932. http://dx.doi.org/10.3390/s8052932 http://www.mdpi.org/sensors/papers/s8052932.pdf.
3. "Wasp Hound". (http://www.sciencentral.com/articles/view.php3? article_id =218392717) Science Central. http://www.sciencentral.com/articles/view.php3?article_id=218392717. Retrieved 23 February 2011.
4. H. M. Hiep et al. "A localized surface plasmon resonance based immunosensor for the detection of casein in milk" Sci. Technol. Adv. Mater. 8 (2007) 331 free download (http://dx.doi.org/10.1016/j.stam.2006.12.010)
5. Gerald A Urban 2009 Meas. Sci. Technol. 20 012001 doi: 10.1088/0957-0233/20/1/012001
6. Fan, F. et al. (2008) Novel Genetically Encoded Biosensors Using Firefly Luciferase. ACS Chem. Biol. 3, 346–51. free download (http://pubs.acs.org/doi/abs/10.1021/cb8000414)
7. S.Q. Lud, M.G. Nikolaides, I. Haase, M. Fischer and A.R. Bausch (2006)."Field Effect of Screened Charges: Electrical Detection of Peptides and Proteins by a Thin Film Resistor" Chem.Phys.Chem. 7(2), 379-384 [1] (http://dx.doi.org/10.1002/cphc.200500484)
8. Vockenroth I, Atanasova P, Knoll W, Jenkins A, Köper I (2005). "Functional tethered bilayer membranes as a biosensor platform". IEEE Sensors 2005 - The 4-th IEEE Conference on Sensors: 608–610.
9. Cornell BA, BraachMaksvytis VLB, King LG, et al. (1997). "A biosensor that uses ion-channel switches". Nature **387** (6633): 580–583.
10. Oh S, Cornell B, Smith D, et al. (2008). "Rapid detection of influenza A virus in clinical samples using an ion channel switch biosensor". Biosensors & Bioelectronics **23** (7): 1161–1165.
11. Krishnamurthy V, Monfared S, Cornell B (2010). ": Ion Channel Biosensors Part I Construction Operation and Clinical Studies". IEEE Transactions on Nanotechnology **9** (3): 313–322.
12. Pohanka M, Skladal P, Kroca M (2007)."Biosensors for biological warfare agent detection". Def. Sci. J. **57**(3):185-93.

13. Pohanka M, Jun D, Kuca K (2007)."Mycotoxin assay using biosensor technology: a review. Drug Chem. Toxicol. **30**(3):253-61.
14. Nanomedicine and its potential in diabetes research and practice. Pickup JC, Zhi ZL, Khan F, Saxl T, Birch DJ. Diabetes Metab Res Rev. 2008 Nov-Dec;24(8):604-10.
15. Entrapment of biomolecules in sol-gel matrix for applications in biosensors: problems and future prospects. Gupta R, Chaudhury NK. Biosens Bioelectron. 2007 May 15;22(11):2387-99.
16. Clark HA, Kopelman R, Tjalkens R, Philbert MA. Optical nanosensors for chemical analysis inside single living cells. 2. Sensors for pH and calcium and the intracellular application of PEBBLE sensors. Anal Chem. 1999 Nov 1;71(21):4837-43.
17. Percutaneous fiber-optic sensor for chronic glucose monitoring *in vivo*. Liao KC, Hogen-Esch T, Richmond FJ, Marcu L, Clifton W, Loeb GE. Biosens Bioelectron. 2008 May 15;23(10):1458-65.
18. http://www.technologyreview.com/biomedicine/21603/?a=f
19. What are biosensors http://www.lsbu.ac.uk/biology/enztech/biosensors.html
20. Biosensor Applications - Drug and explosives detection products from Sweden (http://www.biosensor.se)
21. Scratching at the surface of biosensors – (http://www.rsc.org/Publishing/ChemTech/Volume/2009/02/ biosensors.asp) an *Instant Insight (http://www.rsc.org/Publishing/ChemTech/Instant_insights.asp) discussing how surface chemistry lets porous silicon biosensors fulfil their promise from the Royal Society of Chemistry
22. Biosensor Forum - Social network for researchers and organizations (http://www.biosensorforum.com)

CHAPTER

19 Microscopy

19.1 INTRODUCTION

Microscopy involves use of microscopes for magnifying samples or objects which cannot be seen with the naked eye. The goal is to render these objects visible for research, allowing researchers to study these objects or samples, their structure, composition and how they work. There are a number of different types of microscopes.

Light Microscopy, Bright Field Microscopy, Polarized Light Microscopy, Fluorescence Microscopy, Electron Microscopy, Atomic Force Microscopy, etc. Microscopy which has served as a fundamental technique for centuries remains an invaluable tool in chemistry, biology, health care and forensics. Increasingly, it is being integrated into modern chemical instrumentation and is of value as a powerful tool across scientific disciplines.

Of all the techniques used in biology, microscopy is probably the most important. The vast majority of living organisms are too small to be seen in any detail with the human eye, and cells and their organelles can only be seen with the aid of a microscope. Cells were first seen in 1665 by Robert Hooke (who named them after monks' cells in a monastery), and were studied in more detail by Leeuwenhoek using a primitive microscope.

Units of measurement

metre	m	= 1 m
millimeter	mm	$= 10^{-3}$ m
micrometer	um	$= 10^{-6}$ m
nanometer	nm	$= 10^{-9}$ m
picometer	pm	$= 10^{-12}$ m
angstrom	A	$= 10^{-10}$ m (obsolete)

magnification and resolution

By using more lenses microscopes can magnify by a larger amount, but this doesn't always mean that more detail can be seen. The amount of detail depends on the resolving power of a microscope, which is the smallest separation at which two separate objects can be distinguished (or resolved).

The resolving power of a microscope is ultimately limited by the wavelength of light (400-600 nm for visible light). To improve the resolving power a shorter wavelength of light is needed, and sometimes microscopes have blue filters for this purpose (because blue has the shortest wavelength of visible light).

Magnification is how much bigger a sample appears to be under the microscope than it is in real life.

$$\text{Overall magnification} = \text{Objective lens} \times \text{Eyepiece lens}$$

Resolution is the ability to distinguish between two points on an image, i.e., the amount of detail.

- The resolution of an image is limited by the wavelength of radiation used to view the sample.
- This is because when objects in the specimen are much smaller than the wavelength of the radiation being used, they do not interrupt the waves, and so are not detected.
- The wavelength of light is much larger than the wavelength of electrons, so the resolution of the light microscope is a lot lower.
- Using a microscope with a more powerful magnification will **not** increase this resolution any further. It will increase the size of the image, but objects closer than 200 nm will still only be seen as one point.

19.2 HISTORY OF THE MICROSCOPE

During that historic period known as the Renaissance, after the "dark" Middle Ages, there occurred the inventions of printing, gunpowder and the mariner's compass, followed by the discovery of America. Equally remarkable was the invention of the light microscope: an instrument that enables the human eye, by means of a lens or combination of lenses, to observe enlarged images of tiny objects. It made visible the fascinating details of worlds within worlds.

19.2.1 Invention of Glass Lenses

Long before, in the hazy unrecorded past, someone picked up a piece of transparent crystal thicker in the middle than at the edges, looked through it, and discovered that it made things look larger. Someone also found that such a crystal would focus the sun's rays and set fire to a piece of parchment or cloth. Magnifiers and "burning glasses" or "magnifying glasses" are mentioned in the writings of Seneca and Pliny the Elder, Roman philosophers during the first century A.D., but apparently they were not used much until the invention of spectacles, toward the end of the 13th century. They were named lenses because they are shaped like the seeds of a lentil. The earliest simple microscope was merely a tube with a plate for the object at one end and, at the other, a lens which gave a magnification less than ten diameters – ten times the actual size. These excited general wonder when used to view fleas or tiny creeping things and so were dubbed "flea glasses."

19.2.2 Birth of the Light Microscope

About 1590, two Dutch spectacle makers, Zaccharias Janssen and his son Hans, while experimenting with several lenses in a tube, discovered that nearby objects appeared greatly enlarged. That was the forerunner of the compound microscope and of the telescope. In 1609, Galileo, father of modern physics and astronomy, heard of these early experiments, worked out the principles of lenses, and made a much better instrument with a focusing device. The father of microscopy, Anton van Leeuwenhoek of Holland, started as an apprentice in a dry goods store where magnifying glasses were used to count the threads in cloth. He taught himself new methods for grinding and polishing tiny lenses of great curvature which gave magnification up to 270 diameters, the finest known at that time. These led to the building of his microscopes and the biological discoveries for which he is famous. He was the first to see and describe bacteria, yeast plants, the teeming life in a drop of water, and the circulation of blood corpuscles in capillaries. He used his lenses to make pioneer studies on an extraordinary variety of things, both living and non-living, and reported his findings in over a hundred letters to the Royal Society of England and the French Academy. Robert Hooke, the English father of microscopy, re-confirmed Anton van Leeuwenhoek's discoveries of the existence of tiny living organisms in a drop of water. Hooke made a copy of Leeuwenhoek's light microscope and then improved upon his design. Later, a few major improvements were made until the middle of the 19th century. Then several European countries began to manufacture fine optical equipment but none finer than the marvelous instruments built by the American, Charles A. Spencer, and the industry he founded. Present-day instruments, changed but little, give magnification up to 1250 diameters with ordinary light and up to 5000 with blue light.

19.2.3 Beyond the Light Microscope

A light microscope, even one with perfect lenses and perfect illumination, simply cannot be used to distinguish objects that are smaller than half the wavelength of light. White light has an average wavelength of 0.55 micrometers, half of which is 0.275 micrometers. (One micrometer is a thousandth of a millimeter, and there are about 25,000 micrometers to an inch. Micrometers are also called microns.) Any two lines that are closer together than 0.275 micrometers will be seen as a single line, and any object with a diameter smaller than 0.275 micrometers will be invisible or, at best, show up as a blur. To see tiny particles under a microscope, scientists must bypass light altogether and use a different sort of "illumination," one with a shorter wavelength.

19.3 PRINCIPLES OF MICROSCOPY

The microscope that is available for general use is a sophisticated optical instrument that can provide with high-resolution images of a variety of specimens. Image quality is based largely on ability to use the microscope properly. Some basic information is presented.

Resolution

The magnification of small things is a necessary facet of biological research, but the fine detail in cells and in sub-cellular components requires that any imaging system be capable of providing spatial information across small distances. Resolution is defined as the ability to distinguish two very small and closely-spaced objects as separate entities. Resolution is best when the distance separating the two tiny objects is small. Resolution is determined by certain physical parameters that include the wavelength of light, and the light-gathering power of the objective and condenser lenses. A simple mathematical equation defines the smallest distance (d_{min}) separating the two very small objects:

$$d_{min} = 1.22 \times \text{wavelength}/\text{N.A.}_{\text{objective}} + \text{N.A.}_{\text{condenser}}$$

This is the theoretical resolving power of a light microscope. In practice, specimen quality usually limits d_{min} to something greater than its theoretical lower limit.

N.A. (numerical aperture) is a mathematical calculation of the light-gathering capabilities of a lens. The N.A. of each objective lens is inscribed in the metal tube, and ranges from 0.25-1.4. The higher the N.A, the better the light-gathering properties of the lens, and the better the resolution. Higher N.A. values also mean shorter working distances (you have to get the lens closer to the object). N.A. values above 1.0 also indicate that the lens is used with some immersion fluid, such as immersion oil.

From the equation above, one should be aware that the N.A. of the condenser is as important as the N.A. of the objective lens in determining resolution. It is for this reason that closure of the condenser diaphragm results in loss of resolution. In practice, at full aperture and with good oil immersion lenses (N.A. 1.4 for both the condenser and the objective) it is possible to be able to resolve slightly better than 0.2 μm. From the equation above, it should also be clear that shorter wavelength light (blue light) will provide you with better resolution (smaller d_{min} values). However, there are practical considerations in how short the wavelength can be. In the early 1950s, a UV microscope was designed, but required quartz objectives and a specialized imaging device. The quartz lenses provided slightly better resolution (d_{min} = 0.1 μm), but image quality suffered from an inability on the part of the manufacturers to correct for aberrations caused by the quartz. The human eye is best adapted for green light and our ability to see detail may be compromised somewhat with the use of blue or violet. Most manufacturers of microscopes correct their simplest lenses (achromats) for green light.

Magnification and imaging

Most microscopes in current use are known as compound microscopes, where a magnified image of an object is produced by the objective lens, and this image is magnified by a second lens system (the ocular or eyepiece) for viewing. Thus, final magnification of the microscope is dependent on the magnifying power of the objective times the magnifying power of the ocular. Objective magnification powers range from 4 × to 100 ×. Lower magnification is impractical on a compound microscope stand because of spatial constraints with image correction and illumination. Higher magnification is impractical because of limitations in light gathering ability and shortness of working distances required for very strong lenses. Ocular magnification ranges are typically 8 × ~12 × though 10 × oculars are most common. As a result, a standard microscope will provide you with a final magnification range of ~40 × up to ~1000 ×.

Each objective lens consists of six or more pieces of glass that combine to produce a clear image of an object. The six or more lenses in the objective lens are needed to provide corrections that produce image clarity. The interaction of light with the glass in a lens produce aberrations that result in loss in image quality because light waves will be bent, or refracted, differently in different portions of a lens, and different colors of light will be refracted to different extents by the glass. Spatial aberrations (e.g., spherical aberration) can be corrected by using lenses with different curvature on their surfaces and chromatic (i.e., color) aberrations can be minimized by using multiple kinds of glass in combination. These corrections increase the cost of the lens to the extent that an apochromatic objective lens exhibiting full color correction and extremely high N.A. can cost several thousand dollars. This objective lens is about the size of your thumb.

The objective lenses in most microscopes are achromats, and best suited for imaging with green light. Green filters narrow the bandwidth of the light, and make achromat objectives reasonably effective for most routine uses. The achromat lenses are not suitable for critical high-resolution imaging with white light, because red and blue light do not focus in the same plane as green light. Chromatic aberrations will degrade resolution in color images obtained with achromatic objectives. Color photomicrography aimed at the highest level of resolution and image clarity should be performed with totally corrected apochromatic objective lenses. Fluorite lenses, offer intermediate levels of correction, better than achromats but not as good as apochromats. Fluorite lenses are well suited for fluorescence microscopy because of their high transmittance of shorter wavelength light. Higher levels of correction make objective lenses more expensive.

The oculars in most microscopes are designed to work optimally with the objective lenses from the same manufacturer. Each manufacturer makes some of the color and spatial corrections in the objective and the remainder of the corrections in the ocular. Mixing brands will usually result in a degraded image. In addition, when you look into a microscope, the magnified and corrected image you see through the oculars is actually a virtual image (as opposed to a real image). The ocular, designed to provide a corrected virtual image when viewed by eye, is not suitable for the generation of photographic or video images through the microscope. For photography or video microscopy it is necessary to use a projection lens that generates a corrected real image. Many of the newer microscopes provide total image corrections in the objective lens, thus obviating many of the concerns about matching glass components from the same manufacturer. Nevertheless, it is a good practice not to mix parts from one manufacturer with those of another, because unintended image degradation can result.

Illumination

An essential factor in producing a good image with the light microscope is obtaining adequate levels of light in the specimen, or object plane. It is not only necessary to obtain bright light around the object, but for optimal imaging, the light should be uniform across the field of view. The best way to illuminate the specimen involves the use of yet another lens system, known as a condenser. The front element of the condenser is usually a large, flattened lens that sits directly beneath the specimen. Its placement on a movable rack provides you with the means to focus the light beam coming past the object and maximize the intensity and control the uniformity of illumination. Two apertures in the illumination system allow you to regulate the diameter of the illumination beam by closing or opening iris diaphragms. One of these diaphragms, housed within the bright-field condenser and known as the condenser diaphragm, allow you to increase contrast, but at the cost of worsening resolution. The second of these diaphragms, known as the field aperture diaphragm, does not affect resolution as dramatically and is regularly adjusted for optimal illumination.

Optimal illumination of a specimen with all microscopes currently manufactured is achieved by using a variation of Kohler Illumination, where (for those of you are technophiles) the filament of the light source is in focus at the rear focal plane of the objective lens. Operationally, it is easy to obtain optimal illumination for bright-field (or phase contrast) by first placing any specimen on the stage and focusing on the object. Next, turn the ring for the field aperture diaphragm (the lowest aperture on the microscope) so that its edges obscure the periphery of the field of view. Next, raise or lower the condenser until the edges of the field aperture diaphragm are clearly focused. Do not refocus the objective on the specimen while you are adjusting the condenser. It may be necessary to center the field aperture diaphragm, using the condenser centering screws. When the microscope is properly illuminated, both the object and the edges of the field aperture diaphragm should be in the same plane of focus and the field iris diaphragm should be centered in the field of view.

19.4 TYPES OF MICROSCOPES

19.4.1 Transmitted Light Microscopy

Transmitted light microscopy is the general term used for any type of microscopy where the light is transmitted from a source on the opposite side of the specimen to the objective lens. Usually, the light is passed through a condenser to focus it on the specimen to get maximum illumination. After the light passes through the specimen it goes through the objective lens to magnify the image of the sample and then to the oculars, where the enlarged image is viewed.

In order to get a usable image in the microscope, the specimen must be properly illuminated. The light path of the microscope must be correctly set up for each optical method and the components used for image generation. The condenser was invented to concentrate the light on the specimen in order to obtain a bright enough image to be useful. The optimum set-up for specimen illumination and image generation is known as Köhler illumination after the man who invented it. It is used for most of the optical configurations. The microscope techniques requiring a transmitted light path include bright field, dark field, phase contrast, polarization and differential interference contrast optics.

19.4.2 Bright Field (Köhler illumination) Microscopy

This is "normal" microscopy when no optical contrast technique is employed. It uses transmitted light to view a specimen that contains inherent contrast/color or is stained. In order to get the best image possible from any transmitted light form of microscopy, and it is crucial that the light path be set up properly. The method for doing this is called Köhler illumination after August Köhler; the man who invented it. It is also known as double diaphragm illumination because it employs both a field and an aperture iris diaphragm to set up the illumination. The condenser is used to focus parallel rays of light on the specimen, as if coming from infinity, thereby giving you the advantages of an evenly illuminated field, a bright image without glare and minimum heating the specimen. As most cells and tissues have insufficient contrast in themselves, staining techniques are generally used. Common stains include Papanicolaou's stain, which is used for cervical smears, and toluidine blue, a general stain used for semi-thin sections of all tissue types.

19.4.3 Dark Field Microscopy

The specimen is illuminated obliquely, with no direct light entering the objective. Features in the specimen plane which scatter light can clearly be seen against a dark background. Dark field illumination is provided by either a simple patch stop, a dark field element in a phase contrast condenser or purpose-built darkfield condenser. The latter is required for high-resolution objectives to prevent the oblique rays entering the wide aperture of the objective. Applications include detection of microorganisms in unstained smear preparations and classical diatom studies.

19.4.4 Phase Contrast

Devised by Zernicka, this technique exploits the fact that light slows slightly when passing through biological specimens. The specimen is illuminated by a hollow cone of light coming through a phase annulus in the condenser. Phase contrast objectives must be used, which have a corresponding phase plate. Light rays passing through the specimen are slightly retarded, and further retardation takes place in the phase plate. When these rays combine with rays which have not taken this path, degrees of constructive and destructive interference occur which produce the characteristic light and dark features in the image.

19.4.5 Polarized Light Microscopy

Polarized light microscopy uses plane-polarized light to analyse structures that are birefringent; structures that have two different refractive indices at right angles to one another (e.g., cellulose microfibrils). Normal, unpolarized, light can be thought of as many sine waves, each oscillating at any one of an infinite number of orientations (planes) around the central axis. Plane-polarized light, produced by a polar, only oscillates in one plane because the polar only transmits light in that plane. The polarized light microscope must be equipped with both a polarizer, positioned in the light path somewhere before the specimen, and an analyzer (a second polarizer) placed in the optical pathway after the objective rear aperture. Image contrast arises from the interaction of plane-polarized light with a birefringent (or doubly-refracting) specimen to produce two individual wave components that are each polarized in mutually perpendicular planes. The velocities of these components are different and vary with the propagation direction through the specimen. After exiting the specimen, the light components become out of phase, but are recombined with constructive and destructive interference when they pass through the analyzer. Polarized light microscopy can be used to measure the amount of retardation that occurs in each direction and so give information about the molecular structure of the birefringent object (e.g., orientation).

19.4.6 Differential Interference Contrast

In this complex form of polarized light microscopy two slightly separate, plane polarized beams of light are used to create a 3D-like image with shades of gray. Wollaston prisms situated in the condenser and above the objective produce the effect, and additional elements add colour to the image. Care must be taken to interpreting DIC images as the apparent hills and valleys in the specimen can be misleading. The height of a "hill" (e.g., the nucleus) is a product of both the actual thickness of the feature (i.e., ray path length) and its refractive index. Variations of the DIC system are named after their originators, Nomarski and de Senarmont. Options can be selected to maximise either resolution or contrast.

19.4.7 Fluorescence Microscopy

Fluorescence can be used as a label or tag when preparing specific biological probes. Some biological substances, such as chlorophyll and some oils and waxes, have primary fluorescence (auto-fluorescence). However, most biological molecules do not fluoresce on their own, so they must be linked with fluorescent molecules fluorochromes) in order to create specific fluorescent probes. The fluorochromes emit light of a given wavelength when excited by incident light of a different (shorter) wavelength. Specimens labeled with a fluorochrome such as fluorescein or green fluorescent protein (GFP) are illuminated with the relevant wavelength of light (blue in these examples) and emit the energy as a longer wavelength (green).

The key feature of fluorescence microscopy is that it employs reflected rather than transmitted light, which means transmitted light techniques such as phase contrast and DIC can be combined with fluorescence microscopy. To view the fluorescence in the microscope, several light filtering components are needed. At the heart of the fluorescence microscope is the dichroic mirror cube which comprises three components: a dichroic beam splitter (partial mirror), an excitation filter and a barrier filter. Specific filters are used to isolate the excitation and emission wavelengths for each fluorochrome. The dichroic mirror reflects shorter wavelengths of light and allows longer wavelengths to pass and is required because the objective acts as both the condenser lens (excitation light) and objective lens (emission light); therefore the beam splitter isolates the emitted light from the excitation wavelength. This epi-illumination type of light path is required to create a dark background so that the fluorescence can be easily seen. The wavelength at which a beam splitter allows the longer wavelengths to pass must be set between the excitation and emission wavelengths of any given fluorochrome so that excitation light is reflected and emission light is allowed to pass through it. A bright light source producing the correct wavelengths for excitation is also required for fluorescence microscopy, normally a mercury arc lamp. When using confocal microscopy to view fluorescence, where up to 95% of the emission light is filtered out, specific wavelength lasers are used, as these are extremely bright and monochromatic.

19.4.8 Confocal Microscopy

Although conventional light and fluorescence microscopy allow the examination of both living and fixed specimens, certain problems exist with these techniques. One of the main problems is out-of-focus blur degrading the image by obscuring important structures of interest, particularly in thick specimens. In conventional microscopy, not only is the plane of focus illuminated, but much of the specimen above and below this point is also illuminated resulting in out-of-focus blur from these areas. This out-of-focus light leads to a reduction in image contrast and a decrease in resolution. In the confocal microscope, all out-of-focus structures are suppressed at image formation. This is obtained by an arrangement of diaphragms, which, at optically conjugated points of the path of rays, act as a point source and as a point detector respectively. The detection pinhole does not permit rays of light from out-of-focus points to pass through it. The wavelength of light, the numerical aperture of the objective and the diameter of the diaphragm (wider detection pinhole reduces the confocal effect) affect the depth of the focal plane. To obtain a full image, the point of light is moved across the specimen by scanning mirrors. The emitted/reflected light passing through the detector pinhole is transformed into electrical signals by a photomultiplier and displayed on a computer monitor.

19.5 LIGHT MICROSCOPY

A light microscope (LM) is an instrument that uses visible light and magnifying lenses to examine small objects not visible to the naked eye, or in finer detail than the naked eye allows. Magnification, however, is not the most important issue in microscopy. Mere magnification without added detail is scientifically useless, just as endlessly enlarging a small photograph may not reveal any more detail, but only larger blurs. The usefulness of any microscope is that it produces better resolution than the eye. Resolution is the ability to distinguish two objects as separate entities, rather than seeing them blurred together as a single smudge. The history of microscopy has revolved largely around technological advances that have produced better resolution.

19.5.1 History of the Light Microscope

Light microscopes date at least to 1595, when Zacharias Jansen (1580–1638) of Holland invented a compound light microscope, one that used two lenses, with the second lens further magnifying the image produced by the first. His microscopes were collapsing tubes used like a telescope in reverse, and produced magnifications up to nine times (9 ×).

Antony van Leeuwenhoek (1632–1723) invented a simple (one-lens) microscope around 1670 that magnified up to 200 × and achieved twice the resolution of the best compound microscopes of his day, mainly because he crafted better lenses. While others were making lenses by such methods as squashing molten glass between pieces of wood, Leeuwenhoek made them by carefully grinding and polishing solid glass. He thus became the first to see individual cells, including bacteria, protozoans, muscle cells, and sperm.

Englishman Robert Hooke (1635–1703) further refined the compound microscope, adding such features as a stage to hold the specimen, an illuminator, and coarse and fine focus controls. Until 1800, compound microscopes designed by Hooke and others were limited to magnifications of 30 × to 50 ×, and their images exhibited blurry edges (spherical aberration) and rainbow-like distortions (chromatic aberration). The most significant improvement in microscope optics was achieved in the nineteenth century, when business partners Carl Zeiss (1816–1888) and Ernst Abbe (1840–1905) added the substage condenser and developed superior lenses that greatly reduced chromatic and spherical aberration, while permitting vastly improved resolution and higher magnification.

19.5.2 Tissue Preparation

The advancement of light microscopy also required methods for preserving plant and animal tissues and making their cellular details more visible, methods collectively called histotechnique (from *histo*, meaning "tissue"). In brief, classical histotechnique involves preserving a specimen in a fixative, such as formalin, to prevent decay; embedding it in a block of paraffin and slicing it very thinly with an instrument called a microtome; removing the paraffin with a solvent; and then staining the tissue, usually with two or more dyes. The slices of tissue, called histological sections, are typically thinner than a single cell. The colors of a prepared tissue are not natural colors, but they make the tissue's structural details more visible. A widely used stain combination called hematoxylin and eosin, for example, typically colors cell nuclei violet and the cytoplasm pink. Other methods of histotechnique have been developed for special purposes. One variation is to embed the tissue in special plastics (resins).

A compound light microscope allows for thinner sectioning (Fig. 19.1). Another is the frozen section method, in which a tissue is frozen with compressed carbon dioxide and sectioned with a special cold microtome, eliminating the time-consuming process of paraffin embedding. Some prefer this method for its relative simplicity, and its speed is an asset in hospitals, where a biopsied tissue may need to be examined rapidly and the diagnosis reported to the surgeon while the patient is in the operating room.

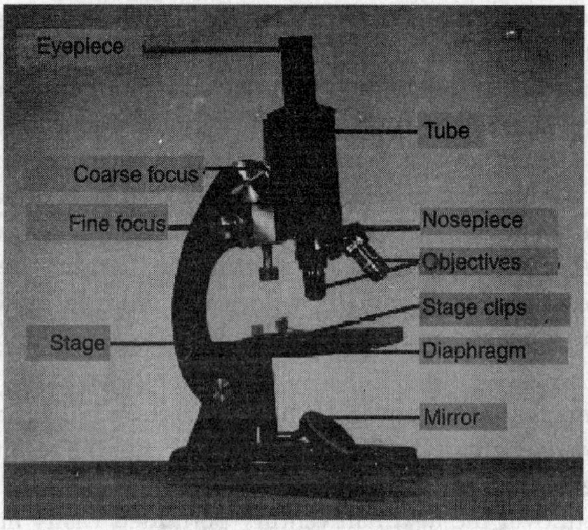

Fig. 19.1 *A compound light microscope.*

19.5.3 Varieties of Light Microscopes

Most compound microscopes today have an illuminator built into the base. A condenser located below the stage has lenses that focus the light on the specimen and a diaphragm that regulates contrast. After passing through the specimen on the stage, the light enters an objective lens. Most light microscopes have three or four objective lenses on a rotating turret. These lenses magnify the image by 4 × to 100 ×. The light then passes up the body tube to an ocular lens that magnifies the image another 10 × to 15 ×. Research-grade microscopes and the better student microscopes have a pair of ocular lenses so that one can view the specimen with both eyes at once. There are many varieties of compound light microscopes for special purposes. For viewing tissue cultures covered with liquid media, biologists can use an inverted light microscope in which the culture is illuminated from above and the objective lenses are positioned below the specimen. The phase contrast microscope can be used to enhance contrast in living specimens, thus avoiding the use of lethal fixatives and stains. The polarizing light microscope is used for analyzing crystals and minerals, among other things. The fluorescence microscope is used to examine structures that bind special fluorescent dyes. It can be used, for example, to identify where a dyetagged hormone binds to its target cell.

Compound light microscopes achieve useful magnifications up to 1200 × and resolutions down to about 0.25 micrometers. That is, two objects in a cell can be as close as 0.25 micrometers and still detected as separate entities. Such resolution is good enough to see most bacteria and some mitochondria and microvilli.

These microscopes generally require thin, transparent, relatively small specimens. They also require that the user adjusts to the phenomenon of optical inversion; if a specimen is moved to the left, it appears under the microscope to move right; when moved up, it appears to move down; and vice versa. The stereomicroscope works at much lower magnification and resolution, but has several advantages: (1) it has two lens systems that view the specimen from slightly different angles, thus giving the specimen a stereoscopic (three-dimensional) appearance; (2) it can use either transmitted or reflected light; and with reflected light, it can be used to view opaque specimens such as rocks, fossils, insects, electronic circuit boards, and so forth; (3) it has a much greater working distance between the specimen and objective lens, allowing for the examination of relatively large objects and for easier manipulation of objects under the microscope; (4) the working distance enables relatively easy dissection of specimens such as insects, allowing hands and instruments to reach the working space while one looks through the microscope; and (5) it does not produce optical inversion; that is, movements to the right appear to go to the right, making dissection and other manipulations much easier.

The utility of light microscopy is governed by its use of visible light, which limits resolution. The shorter the wavelength of the illumination, the better the resolution. Electron beams have shorter wavelengths than photons. The invention of the electron microscope in the late 1930s and its refinement over the next half century permitted vastly improved visualization of cell and tissue fine structure.

19.6 BRIGHT FIELD MICROSCOPY

With a conventional bright field microscope, light from an incandescent source is aimed toward a lens beneath the stage called the condenser, through the specimen, through an objective lens, and to the eye through a second magnifying lens, the ocular or eyepiece. We see objects in the light path because natural pigmentation or stains absorb light differentially, or because they are thick enough to absorb a significant amount of light despite being colorless. A *Paramecium* should show up fairly well in a bright field microscope, although it will not be easy to see cilia or most organelles. Living bacteria won't show up at all unless the viewer hits the focal plane by luck and distorts the image by using maximum contrast.

A good quality microscope has a built-in illuminator, adjustable condenser with aperture diaphragm (contrast) control, mechanical stage, and binocular eyepiece tube. The condenser is used to focus light on the specimen through an opening in the stage. After passing through the specimen, the light is displayed to the eye with an apparent field that is much larger than the area illuminated. The magnification of the image is simply the objective lens magnification (usually stamped on the lens body) times the ocular magnification.

Students are usually aware of the use of the coarse and fine focus knobs, used to sharpen the image of the specimen. They are frequently unaware of adjustments to the condenser that can affect resolution and contrast. Some condensers are fixed in position; others are focusable, so that the quality of light can be adjusted. Usually, the best position for a focusable condenser is as close to the stage as possible. The bright field condenser usually contains an aperture diaphragm, a device that controls the diameter of the light beam coming up through the condenser, so that when the diaphragm is stopped down (nearly closed) the light comes straight up through the center of the condenser lens and the contrast is high. When the diaphragm is wide open the image is brighter and the contrast is low.

A disadvantage of having to rely solely on an aperture diaphragm for contrast is that beyond an optimum point the more contrast you produce the more you distort the image. With a small, unstained, unpigmented specimen, you are usually past optimum contrast when you begin to see the image.

Using a bright field microscope

First, think about what you want to do with the microscope. What is the maximum magnification you will need? Are you looking at a stained specimen? How much contrast/resolution do you require? Next, start setting up for viewing.

Mount the specimen on the stage

The cover slip must be up if there is one. High magnification objective lenses can't focus through a thick glass slide; they must be brought close to the specimen, that is why cover slips are so thin. The stage may be equipped with simple clips (less expensive microscopes), or with some type of slide holder. The slide may require manual positioning, or there may be a mechanical stage (preferred) that allows precise positioning without touching the slide.

Optimize the lighting

A light source should have a wide dynamic range, to provide high intensity illumination at high magnifications, and lower intensities so that the user can view comfortably at low magnifications. Better microscopes have a built-in illuminator, and the best microscopes have controls over light intensity and shape of the light beam. If your microscope requires an external light source, make sure that the light is aimed toward the middle of the condenser. Adjust illumination so that the field is bright without hurting the eyes.

Adjust the condenser

To adjust and align the microscope, start by reading the manual. If no manual is available, try using these guidelines. If the condenser is focusable, position it with the lens as close to the opening in the stage as you can get it. If the condenser has selectable options, set it to bright field. Start with the aperture diaphragm stopped down (high contrast). You should see the light that comes up through the specimen change brightness as you move the aperture diaphragm lever.

Think about what you are looking for

It is a lot harder to find something when you have no expectations as to its appearance. How big is it? Will it be moving? Is it pigmented or stained, and if so what is its color? Where do you expect to find it on a slide? For example, students typically have a lot of trouble finding stained bacteria because with the unaided eye and at low magnifications the stuff looks like dirt. It helps to know that as smears dry down they usually leave rings so that the edge of a smear usually has the densest concentration of cells.

Focus, locate, and center the specimen

Start with the lowest magnification objective lens, to home in on the specimen and/or the part of the specimen you wish to examine. It is rather easy to find and focus on sections of tissues, especially if they are fixed and stained, as with most prepared slides. However, it can be very difficult to locate living, minute specimens such as bacteria or unpigmented protists. A suspension of yeast cells makes a good practice specimen for finding difficult objects.

- Use dark field mode (if available) to find unstained specimens. If not, start with high contrast (aperture diaphragm closed down).
- Start with the specimen out of focus so that the stage and objective must be brought closer together. The first surface to come into focus as you bring stage and objective together is the top of the cover slip. With smears, a cover slip is frequently not used, so the first thing you see is the smear itself.
- If you are having trouble, focus on the edge of the cover slip or an air bubble, or something that you can readily recognize. The top edge of the cover slip comes into focus first, then the bottom, which should be in the same plane as your specimen.
- Once you have found the specimen, adjust contrast and intensity of illumination, and move the slide around until you have a good area for viewing.

Adjust eyepiece separation, focus

With a single ocular, there is nothing to do with the eyepiece except to keep it clean. With a binocular microscope (preferred) you need to adjust the eyepiece separation just like you do a pair of binoculars. Binocular vision is much more sensitive to light and detail than monocular vision, so if you have a binocular microscope, take advantage of it.

One or both of the eyepieces may be a telescoping eyepiece, that is, you can focus it. Since very few people have eyes that are perfectly matched, most of us need to focus one eyepiece to match the other image. Look with the appropriate eye into the fixed eyepiece and focus with the microscope focus knob. Next, look into the adjustable eyepiece (with the other eye of course), and adjust the eyepiece, not the microscope.

Select an objective lens for viewing

The lowest power lens is usually 3.5 or 4 ×, and is used primarily for initially finding specimens. We sometimes call it the scanning lens for that reason. The most frequently used objective lens is the 10 × lens, which gives a final magnification of 100 × with a 10 × ocular lens. For very small protists and for details in prepared slides such as cell organelles or mitotic figures, you will need a higher magnification. Typical high magnification lenses are 40 × and 97 × or 100 ×. The latter two magnifications are used exclusively with oil in order to improve resolution.

Move up in magnification by steps. Each time you go to a higher power objective, refocus and re-center the specimen. Higher magnification lenses must be physically closer to the specimen itself, which poses the risk of jamming the objective into the specimen. Be very cautious when focusing. By the way, good quality sets of lenses are parfocal, that is, when you switch magnifications the specimen remains in focus or close to focused.

Bigger is not always better. All specimens have three dimensions, and unless a specimen is extremely thin you will be unable to focus with a high magnification objective. The higher the magnification, the harder it is to "chase" a moving specimen.

Adjust illumination for the selected objective lens

The apparent field of an eyepiece is constant regardless of magnification used. So it follows that when you raise magnification the area of illuminated specimen you see is smaller. Since you are looking at a smaller area, less light reaches the eye, and the image darkens. With a low power objective you may have to cut down on illumination intensity. With a high power you need all the light you can get, especially with less expensive microscopes.

When to use bright field microscopy

Bright field microscopy is best suited to viewing stained or naturally pigmented specimens such as stained prepared slides of tissue sections or living photosynthetic organisms. It is useless for living specimens of bacteria, and inferior for non-photosynthetic protists or metazoans, or unstained cell suspensions or tissue sections. Here is a not-so-complete list of specimens that might be observed using bright-field microscopy, and appropriate magnifications (preferred final magnifications are emphasized).

- Prepared slides, stained - bacteria (1000 ×), thick tissue sections (100 ×, 400 ×), thin sections with condensed chromosomes or specially stained organelles (1000 ×), large protists or metazoans (100 ×).
- Smears, stained - blood (400 ×, 1000 ×), negative stained bacteria (400 ×, 1000 ×).
- Living preparations (wet mounts, unstained) - pond water (40 ×, 100 ×, 400 ×), living protists or metazoans (40 ×, 100 ×, 400 × occasionally), algae and other microscopic plant material (40 ×, 100 ×, 400 ×). Smaller specimens will be difficult to observe without distortion, especially if they have no pigmentation.

Care of the microscope

- Everything on a good quality microscope is unbelievably expensive, so be careful.
- Hold a microscope firmly by the stand, only. Never grab it by the eyepiece holder, for example.
- Hold the plug (not the cable) when unplugging the illuminator.
- Since bulbs are expensive, and have a limited life, turn the illuminator off when you are done.
- Always make sure the stage and lenses are clean before putting away the microscope.
- Never use a paper towel, a kimwipe, your shirt, or any material other than good quality lens tissue or a cotton swab (must be 100% natural cotton) to clean an optical surface. Be gentle. You may use an appropriate lens cleaner or distilled water to help remove dried material. Organic solvents may separate or damage the lens elements or coatings.
- Cover the instrument with a dust jacket when not in use.
- Focus smoothly, don't try to speed through the focusing process or force anything. For example if you encounter increased resistance when focusing then you've probably reached a limit and you are going in the wrong direction.

19.7 POLARIZED LIGHT MICROSCOPY

19.7.1 Introduction

Polarized light is a contrast-enhancing technique that improves the quality of the image obtained with birefringent materials when compared to other techniques such as dark field and bright field illumination, differential interference contrast, phase contrast, Hoffman modulation contrast, and fluorescence. Polarized light microscopes have a high degree of sensitivity and can be utilized for both quantitative and qualitative studies targeted at a wide range of anisotropic specimens. Qualitative polarizing microscopy is very popular in practice, with numerous volumes dedicated to the subject. In contrast, the quantitative aspects of polarized light microscopy, which is primarily employed in crystallography, represent a far more difficult subject that is usually restricted to geologists, mineralogists, and chemists. However, steady advances made over the past few years have enabled biologists to study the birefringent character of many anisotropic sub-cellular assemblies (Fig. 19.2).

The polarized light microscope is designed to observe and photograph specimens that are visible primarily due to their optically anisotropic character. In order to accomplish this task, the microscope must be equipped with both a polarizer, positioned in the light path somewhere before the specimen, and an analyzer, placed in the optical pathway between the objective rear aperture and the observation tubes or camera port. Image contrast arises from the interaction of plane-polarized light with a birefringent (or doubly-refracting) specimen to produce two individual wave components that are each polarized in mutually perpendicular planes. The velocities of these components, which are termed the ordinary and the extraordinary wavefronts, are different and vary with the propagation direction through the specimen. After exiting the specimen, the light components become out of phase, but are recombined with constructive and destructive interference when they pass through the analyzer. These concepts are outlined for the wavefront field generated by a hypothetical birefringent specimen. In addition, the critical optical and mechanical components of a modern polarized light microscope are illustrated in the Fig. 19.2.

Polarized light microscopy is capable of providing information on absorption color and optical path boundaries between minerals of differing refractive indices, in a manner similar to bright field illumination, but the technique can also distinguish between isotropic and anisotropic substances. Furthermore, the contrast-enhancing technique exploits the optical properties specific to anisotropy and reveals detailed information concerning the structure and composition of materials that are invaluable for identification and diagnostic purposes.

Isotropic materials, which include a variety of gases, liquids, unstressed glasses and cubic crystals, demonstrate the same optical properties when probed in all directions. These materials have only one refractive index and no restriction on the vibration direction of light passing through them. In contrast, anisotropic materials, which include 90 percent of all solid substances, have optical properties that vary with the orientation of incident light with the crystallographic axes. They demonstrate a range of refractive indices depending both on the propagation direction of light through the substance and on the vibrational plane coordinates.

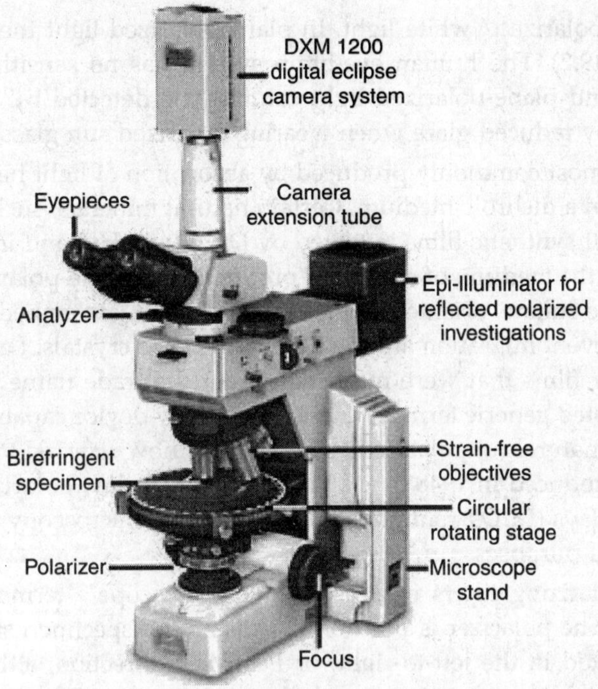

Fig. 19.2 *Polarized light microscope configuration.*

More importantly, anisotropic materials act as beam splitters and divide light rays into two orthogonal components as illustrated above. The technique of polarizing microscopy exploits the interference of the split light rays, as they are reunited along the same optical path to extract information about anisotropic materials.

Polarized light microscopy is perhaps best known for its applications in the geological sciences, which focus primarily on the study of minerals in rock thin sections. However, a wide variety of other materials can readily be examined in polarized light, including both natural and industrial minerals, cement composites, ceramics, mineral fibers, polymers, starch, wood, urea, and a host of biological macromolecules and structural assemblies. The technique can be used both qualitatively and quantitatively with success, and is an outstanding tool for the materials sciences, geology, chemistry, biology, metallurgy, and even medicine.

Although an understanding of the analytical techniques of polarized microscopy may be perhaps more demanding than other forms of microscopy, it is well worth pursuing, simply for the enhanced information that can be obtained over brightfield imaging. An awareness of the basic principles underlying polarized light microscopy is also essential for the effective interpretation of differential interference contrast (DIC).

19.7.2 Basic of Polarized Light

The wave model of light describes light waves vibrating at right angles to the direction of propagation with all vibration directions being equally probable. This is referred to as "common" or "non-polarized" white light. In plane-polarized light there is only one vibration direction (Fig. 19.2). The human eye-brain system has no sensitivity to the vibration directions of light, and plane-polarized light can only be detected by an intensity or color effect, for example, by reduced glare when wearing polarized sun glasses.

Polarized light is most commonly produced by absorption of light having a set of specific vibration directions in a dichroic medium. Certain natural minerals, such as tourmaline, possess this property, but synthetic films invented by Dr. Edwin H. Land in 1932 soon overtook all other materials as the medium of choice for production of plane-polarized light. Tiny crystallites of iodoquinine sulfate, oriented in the same direction, are embedded in a transparent polymeric film to prevent migration and reorientation of the crystals. Land developed sheets containing polarizing films that were marketed under the trade name of Polaroid®, which has become the accepted generic term for these sheets. Any device capable of selecting plane-polarized light from natural (unpolarized) white light is now referred to as a polar or polarizer, a name first introduced in 1948 by A. F. Hallimond. Today, polarizers are widely used in liquid crystal displays (LCDs), sunglasses, photography, microscopy, and for a myriad of scientific and medical purposes.

There are two polarizing filters in a polarizing microscope - termed the polarizer and analyzer (Fig. 19.2). The polarizer is positioned beneath the specimen stage usually with its vibration azimuth fixed in the left-to-right, or East-West direction, although most of these elements can be rotated through 360 degrees. The analyzer, usually aligned with a vibration direction oriented North-South, but again rotatable on some microscopes, is placed above the objectives and can be moved in and out of the light path as required. When both the analyzer and polarizer are inserted into the optical path, their vibration azimuths are positioned at right angles to each other. In this configuration, the polarizer and analyzer are said to be crossed, with no light passing through the system and a dark viewfield present in the eyepieces.

For incident light polarized microscopy, the polarizer is positioned in the vertical illuminator and the analyzer is placed above the half mirror. Most rotatable polarizers are graduated to indicate the rotation angle of the transmission azimuth, while analyzers are usually fixed into position (although advanced models can be rotated either 90 or 360 degrees). The polarizer and analyzer are the essential components of the polarizing microscope, but other desirable features include:

- **Specialized Stage:** A 360-degree circular rotating specimen stage to facilitate orientation studies with centration of the objectives and stage with the microscope optical axis to make the center of rotation coincide with the center of the field of view. Many stages designed for polarized light microscopy also contain a vernier scale so that rotation angle can be measured to an accuracy of 0.1 degree. For advanced studies of conoscopic images, a universal stage having multiple axes of rotation can also be employed to enable observation of the specimen from any direction.
- **Strain Free Objectives:** Stress introduced into the glass of an objective during assembly can produce spurious optical effects under polarized light, a factor that

could compromise performance. Objectives designed for polarized light observation are distinguished from ordinary objectives with the inscription P, PO, or Pol on the barrel. The performance of an objective is limited by several factors, including the anti-reflection coatings used on lens surfaces, and the refractive properties due to angle of incident light on the front lens. In addition, lens strain can be introduced at the cement junction between elements in a lens group or from a single or group of lenses that has been mounted too tightly in the frame.

- **Centerable Revolving Nosepiece:** Because the objective optical axis position varies from one assembly to another, many polarized light microscopes are equipped with a specialized nosepiece that contains a centering mechanism for individual objectives. This enables each objective to be centered with respect to the stage and microscope optical axis so that specimen features remain in the center of the viewfield when the stage is rotated through 360 degrees.
- **Strain-Free Condenser:** Condensers designed for polarized light microscopy have several features in common, including the use of strain free lenses. Some condensers are equipped with a receptacle for the polarizer or have the polarizing element mounted directly into the condenser, beneath the aperture diaphragm. Many polarized light condensers have a top lens that can be removed (a swing-lens condenser) from the light path to generate nearly parallel illumination wavefronts for low magnification and birefringence observations.
- **Eyepieces:** Polarized light microscope eyepieces are fitted with a cross wire reticle (or graticule) to mark the center of the field of view. Often, the cross wire reticle is substituted for a photomicrography reticle that assists in focusing the specimen and composing images with a set of frames bounding the area of the viewfield to be captured either digitally or onto film. Orientation of the eyepiece with respect to the polarizer and analyzer is guaranteed by a point pin that slides into the observation tube sleeve.
- **Bertrand Lens:** A specialized lens mounted in an intermediate tube or within the observation tubes, a Bertrand lens projects an interference pattern formed at the objective rear focal plane into focus at the microscope image plane. The lens is designed to enable easy examination of the objective rear focal plane, to allow accurate adjustment of the illuminating aperture diaphragm and to view interference
- **Compensator and Retardation Plates:** Many polarized light microscopes contain a slot to allow the insertion of compensators and/or retardation plates between the crossed polarizers, which are used to enhance optical path differences in the specimen. In most modern microscope designs, this slot is placed either in the microscope nosepiece or an intermediate tube positioned between the body and eyepiece tubes. Compensation plates inserted into the slot are then situated between the specimen and the analyzer.

Polarized light microscopy can be used both with reflected (incident or epi) and transmitted light. Reflected light is useful for the study of opaque materials such as ceramics, mineral oxides and sulfides, metals, alloys, composites, and silicon wafers reflected light techniques require a dedicated set of objectives that have not been corrected for viewing through the cover glass, and those for polarizing work should also be strain free.

Polarized light photomicrographs of typical specimen's image are utilizing this technique. On the left is a digital image revealing surface features of a microprocessor integrated circuit. Birefringent elements employed in the fabrication of the circuit are clearly visible in the image, which displays a portion of the chip's arithmetic logic unit. The blemished surface of a ceramic is superconducting crystal (bismuth base) which shows birefringent crystalline areas with interference colors interspersed with grain boundaries. Metallic thin films are also visible with reflected polarized light. A blister is that form imperfections in an otherwise confluent thin film of copper (about 0.1 micron thick) sandwiched over a nickel/sodium chloride substrate to form a metallic super lattice assembly.

Careful specimen preparation is essential for good results in polarized light microscopy. The method chosen will depend on the type of material studied. In geological applications, the standard thickness for rock thin sections is 25-30 micrometers. Specimens can be ground down with diamond impregnated wheels and then hand finished to the correct thickness using abrasive powders of successively decreasing grit size. The final specimen should have a cover glass cemented with an optically transparent adhesive. Softer materials can be prepared in a manner similar to biological samples using a microtome. Slices between one and 40 micrometers thick are used for transmitted light observations. These should be strain-free and free from any knife marks. Biological and other soft specimens are mounted between the slide and the cover glass using a mounting medium whose composition will depend on the chemical and physical nature of the specimen. This is particularly significant in the study of synthetic polymers where some media can chemically react with the material being studied and cause degrading structural changes (artifacts).

19.7.3 Applications of Polarized Light Microscopy

As described above, polarized light microscopy is utilized in a broad range of disciplines, including medicine, biology, geology, materials science, and the food industry. The specimens that are readily examined between crossed polarizers originate from a variety of natural and synthetic sources and include gout crystals, amyloid, muscle tissue, teeth, minerals, solid crystals, liquid crystals, fibers, fats, glasses, ceramics, metals, alloys, among others.

19.7.3.1 Identification of gout crystals

One of the most common medical applications for polarized light microscopy is the identification of gout crystals (monosodium urate) with a first order retardation plate. This practice is so common that many microscope manufacturers offer a gout kit attachment for their laboratory brightfield microscopes that can be purchased by physicians. Gout is an acute, recurrent disease caused by precipitation of urate crystals and characterized by painful inflammation of the joints, primarily in the feet and hands. In practice, several drops of fresh synovial fluid are sandwiched between a microscope slide and cover glass and sealed with nail polish to prevent drying. After the specimen has been prepared, it is examined between crossed polarizers with a first order retardation plate inserted into the optical path.

19.7.3.2 Identification of asbestos fibers

Asbestos is a generic name for a group of naturally occurring mineral fibers, which have been widely used as insulating materials, brake pads, and to reinforce concrete. These materials can be harmful to the health when inhaled and it is important that their presence in the environment be easily identified. Specimens are commonly screened using scanning electron microscopy and X-ray microanalysis, but polarizing microscopy provides a quicker and easier alternative that can be utilized to distinguish between asbestos and other fibers and between the major types asbestos, including chrysotile, crocidolite, and amosite. From a healthcare point of view, it is believed that the amphibole asbestos derivatives (crocidolite and amosite) are more harmful than the serpentine, chrysotile.

Plane-polarized light provides information about gross fiber morphology, color, pleochroism, and refractive index. Glass fibers and others that are isotropic will be unaffected by rotation under plane-polarized light while asbestos fibers will display some pleochroism. Chrysotile asbestos fibrils may appear crinkled, like permed or damaged hair, under plane-polarized light, whereas crocidolite and amosite asbestos are straight or slightly curved. Chrysotile has a refractive index of about 1.550, while that of amosite is 1.692, and crocidolite has the highest, with a value of 1.695. Note that the refractive index value of the amphibole asbestos products is much higher than chrysotile. Crocidolite displays blue colors, pleochroism, and murky brown polarization colors. The fast vibration for this fiber is parallel with the long axis. In summary, identification of the three asbestos fiber types depends on shape, refractive indices, pleochroism, birefringence, and fast and slow vibration directions.

19.7.3.3 Uncovering the history of rock formation

Phyllite as well as providing information on component minerals, an examination of geological thin sections using polarizing microscopy can reveal a great deal about how the rock was formed. Phyllite, a metamorphic rock, clearly shows the alignment of crystals under the effects of heat and stress. Small-scale folds are visible in the plane-polarized image and more clearly defined under crossed polarizers with and without the first order retardation plate. The crossed polarizers image reveals that there are several minerals present, including quartz in gray and whites and micas in higher order colors. The alignment of the micas is clearly apparent. Addition of the first order retardation plate improves contrast for clear definition in the image.

Oolite a light gray rock composed of siliceous oolites cemented in compact silica, is formed in the sea. The mineral's name is derived from its structural similarity to fish roe, better known as caviar. Oolite forms in the sea when sand grains are rolled by gentle currents over beds of calcium carbonate or other minerals. These minerals build up around the sand grains and subsequent cementation transforms the grains into coherent rock. The thin sections show the original quartz nuclei on which the buildup of carbonate mineral occurred.

In plane-polarized light the quartz is virtually invisible having the same refractive index as the cement, while the carbonate mineral, with a different refractive index, shows high contrast. The crossed polarizer image reveals quartz grains in grays and whites and the calcium carbonate in the characteristic biscuit colored, high order whites. The groups of quartz grains in some of the cores reveal that these are polycrystalline and are metamorphic quartzite particles. When a first order retardation plate is inserted into the optical path, optical path differences become apparent in the specimen, and contrast is enhanced.

19.7.3.4 Natural and synthetic polymers

During the solidification of polymer melts there may be some organization of the polymer chains, a process that is often dependent upon the annealing conditions. When nucleation occurs, the synthetic polymer chains often arrange themselves tangentially and the solidified regions grow radially. These can be seen in crossed polarized illumination as white regions, termed spherulites, with the distinct black extinction crosses. When these spherulites impinge, their boundaries become polygonal. This can be clearly seen in crossed polarizers but not under plane-polarized light.

The addition of the first order retardation plate confirms the tangential arrangement of the polymer chains. The banding occurring in these spherulites indicates slow cooling of the melt allowing the polymer chains to grow out in spirals. This information on thermal history is almost impossible to collect by any other technique. Nucleation in polymer melts can take place as the result of accidental contamination or contact with a nucleating surface and can lead to substantial weakening of the product. Identification of nucleation can be a valuable aid for quality control.

19.7.3.5 Nylon fibers

Observations under plane-polarized light reveal refractive index differences between a nylon fiber and the mounting medium, and the presence of opacifying titanium dioxide particles. The image under crossed polarizers reveals second and third order polarization colors and their distribution across the fibers indicate that this is a cylindrical and not a lobate fiber useful in predicting mechanical strength. The use of the quartz wedge enables the determination of optical path differences for birefringence measurements.

In summary, polarizing microscopy provides a vast amount of information about the composition and three-dimensional structure of a variety of samples. Virtually unlimited in its scope, the technique can reveal information about thermal history and the stresses and strains to which a specimen was subjected during formation. Useful in manufacturing and research, polarizing microscopy is a relatively inexpensive and accessible investigative and quality control tool, which can provide information unavailable with any other technique.

19.8 FLUORESCENCE MICROSCOPY

In certain classes of atoms and molecules, electrons absorb light, become energized, and then rapidly lose this energy in the form of heat and light emission. If the electron keeps its spin, the electron is said to enter a singlet state, and the kind of light that is emitted as the electron returns to ground state is called fluorescence. If the electron changes its spin when excited, it enters the triplet state, and the kind of light that is emitted as the electron returns to ground state is known as phosphorescence. Phosphorescence is much longer-lived than fluorescence. Both fluorescence and phosphorescence emissions are of particular wavelengths for specific excited electrons. Both types of emission are dependent on specific wavelengths of excitation light, and for both types of emission, the energy of excitation is greater than the energy of emission. Described another way, λ of excitation light is shorter than λ of emission light. In biology, we can utilize fluorescence in localization reactions, to identify particular molecules in complex mixtures or in cells. Fluorescence has the advantage of providing a very high signal-to-noise ratio, which enables us to distinguish spatial distributions of rare molecules. To utilize fluorescence, we need to label the specimen (a cell, a tissue, or a gel) with a suitable molecule (a fluorochrome) whose distribution will become evident after illumination. The fluorescence microscope is ideally suited for the detection of particular fluorochromes in cells and tissues.

The fluorescence microscope that is in wide use today follows the basic "incident-light" design of Ploem, who employed a novel arrangement of filters with a chromatic beam splitter (often wrongly called a dichroic filter both by biologists and microscope sales people). With the incident light fluorescence microscope, the object is illuminated with fluorescence excitation light through the objective lens. The object emits longer-λ fluorescence in response to the shorter- λ excitation light. The objective lens then serves both for illumination and imaging. The chromatic beam splitter transmits or reflects light, depending on its color. For this application, shorter λ light is reflected and longer λ light is transmitted by the splitter. Ploem placed the chromatic splitter in the optical path between the objective lens and the ocular, at a 45° angle, so that it would reflect shorter λ light downward toward the objective. The longer- λ fluorescence emission light would be transmitted through the chromatic beam splitter toward the ocular.

The microscopes operate in the same general fashion as other microscopes. Light beams pass through a condenser lens system and provide illumination of an object at many points simultaneously. For incident excitation light beam the objective lens also acts as a condenser for the excitation light beam. In its interaction with the object, some of this light is absorbed, some of this light is scattered, some of this light is reflected, and some of this light is slowed or retarded (relative to a beam of light that does not pass through the object). A portion of the light that has interacted with the object then passes through the imaging lens system of the microscope where it provides with visual or pictorial image information about the object. Like the process of illumination, the process of image generation operates in a parallel fashion, where large numbers of light beams contribute to the image simultaneously. Resolution is limited by the closeness of overlapping points of brightness or darkness. In a practical sense, the limit of resolution is 0.18-0.2 µm with the best available objective lenses and a good specimen.

To observe cells with the fluorescence microscope, it is important to know the spectral characteristics of the fluorochrome that has been employed. In order to excite the fluorochrome properly and then observe its fluorescence emission, the appropriate filter packages must be present in the microscope. The fluorochrome may not fluoresce at all if the cells are illuminated with the inappropriate filter pack present in the optical path. Finally, for any kind of fluorescence localizations to be performed, it is essential to have the appropriate controls, to be sure that the cells do not exhibit excessive autofluorescence (that is, they do not glow in the absence of the fluorochrome), and that the fluorochrome is responsible for the localization pattern observed. In the laboratory, we have several microscopes equipped for incident light fluorescence microscopy.

19.9 ELECTRON MICROSCOPY

19.9.1 What is Electron Microscopy?

The electron microscope is a type of microscope that uses a beam of electrons to create an image of the specimen. It is capable of much higher magnifications and has a greater resolving power than a light microscope, allowing it to see much smaller objects in finer detail. They are large, expensive pieces of equipment, generally standing alone in a small, specially designed room and requiring trained personnel to operate them.

19.9.2 History of Electron Microscopy

By the middle of the 19th century, microscopists had accepted that it was simply not possible to resolve structures of less than half a micrometre with a light microscope because of the Abbe's formula, but the development of the cathode ray tube was literally about to change the way they looked at things; by using electrons instead of light! Hertz (1857-94) suggested that cathode rays were a form of wave motion and Weichert, in 1899, found that these rays could be concentrated into a small spot by the use of an axial magnetic field produced by a long solenoid. But it was not until 1926 that Busch showed theoretically that a short solenoid converges a beam of electrons in the same way that glass can converge the light of the sun, that a direct comparison was made between light and electron beams. Busch should probably therefore be known as the father of electron optics.

In 1931 the German engineers Ernst Ruska and Maximillion Knoll succeeded in magnifying and electron image. This was, in retrospect, the moment of the invention of the electron microscope but the first prototype was actually built by Ruska in 1933 and was capable of resolving to 50 nm. Although it was primitive and not really fit for practical use, Ruska was recognised some 50 years later by the award of a Nobel Prize. The first commercially available electron microscope was built in England by Metropolitan Vickers for Imperial College, London, and was called the EM1, though it never surpassed the resolution of a good optical microscope. The early electron microscopes did not excite the optical microscopists because the electron beam, which had a very high current density, was concentrated into a very small area and was very hot and therefore charred any non-metallic specimens that were examined. When it was found that you could successfully examine biological specimens in the electron microscope after treating them with osmium and cutting very thin slices of the sample, the electron microscope began to appear as a viable proposition. At the University of Toronto, in 1938, Eli Franklin Burton and students Cecil Hall, James Hillier and Albert Prebus constructed the first electron microscope in the New World. This was an effective, high-resolution instrument, the design of which eventually led to what was to become known as the RCA (Radio Corporation of America) range of very successful microscopes.

Unfortunately, the outbreak of the Second World War in 1939 held back their further development somewhat, but within 20 years of the end of the war routine commercial electron microscopes were capable of 1 nm resolution.

19.9.3 Electron Microscopy

19.9.3.1 Concept

Electron microscopes are scientific instruments that use a beam of highly energetic electrons to examine objects on a very fine scale. This examination can yield information about the topography (surface features of an object), morphology (shape and size of the particles making up the object), composition (the elements and compounds that the object is composed of and the relative amounts of them) and crystallographic information (how the atoms are arranged in the object).

Electron microscopes were developed due to the limitations of light microscopes which are limited by the physics of light to 500 × or 1000 × magnification and a resolution of 0.2 micrometers. In the early 1930s this theoretical limit had been reached and there was a scientific desire to see the fine details of the interior structures of organic cells (nucleus, mitochondria, etc.). This required 10,000 × plus magnification which was just not possible using light microscopes. The Transmission Electron Microscope (TEM) was the first type of electron microscope to be developed and is patterned exactly on the light transmission microscope except that a focused beam of electrons is used instead of light to "see through" the specimen. It was developed by Max Knoll and Ernst Ruska in Germany in 1931.

The first Scanning Electron Microscope (SEM) debuted in 1942 with the first commercial instruments around 1965. Its late development was due to the electronics involved in "scanning" the beam of electrons across the sample.

Electron microscopes (EMs) function exactly as their optical counterparts except that they use a focused beam of electrons instead of light to "image" the specimen and gain information as to its structure and composition. The basic steps involved in all EMs are the following: A stream of electrons is formed in high vacuum (by electron guns). This stream is accelerated towards the specimen (with a positive electrical potential) is confined and focused using metal apertures and magnetic lenses into a thin, focused, monochromatic beam.

The sample is irradiated by the beam and interactions occur inside the irradiated sample, affecting the electron beam. These interactions and effects are detected and transformed into an image. The above steps are carried out in all EMs regardless of type. A more specific treatment of the working of two different types of EMs (SEM, TEM) as well as the function of an electron gun and the theory of electron- specimen interaction is described in more detail below.

19.9.4 Electron Gun

The first and basic part of the microscopes is the source of electrons. It is usually a V-shaped filament made of LaB6 or W (tungsten) that is wreathed with Wehnelt electrode (Wehnelt Cap). Due to negative potential of the electrode, the electrons are emitted from a small area of the filament (point source). A point source is important because it emits monochromatic electrons (with similar energy). The two usual types of electron guns are the conventional electron guns and the field emission guns (FEG). Figure 19.3 illustrates the geometry of an electron gun.

In conventional electron guns, a positive electrical potential is applied to the anode, and the filament (cathode) is heated until a stream of electrons is produced. The electrons are accelerated by the positive potential down the column, and because of the negative potential of cap, all electrons are repelled towards the optic axis. A collection of electrons occurs in the space between the filament tip and Cap, which is called a space charge. Those electrons at the bottom of the space charge (nearest to the anode) can exit the gun area through the small (<1 mm) hole in the Whenelt Cap and then move down the column to be later used in imaging.

A field emission gun consists of a sharply pointed tungsten tip held at several kilovolts negative potential relative to a nearby electrode, so that there is a very high potential gradient at the surface of the tungsten tip. The result of this is that the potential energy of an electron as a function of distance from the metal surface has a sharp peak (from the work function), then drops off quickly (due to electron charge traveling through an electric field). Because electrons are quantum particles and have a probability distribution to their location, a certain number of electrons that are nominally at the metal surface will find themselves at some distance from the surface, such that they can reduce their energy by moving further away from the surface. This transport-via-delocalization is called 'tunneling', and is the basis for the field emission effect. FEGs produce much higher source brightness than in conventional guns (electron current > 1000 times), better monochromaticity, but requires a very good vacuum (~10-7 Pa) (Fig. 19.3).

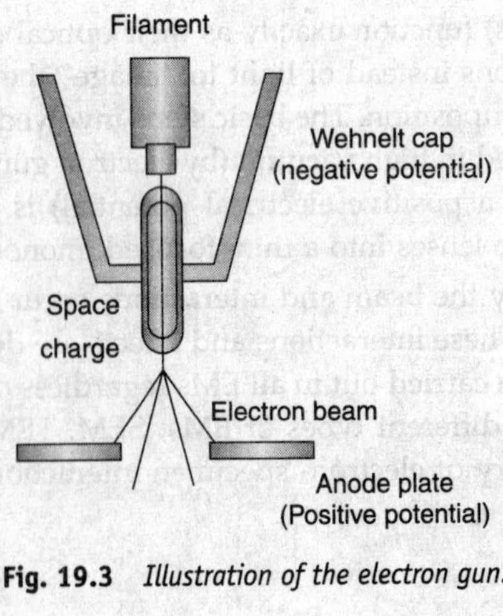

Fig. 19.3 *Illustration of the electron gun.*

19.9.5 Electron-Specimen Interactions

When an electron beam interacts with the atoms in a sample, individual incident electrons undergo two types of scattering – elastic and inelastic (Fig. 19.4). In the former, only the trajectory changes and the kinetic energy and velocity remain constant. In the case of inelastic scattering, some incident electrons will actually collide with and displace electrons from their orbits (shells) around nuclei of atoms comprising the sample. This interaction places the atom in an excited (unstable) state. Specimen interaction is what makes electron microscopy possible. The interactions (inelastic) noted on the top side of the diagram are utilized when examining thick or bulk specimens (Scanning Electron Microscopy, SEM) while on the bottom side are those examined in thin or foil specimens (Transmission Electron Microscopy, TEM) (Fig. 19.4).

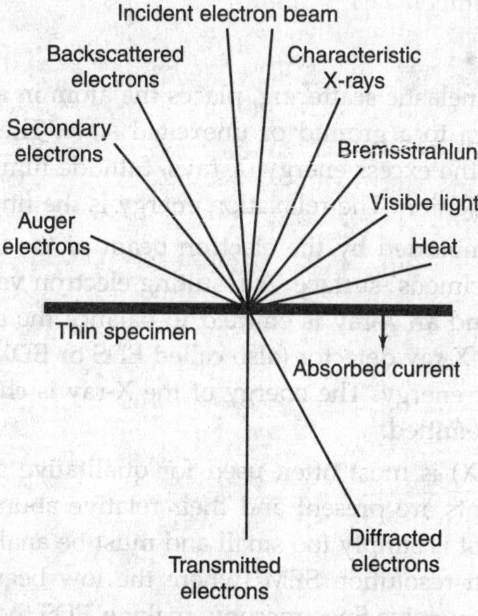

Fig. 19.4 *Effects produced by electron bombardment of a material.*

19.9.6 Reactions Exploited in SEM

Secondary Electrons

When a sample is bombarded with electrons, the strongest region of the electron energy spectrum is due to secondary electrons. The secondary electron yield depends on many factors, and is generally higher for high atomic number targets, and at higher angles of incidence. Secondary electrons are produced when an incident electron excites an electron in the sample and loses most of its energy in the process. The excited electron moves towards the surface of the sample undergoing elastic and inelastic collisions until it reaches the surface, where it can escape if it still has sufficient energy.

Production of secondary electrons is very topography related. Due to their low energy (5 eV) only secondary's that are very near the surface (<10 nm) can exit the sample and be examined. Any changes in topography in the sample that are larger than this sampling depth will change the yield of secondary's due to collection efficiencies. Collection of these electrons is aided by using a "collector" in conjunction with the secondary electron detector.

Backscattered Electrons

Backscattered electrons consist of high-energy electrons originating in the electron beam that are reflected or back-scattered out of the specimen interaction volume. The production of backscattered electrons varies directly with the specimen's atomic number. This differing production rate causes higher atomic number elements to appear brighter than lower atomic number elements. This interaction is utilized to differentiate parts of the specimen that have different average atomic number.

Relaxation of excited atoms

As was mentioned above, inelastic scattering, places the atom in an excited (unstable) state. The atom "wants" to return to a ground or unexcited state. Therefore, at a later time the atoms will relax giving off the excess energy. X-rays, cathode-luminescence and Auger electrons are three ways of relaxation. The relaxation energy is the fingerprint of each element.

When the sample is bombarded by the electron beam of the SEM, electrons are ejected from the atoms on the specimens' surface. A resulting electron vacancy is filled by an electron from a higher shell, and an X-ray is emitted to balance the energy difference between the two electrons. The EDS X-ray detector (also called EDS or EDX) measures the number of emitted X-rays versus their energy. The energy of the X-ray is characteristic of the element from which the X-ray was emitted.

In practice, EDS (or EDX) is most often used for qualitative elemental analysis, simply to determine which elements are present and their relative abundance. In some instances, however, the area of interest is simply too small and must be analyzed by TEM (where EDS is the only option) or high resolution SEM (where the low beam currents used preclude WDS-Wavelength X-ray Dispersive Spectroscopy, making EDS the only option).

Cathodoluminescence (CL) is the emission of photons of characteristic wavelengths from a material that is under high-energy electron bombardment. The electron beam is typically produced in an electron microprobe (EPMA) or scanning electron microscope (SEM-CL).

Auger electrons are electrons ejected by radiationless excitation of a target atom by the incident electron beam. When an electron from the L shell drops to fill a vacancy formed by K-shell ionization, the resulting X-ray photon with energy EK - EL may not be emitted from the atom. If the photon strikes a lower energy electron (e.g., an M-shell electron),then outer electron may be ejected as a low energy Auger electron. Auger electrons are characteristic of the fine structure of the atom and have energies between 280 eV (carbon) and 2.1 keV (sulfur). By discriminating between Auger electrons of various energies, a chemical analysis of the specimen surface can be made. Auger electrons are exploited in Auger Electron Spectroscopy tools (AES).

The volume inside the specimen in which interactions occur while being struck with an electron beam is called specimen interaction volume.

19.9.7 Reactions Exploited in TEM

TEM exploits three different interactions of electron beam-specimen: unscattered electrons (transmitted beam), elastically scattered electrons (diffracted beam) and inelastically scattered electrons.

When incident electrons are transmitted through the thin specimen without any interaction occurring inside the specimen, then the beam of these electrons is called transmitted. The transmission of unscattered electrons is inversely proportional to the specimen thickness. Areas of the specimen that are thicker will have fewer transmitted unscattered electrons and so will appear darker, conversely the thinner areas will have more transmitted and thus will appear lighter.

Another part of the incident electrons, are scattered (deflected from their original path) by atoms in the specimen in an elastic fashion (no loss of energy). These scattered electrons are then transmitted through the remaining portions of the specimen. All electrons follow Bragg's law and thus are scattered according to

$$n.\lambda = 2.d.\sin(\theta)$$

where:

λ is the wavelength of the rays

θ is the angle between the incident rays and the surface of the crystal and

d is the spacing between layers of atoms.

All incident electrons have the same energy (thus wavelength) and enter the specimen normal to its surface. All incidents that are scattered by the same atomic spacing will be scattered by the same angle. These scattered electrons can be collated using magnetic lenses to form a pattern of spots; each spot corresponding to a specific atomic spacing (a plane). This pattern can then yield information about the orientation, atomic arrangements and phases present in the area being examined.

Finally, another way that incident electrons can interact with the specimen is inelastically. Incident electrons that interact with specimen atoms is an inelastic fashion, loosing energy during the interaction. These electrons are then transmitted through the rest of the specimen. Inelastically scattered electrons can be utilized in two ways; Electron Energy Loss Spectroscopy (EELS) and Kikuchi Bands.

Elemental composition and atomic bonding state can be determined by analyzing the energy with the spectroscope attached under the electron microscope (Electron Energy Loss Spectroscopy). Because the analyzing region can be selected from a part of the enlarged electron microscopic image, one can analyze very small region. Moreover, by selecting electrons with a specific loss energy by a slit so as to image them, element distribution in specimen can be visualized (elemental mapping)

Kikuchi lines appear in transmission electron diffraction patterns of relatively thick crystals due to Bragg reflection of the inelastically scattered electrons. They are alternating light and dark lines that are related to atomic spacings in the specimen. These bands can be either measured (their width is inversely proportional to atomic spacing) or "followed" like a roadmap to the "real" elasticity scattered electron pattern.

19.10 TYPES OF ELECTRON MICROSCOPES

All electron microscopes use electromagnetic and/or electrostatic lenses to control the path of electrons. Glass lenses, used in light microscopes, have no effect on the electron beam. The basic design of an electromagnetic lens is a solenoid (a coil of wire around the outside of a tube) through which one can pass a current, thereby inducing an electromagnetic field. The electron beam passes through the centre of such solenoids on its way down the column of the electron microscope towards the sample. Electrons are very sensitive to magnetic fields and can therefore be controlled by changing the current through the lenses.

The faster the electrons travel, the shorter their wavelength. The resolving power of a microscope is directly related to the wavelength of the irradiation used to form an image. Reducing wavelength increases resolution. Therefore, the resolution of the microscope is increased if the accelerating voltage of the electron beam is increased. The accelerating voltage of the beam is quoted in kilovolts (kV). It is now possible to purchase a 1000 kV electron microscope, though this is not commonly found.

Although modern electron microscopes can magnify objects up to about two million times, they are still based upon Ruska's prototype and the correlation between wavelength and resolution. The electron microscope is an integral part of many laboratories such as the John Innes Centre. Researchers can use it to examine biological materials (such as microorganisms and cells), a variety of large molecules, medical biopsy samples, metals and crystalline structures, and the characteristics of various surfaces. Nowadays, electron microscopes have many other uses outside research. They can be used as part of a production line, such as in the fabrication of silicon chips, or within forensic laboratories for looking at samples such as gunshot residues. In the arena of fault diagnosis and quality control, they can be used to look for stress lines in engine parts or simply to check the ratio of air to solids in ice cream.

19.10.1 Transmission Electron Microscope (TEM)

The original form of electron microscopy, transmission electron microscopy (TEM) involves a high voltage electron beam emitted by a cathode and formed by magnetic lenses. The electron beam that has been partially transmitted through the very thin (and so semitransparent for electrons) specimen carries information about the structure of the specimen. The spatial variation in this information (the "image") is then magnified by a series of magnetic lenses until it is recorded by hitting a fluorescent screen, photographic plate, or light sensitive sensor such as a CCD (charge-coupled device) camera. The image detected by the CCD may be displayed in real time on a monitor or computer.

Transmission electron microscopes produce two-dimensional, black and white images.

Resolution of the TEM is also limited by spherical and chromatic aberration, but a new generation of aberration correctors has been able to overcome or limit these aberrations. Software correction of spherical aberration has allowed the production of images with sufficient resolution to show carbon atoms in diamond separated by only 0.089 nm and atoms in silicon at 0.078 nm at magnifications of 50 million times. The ability to determine the positions of atoms within materials has made the TEM an indispensable tool for nanotechnologies research and development in many fields, including heterogeneous catalysis and the development of semiconductor devices for electronics and photonics. In the life sciences, it is still mainly the specimen preparation which limits the resolution of what we can see in the electron microscope, rather than the microscope itself.

At JIC we have a high voltage (200kV) TEM, which was installed in 2008. We have two digital cameras on it, one is higher resolution than the other, so that the need for developing and printing film has been negated. Our TEM is designed for use with biological samples and is capable of resolving to better than 1 nm. It is also capable of 3-D tomography which involves taking a succession of images whilst tilting the specimens through increasing angles, which can then be combined to form a three-dimensional image of the specimen.

19.10.2 Scanning Electron Microscope (SEM)

Unlike the TEM, where the electrons in the primary beam are transmitted through the sample, the Scanning Electron Microscope (SEM) produces images by detecting secondary electrons which are emitted from the surface due to excitation by the primary electron beam. In the SEM, the electron beam is scanned across the surface of the sample in a raster pattern, with detectors building up an image by mapping the detected signals with beam position.

TEM resolution is about an order of magnitude better than the SEM resolution. Our TEM can easily resolve details of 0.2 nm. Our two SEMs at JIC are both relatively recent acquisitions and are high-resolution instruments capable of about 2 nm resolution on biological samples. Because the SEM image relies on electron interactions at the surface rather than transmission it is able to image bulk samples and has a much greater depth of view, and so can produce images that are a good representation of the 3D structure of the sample. SEM images are therefore considered to provide us with 3D, topographical information about the sample surface but will still always be only in black and white.

In the SEM, we use much lower accelerating voltages to prevent beam penetration into the sample since what we require is generation of the secondary electrons from the true surface structure of a sample. Therefore, it is common to use low kV, in the range 1-5 kV for biological samples, even though our SEMs are capable of up to 30 kV.

At JIC we currently have two SEMs, both with high-resolution capabilities, digital imaging facilities and cryo-systems which enable them to be used for looking at frozen-hydrated specimens.

19.10.3 Sample Preparation

Materials to be viewed in an electron microscope generally require processing to produce a suitable sample. This is mainly because the whole of the inside of an electron microscope is under high vacuum in order to enable the electron beam to travel in straight lines. The technique required varies depending on the specimen, the analysis required and the type of microscope.

Cryofixation — Freezing a specimen rapidly, typically to liquid nitrogen temperatures or below, that the water forms ice. This preserves the specimen in a snapshot of its solution state with the minimal of artefacts. An entire field called cryo-electron microscopy has branched from this technique. With the development of cryo-electron microscopy, it is now possible to observe virtually any biological specimen close to its native state.

Fixation — A general term used to describe the process of preserving a sample at a moment in time and to prevent further deterioration so that it appears as close as possible to what it would be like in the living state, although it is now dead. In chemical fixation for electron microscopy, glutaraldehyde is often used to crosslink protein molecules and osmium tetroxide to preserve lipids.

Dehydration — Removing water from the samples. The water is generally replaced with organic solvents such as ethanol or acetone as a stepping stone towards total drying for SEM specimens or infiltration with resin and subsequent embedding for TEM specimens.

Embedding — Infiltration of the tissue with wax (for light microscopy) or a resin (for electron microscopy) such as araldite or LR White, which can then be polymerised into a hardened block for subsequent sectioning.

Sectioning — The production of thin slices of the specimen. For light microscopy, the sections can be a few micrometres thick but for electron microscopy they must be very thin so that they are semitransparent to electrons, typically around 90 nm. These ultra-thin sections for electron microscopy are cut on an ultramicrotome with a glass or diamond knife. Glass knives can easily be made in the laboratory and are much cheaper than diamond, but they blunt very quickly and therefore need replacing frequently.

Staining — Uses heavy metals such as lead and uranium to scatter imaging electrons and thus give contrast between different structures, since many (especially biological) materials are nearly "transparent" to the electron beam. By staining the samples with heavy metals, we add electron density to it which results in there being more interactions between the electrons in the primary beam and those of the sample, which in turn provides us with contrast in the resultant image. In biology, specimens can be stained "en bloc" before embedding and/or later, directly after sectioning, by brief exposure of the sections to solutions of the heavy metal stains.

Freeze-fracture and **freeze-etch** — A preparation method particularly useful for examining lipid membranes and their incorporated proteins in "face on" view. The fresh tissue or cell suspension is frozen rapidly (cryofixed), then fractured by simply breaking or by using a microtome while maintained at liquid nitrogen temperature. The cold, fractured surface is generally "etched" by increasing the temperature to about −95°C for a few minutes to let some surface ice sublime to reveal microscopic details. For the SEM, the sample is now ready for imaging. For the TEM, it can then be rotary-shadowed with evaporated platinum at low angle (typically about 6°) in a high vacuum evaporator. A second coat of carbon, evaporated perpendicular to the average surface plane is generally performed to improve stability of the replica coating. The specimen is returned to room temperature and pressure, and then the extremely fragile "shadowed" metal replica of the fracture surface is released from the underlying biological material by careful chemical digestion with acids, hypochlorite solution or SDS detergent. The floating replica is thoroughly washed from residual chemicals, carefully picked up on an EM grid, dried then viewed in the TEM.

Sputter Coating — An ultra-thin coating of electrically-conducting material, deposited by low vacuum coating of the sample. This is done to prevent charging of the specimen which would occur because of the accumulation of static electric fields due to the electron irradiation required during imaging. It also increases the amount of secondary electrons that can be detected from the surface of the sample in the SEM and therefore increases the signal to noise ratio. Such coatings include gold, gold/palladium, platinum, chromium, etc. Resolution is given in Table 19.1

19.10.4 Resolution of an Electron Microscope

Table 19.1

Bibliographic Entry	*Result (w/surrounding text)*	*Standardized Result*
Curtis, Helena. Biology: 5th Edition. New York: Worth, 1989: 96.	"TEM at present afforts a resolving power of about 0.2 nm .... Although the resolving power of the STM is only about 10 nm, this instrument has become a valuable tool for biologists."	0.2 nm (TEM) 10 nm (STM)
Slayter, Elizabeth M. "Electron Microscope." Grolier Multimedia Encyclopedia. Grolier, 1997.	"... [TEM] resolutions of 2 nm are common .... Although [STM] resolution is limited to 10 nm, readily interpretable images of surface topography emerge."	2 nm (TEM) 10 nm (STM)
Macmillan Encyclopedia of Physics. New York: Simon & Schuster, 1996: 452-454.	"The smallest distance that can be resolved with a TEM is approximately 0.2-0.5 nm .... A typical [STM] resolution of several tenths of a nm can be achieved."	0.2–0.5 nm (TEM) ~ 0.1 nm (STM)
Kane, Joseph & Morton Sternheim. Physics. New York: Wiley, 1978: 620.	"In practice, the resolution of the TEM is limited to about 0.2 nm."	0.2 nm (TEM)
Robertson, Brian. What are Electron Microscopes? Center for Materials Research and Analysis. University of Nebraska at Lincoln.	"Electron Microscopes were developed due to the limitations of Light Microscopes which are limited by the physics of light to 500x or 1000x magnification and a resolution of 0.2 micrometers."	< 200 nm

An electron microscope is an instrument that uses electrons instead of light for the imaging of objects. The development of the transmission electron microscope was based on theoretical work done by Louis de Broglie, who found that wavelength is inversely proportional to momentum. In 1926, Hans Busch discovered that magnetic fields could act as lenses by causing electron beams to converge to a focus. A few years later, Max Knoll and Ernst Ruska made the first modern prototype of an electron microscope.

In a TEM, a monochromatic beam of electrons is accelerated through a potential of 40 to 100 kilovolts (kV) and passed through a strong magnetic field that acts as a lens. The resolution of a modern TEM is about 0.2 nm. This is the typical separation between two atoms in a solid. This resolution is 1,000 times greater than a light microscope and about 500,000 times greater than that of a human eye. The STM is similar to the TEM except for the fact that it causes an electron beam to scan rapidly over the surface of the sample and yields an image of the topography of the surface. The resolution of a STM is about 10 nm. The resolution is limited by the width of the exciting electron beam and by the interaction volume of electrons in a solid.

Resolution is the finest detail that can be distinguished in an image. The resolving power of a microscope is quite different from its magnification. You can enlarge a photograph indefinitely using more powerful lenses, but the image will blur together and be unreadable. Therefore, increasing the magnification will not improve resolution. The minimum separation (*d*) that can be resolved by any kind of a microscope is given by the following formula:

$$d = \lambda/(2n \sin\theta)$$

where n is the refractive index (which is 1 in the vacuum of an electron microscope) and λ is the wavelength. Since resolution and d are inversely proportional, this formula suggests that they way to improve resolution is to use shorter wavelengths and media with larger indices of refraction. The electron microscope exploits these principles by using extremely short wavelengths of accelerated electrons to form high-resolution images.

Today, electron microscopy is widely used in metallurgy, biology, material science, physics, chemistry, and many other technological fields. It has been an integral part in the understanding of the complexities of cellular structure, the fine structure of metals and crystalline materials as well as numerous other areas of the microscopic world.

19.10.5 Disadvantages of Electron Microscopy

Electron microscopes are very expensive to buy and maintain. They are dynamic rather than static in their operation: requiring extremely stable high voltage supplies, extremely stable currents to each electromagnetic coil/lens, continuously-pumped high/ultra-high vacuum systems and a cooling water supply circulation through the lenses and pumps. As they are very sensitive to vibration and external magnetic fields, microscopes aimed at achieving high resolutions must be housed in buildings with special services.

A significant amount of training is required in order to operate an electron microscope successfully and electron microscopy is considered a specialized skill.

The samples have to be viewed in a vacuum, as the molecules that make up air would scatter the electrons. This means that the samples need to be specially prepared by sometimes lengthy and difficult techniques to withstand the environment inside an electron microscope. Recent advances have allowed some hydrated samples to be imaged using an environmental scanning electron microscope, but the applications for this type of imaging are Still Limited.

19.11 ATOMIC FORCE MICROSCOPY

Scanning probe microscopes (SPM) define a broad group of instruments used to image and measure properties of material, chemical, and biological surfaces. SPM images are obtained by scanning a sharp probe across a surface while monitoring and compiling the tip–sample interactions to provide an image. The two primary forms of SPM are scanning tunneling microscopy (STM) and atomic force microscopy (AFM). STM was first developed in 1982 at IBM in Zurich by Binnig, et al. The invention of the scanning tunneling microsope (for which Binnig and Rohrer were awarded the Nobel Prize in Physics in 1986) has had a great impact on the technical community by providing a new and unique tool to advance fundamental science and technology. Although the ability of the STM to image and measure material surface morphology with atomic resolution has been well documented, only good electrical conductors are candidates for this technique. This significantly limits the materials that can be studied using STM and led to the development, in 1986, of the atomic force microscope by Binnig, Quate, and Gerber. This enabled the detection of atomic scale features on a wide range of insulating surfaces that include ceramic materials, biological samples, and polymers.

Prior to the invention and commercial availability of SPMs, researchers traditionally used (and still use) a variety of microscopes to image surfaces and measure surface morphology on a microscale. Optical microscopes are the most common instrument available to image any sample that is not completely optically transparent. Resolution is limited to about 1μm and only images and size measurements from features lying in the surface (*x-y*) plane are obtainable. Also, optical microscopy has a relatively small depth of field. A more advanced technique, scanning electron microscopy (SEM), has been widely used, since its inception in the mid-1900s, to image microscopic features on sample surfaces. SEMs provide much greater resolution (~5nm) than optical microscopes and have a relatively large depth of field. Because a beam of electrons must travel to the sample to provide an image, the samples must be vacuum compatible and either electrically conductive or coated with a conductive layer to avoid charge buildup.

AFM provides a number of advantages over conventional microscopy techniques. AFMs probe the sample and make measurements in three dimensions, *x, y,* and *z* (normal to the sample surface), thus enabling the presentation of three-dimensional images of a sample surface. This provides a great advantage over any microscope available previously. With good samples (clean, with no excessively large surface features), resolution in the *x-y* plane ranges from 0.1 to 1.0 nm and in the *z* direction is 0.01 nm (atomic resolution). AFMs require neither a vacuum environment nor any special sample preparation, and they can be used in either an ambient or liquid environment. With these advantages AFM has significantly impacted the fields of materials science, chemistry, biology, physics, and the specialized field of semiconductors (Fig. 19.5).

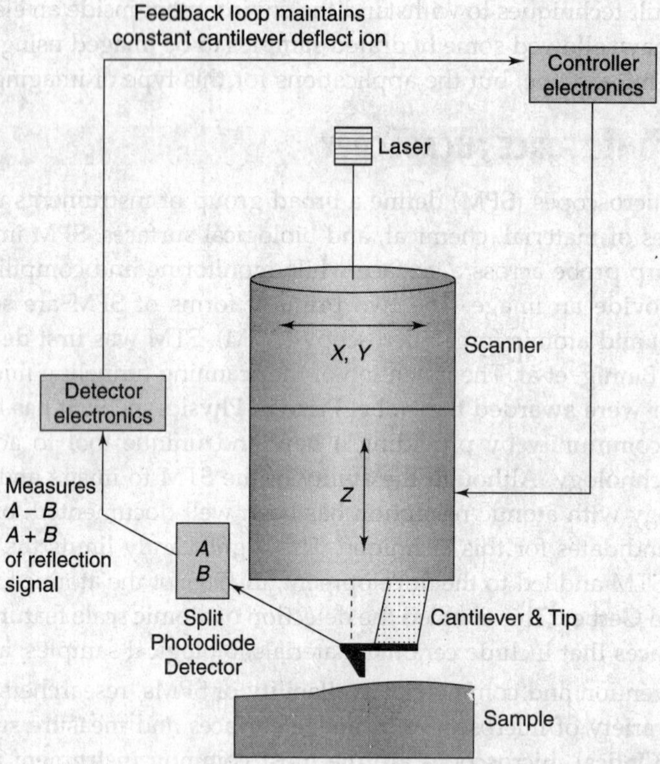

Fig. 19.5 *Operating principles of the AFM in the contact mode.*

Contact mode AFM is one of the more widely used scanning probe modes, and operates by rastering a sharp tip (made either of silicon or Si3N4 attached to a low spring constant cantilever) across the sample. An extremely low force (~10-9 N, interatomic force range) is maintained on the cantilever, thereby pushing the tip against the sample as it rasters. Either the repulsive force between the tip and sample or the actual tip deflection is recorded relative to spatial variation and then converted into an analogue image of the sample surface.

The principle behind the operation of an AFM in the contact mode is shown in Fig. 1.4. The AFM tip is first brought (manually) close to the sample surface, and then the scanner makes a final adjustment in tip-sample distance based on a set point determined by the user. The tip, now in contact with the sample surface through any adsorbed gas layer, is then scanned across the sample under the action of a piezoelectric actuator, either by moving the sample or the tip relative to the other. A laser beam aimed at the back of the cantilever–tip assembly reflects off the cantilever surface to a split photodiode, which detects the small cantilever deflections. A feedback loop, shown schematically in Fig. 1.4, maintains constant tip-sample separation by moving the scanner in the *z* direction to maintain the setpoint deflection.

Without this feedback loop, the tip would "crash" into a sample with even small topographic features (although this phenomenon can happen even with careful AFM operation). By maintaining a constant tip-sample separation and using Hooke's Law ($F = - kx$ where F is force, k is the spring constant, and x is the cantilever deflection), the force between the tip and the sample is calculated. Finally, the distance the scanner moves in the z direction is stored in the computer relative to spatial variation in the x-y plane to generate the topographic image of the sample surface.

The laboratory setup for an AFM (specifically the Dimension1 3000 SPM), includes the scanner, sample display screen, control screen, processor, and image display screen. A special table to isolate mechanical and acoustical vibrations is also usually necessary to perform high-resolution (atomic scale) work.

Examples of images generated using contact mode AFM. A freshly cleaved surface of mica was imaged with a BioScope1 AFM to reveal the atomic structure of mica, use of the wet cell in contact mode (Imaged with a NanoScope1 SPM), which is a great advantage when studying biological materials, to image bovine bone. The hard hydroxyapatite mineral phase appears to be raised above the interconnecting soft collagen phase. Recent advances in the development of AFM technology have led to a number of useful imaging modes including TappingMode and LiftMode AFM. Although operating in the contact mode has proven successful, it suffers from a number of drawbacks that limit its use on a number of sample types. First, the constant downward force on the tip often damages (and thus changes) many softer surfaces (polymers and biological samples) and even some hard surfaces such as silicon. Also, many samples, such as small particles or biological samples like DNA and cells, must be placed on a substrate for imaging purposes. In contact mode, the sample is often destroyed or even pushed out of the field of view by the rastering tip. These complications have been addressed through the development of Tapping Mode AFM. In the TappingMode, the AFM tip–cantilever assembly oscillates at the sample surface while the tip is scanned; thus, the tip lightly taps the sample surface while rastering and only touches the sample at the bottom of each oscillation. This prevents damage to soft specimens and avoids the "pushing" of specimens around on the substrate. By using a constant oscillation amplitude, a constant tip-sample distance is maintained until the scan is complete. Tapping Mode AFM can be performed on both wet and dry sample surfaces.

Resolution of a d-spacing (the spacing between crystalline lattice planes) of 70 nm confirms measurements made with a transmission electron microscope; the latter requires tedious and time consuming specimen preparation versus the minimal sample preparation for the AFM. Sample preparation consisted solely of a simple smear on a glass slide. Note the sickled cell has indented a normal (softer) red blood cell.

Lift Mode AFM provides the operator with a tool to record dual information about a sample surface at one location, such as topography and magnetic gradients (obtained in the magnetic force microscopy or MFM mode), thereby allowing the useful association of the two images. Lift Mode AFM operates by first scanning a line on the sample surface in Tapping Mode to obtain the topographical information. Then, the tip is lifted to a distance above the sample set by the operator and the same line retraced in a non-contact mode to obtain, for example, near surface magnetic field information. The process is repeated until the scan is complete and both images are saved. To perform MFM, a ferromagnetic tip and a ferromagnetic or paramagnetic sample are required. An example of coordinating images generated using LiftMode AFM. The image on the left shows the surface topography of a hard disc with an 800 bit test track while the image on the right is an MFM image of the same area showing the magnetic force gradients. Note the MFM image shows clear resolution of the bit transition gradients without contamination from the surface topography. Lift Mode AFM can also be used to record topography and electric fields or phase imaging data.

In summary, the continuing development of AFM technology provides scientists with a powerful tool to characterize a variety of sample surfaces. Minimal sample preparation, use in ambient conditions, and the ability to image non-conducting specimens at the atomic scale (in some cases) makes AFM an extremely versatile and useful form of microscopy. Recent advances in AFM have allowed the successful imaging of soft polymeric and biological samples and the imaging of magnetic microstructures.

19.11.1 Atomic Force Microscopy – Advances

The Atomic Force Microscope was developed to overcome a basic drawback with STM that it can only image conducting or semiconducting surfaces. The AFM, however, has the advantage of imaging almost any type of surface, including polymers, ceramics, composites, glass, and biological samples.

Binnig, Quate, and Gerber invented the Atomic Force Microscope in 1985. Their original AFM consisted of a diamond shard attached to a strip of gold foil. The diamond tip contacted the surface directly, with the interatomic van der Waals forces providing the interaction mechanism. Detection of the cantilever's vertical movement was done with a second tip—an STM placed above the cantilever.

AFM probe deflection

Today, most AFMs use a laser beam deflection system, introduced by Meyer and Amer, where a laser is reflected from the back of the reflective AFM lever and onto a position-sensitive detector. AFM tips and cantilevers are microfabricated from Si or Si_3N_4. Typical tip radius is from a few to 10s of nm.

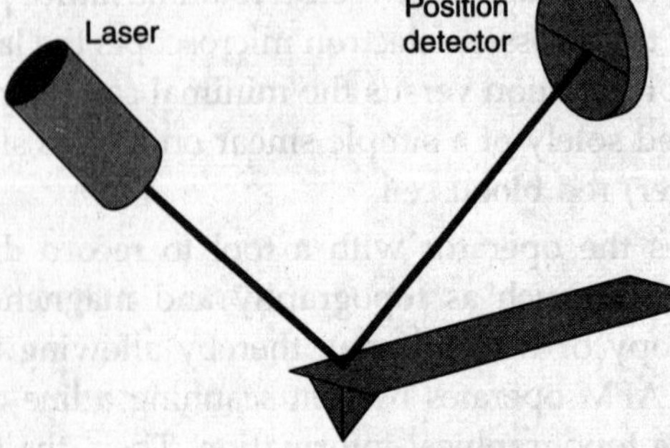

Beam deflection system, using a laser and photodector to measure the beam position.

Measuring forces

Because the atomic force microscope relies on the forces between the tip and sample, knowing these forces is important for proper imaging. The force is not measured directly, but calculated by measuring the deflection of the lever, and knowing the stiffness of the cantilever. Hooke's law gives $F = -kz$, where F is the force, k is the stiffness of the lever, and z is the distance the lever is bent.

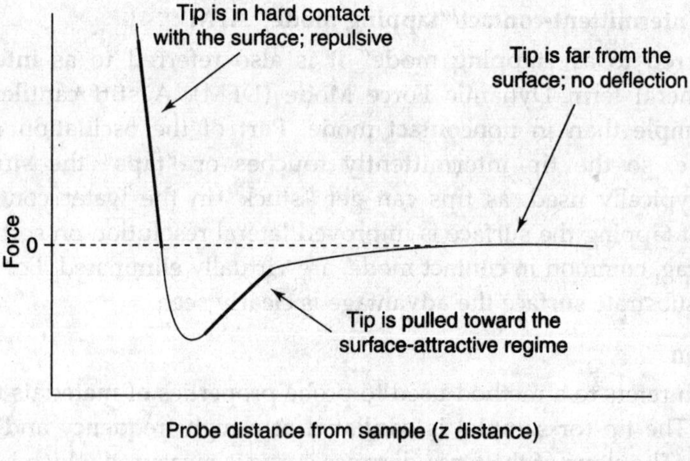

AFM Modes of operation

Because of AFM's versatility, it has been applied to a large number of research topics. The Atomic Force Microscope has also gone through many modifications for specific application requirements.

Contact Mode

The first and foremost mode of operation, contact mode is widely used. As the tip is raster-scanned across the surface, it is deflected as it moves over the surface corrugation. In constant force mode, the tip is constantly adjusted to maintain a constant deflection, and therefore constant height above the surface. It is this adjustment that is displayed as data. However, the ability to track the surface in this manner is limited by the feedback circuit. Sometimes the tip is allowed to scan without this adjustment, and one measures only the deflection. This is useful for small, high-speed atomic resolution scans, and is known as variable-deflection mode.

Because the tip is in hard contact with the surface, the stiffness of the lever needs to be less that the effective spring constant holding atoms together, which is on the order of 1 – 10 nN/nm. Most contact mode levers have a spring constant of < 1 N/m.

Lateral Force Microscopy

LFM measures frictional forces on a surface. By measuring the "twist" of the cantilever, rather than merely its deflection, one can qualitatively determine areas of higher and lower friction.

Noncontact mode

Noncontact mode belongs to a family of AC modes, which refers to the use of an oscillating cantilever. A stiff cantilever is oscillated in the attractive regime, meaning that the tip is quite close to the sample, but not touching it (hence, "noncontact"). The forces between the tip and sample are quite low, on the order of pN (10^{-12} N). The detection scheme is based on measuring changes to the resonant frequency or amplitude of the cantilever.

Dynamic Force/Intermittent-contact/"tapping mode" AFM

Commonly referred to as "tapping mode" it is also referred to as intermittent-contact or the more general term Dynamic Force Mode (DFM). A stiff cantilever is oscillated closer to the sample than in noncontact mode. Part of the oscillation extends into the repulsive regime, so the tip intermittently touches or "taps" the surface. Very stiff cantilevers are typically used, as tips can get "stuck" in the water contamination layer. The advantage of tapping the surface is improved lateral resolution on soft samples. Lateral forces such as drag, common in contact mode, are virtually eliminated. For poorly adsorbed specimens on a substrate surface the advantage is clearly seen.

Force Modulation

Force modulation refers to a method used to probe properties of materials through sample/tip interactions. The tip (or sample) is oscillated at a high frequency and pushed into the repulsive regime. The slope of the force-distance curve is measured which is correlated to the sample's elasticity. The data can be acquired along with topography, which allows comparison of both height and material properties.

Phase Imaging

In phase mode imaging, the phase shift of the oscillating cantilever relative to the driving signal is measured. This phase shift can be correlated with specific material properties that effect the tip/sample interaction. The phase shift can be used to differentiate areas on a sample with such differing properties as friction, adhesion, and viscoelasticity. The techniques is used simultaneously with DFM mode, so topography can be measured as well.

Examples of atomic force microscope systems

The Nanosurf AFM systems are designed to be easy to use, and ideal for those just getting started with AFM. There are several models for a very wide range of applications. Here you can find a list of various atomic force microscopes.

19.11.2 Applications of AFM

The number of applications for AFM has exploded since it was invented in 1986 and nowadays this technique is involved in many fields of nanoscience and nanotechnology. The remarkable feature of STM and AFM instruments is their ability to examine samples not only in an ultrahigh vacuum but also at ambient conditions or even in liquids.

One of the advantages of AFM is that it can image the non-conducting surfaces, and therefore it is very suitable for biological systems.

The AFM is capable of measuring nanometer scale images of insulating surfaces with little or no sample preparation as well as measuring three-dimensional images of surfaces and studying the topography.

Some possible applications of AFM are:

- Substrate roughness analysis.
- Step formation in thin film epitaxial deposition.
- Pin-holes formation or other defects in oxides growth.
- Grain size analysis.
- Phase mode is very sensitive to variations in material properties, including surface stiffness, elasticity and adhesion.
- Comparing the tip-samples forces curves for materials to study the ratio of Young´s modulus (graphite as a reference for measure of the indentation).
- Obtaining information of what is happening under indentation at very small loads.
- By *in situ* AFM analysis with changes in temperature we can study changes in the structure.

19.13 REFERENCES

1. Ernst Ruska (1986). "Ernst Ruska Autobiography". (http://nobelprize.org/nobel_prizes/physics/laureates/1986/ruska-autobio.html.) Nobel Foundation. http://nobelprize.org/nobel_prizes/physics/laureates/1986/ruska-autobio.html. Retrieved 2010-01-31.
2. Kruger DH, Schneck P, Gelderblom HR (May 2000). "Helmut Ruska and the visualisation of viruses". (http://linkinghub.elsevier.com/retrieve/pii/S0140673600022509) . Lancet **355** (9216): 1713–7. doi:10.1016/S0140-6736(00)02250-9. PMID 10905259. http://www.ncbi.nlm.nih.gov/pubmed/10905259 http://linkinghub.elsevier.com/retrieve/pii/S0140673600022509.
3. M von Ardenne and D Beischer (1940). "Untersuchung von metalloxyd-rauchen mit dem universal-elektronenmikroskop" (in German). Zeitschrift Electrochemie **46**: 270–277.
4. "James Hillier". (http://web.mit.edu/Invent/iow/hillier.html) Inventor of the Week: Archive. 2003-05-01. http://web.mit.edu/Invent/iow/hillier.html. Retrieved 2010-01-31.
5. Erni, Rolf; Rossell, MD; Kisielowski, C; Dahmen, U (2009). "Atomic-Resolution Imaging with a Sub-50-pm Electron Probe". Physical Review Letters **102** (9): 096101. doi:10.1103/PhysRevLett.102.096101. (http://dx.doi.org/10.1103/PhysRevLett.102.096101). PMID 19392535 http://www.ncbi.nlm.nih.gov/pubmed/19392535
6. "The Scale of Things". http://www.sc.doe.gov/bes/scale_of_things.html Office of Basic Energy Sciences, U.S. Department of Energy. 2006-05-26. http://www.sc.doe.gov/bes/scale_of_things.html. Retrieved 2010-01-31.

7. O'Keefe MA, Allard LF (pdf). Sub-Ångstrom Electron Microscopy for Sub-Ångstrom Nano-Metrology. http://www.osti.gov/bridge/servlets/purl/821768-E3YVgN/native/821768.pdf. Information Bridge: DOE Scientific and Technical Information - Sponsored by OSTI. http://www.osti.gov/bridge/servlets/purl/821768-E3YVgN/native/821768.pdf. Retrieved 2010-01-31.
8. McMullan D (1993). "Scanning Electron Microscopy, 1928 - 1965". http://www-g.eng.cam.ac.uk/125/achievements/mcmullan/mcm.htm 51st Annual Meeting of the Microscopy Society of America. Cincinnati, OH. http://www-g.eng.cam.ac.uk/125/achievements/mcmullan/mcm.htm. Retrieved 2010-01-31.
9. "SPLEEM". http://ncem.lbl.gov/frames/spleem.html National Center for Electron Microscopy (NCEM). http://ncem.lbl.gov/frames/spleem.html. Retrieved 2010-01-31.
10. Nebesářová1, Jana; Vancová, Marie (2007). "How to Observe Small Biological Objects in Low-Voltage Electron Microscope". http://journals.cambridge.org/abstract_S143192760708124X Microscopy and Microanalysis **13** (3): 248–249. http://journals.cambridge.org/abstract_S143192760708124X.
11. Drummy, Lawrence, F.; Yang, Junyan; Martin, David C. (2004). "Low-voltage electron microscopy of polymer and organic molecular thin films". Ultramicroscopy **99** (4): 247–256. doi:10.1016/j.ultramic.2004.01.011. (http://dx.doi.org/10.1016/j.ultramic.2004.01.011) PMID 15149719. http://www.ncbi.nlm.nih.gov/pubmed/15149719.
12. Adrian, Marc; Dubochet, Jacques; Lepault, Jean; McDowall, Alasdair W. (1984). "Cryo-electron microscopy of viruses". Nature **308** (5954): 32–36. doi:10.1038/308032a0. (http://dx.doi.org/10.1038/308032a0) PMID 6322001. (http://www.ncbi.nlm.nih.gov/pubmed/6322001).
13. Sabanay, I.; Arad, T.; Weiner, S.; Geiger, B. (1991). "Study of vitrified, unstained frozen tissue sections by cryoimmunoelectron microscopy". (http://jcs.biologists.org/cgi/content/abstract/100/1/227) Journal of Cell Science **100** (1): 227–236. PMID 1795028. (http://www.ncbi.nlm.nih.gov/pubmed/1795028) http://jcs.biologists.org/cgi/content/abstract/100/1/227.
14. Kasas, S.; Dumas, G.; Dietler, G.; Catsicas, S.; Adrian, M. (2003). "Vitrification of cryoelectron microscopy specimens revealed by high-speed photographic imaging". Journal of Microscopy **211** (1): 48–53. doi:10.1046/j.1365-2818.2003.01193.x. (http://dx.doi.org/10.1046/j.1365-2818.2003.01193.x).
15. Bradbury, Savile, and Brian Bracegirdle. Introduction to Light Microscopy. New York: Springer-Verlag, 1998.
16. Jones, Thomas E. History of the Light Microscope. <http://www.utmem.edu/~thjones/hist/hist_mic.htm> .
17. Levine, S., and L. Johnstone. The Microscope Book. New York: Sterling Publishing Co., 1996.

18. Nachtigall, Werner. Exploring with the Microscope: A Book of Discovery and Learning. London: Sterling Publications, 1997.
19. Rogers, K. The Usborne Complete Book of the Microscope. Tulsa, OK: EDC Publishing, 1999.
20. WWW Virtual Library: Microscopy. <http://www.ou.edu/research/electron/mirror/>
21. William R. Herguth, President, Guy Nadeau.Applications of Scanning Electron Microscopy and Energy Dispersive Spectroscopy (SEM/EDS) To Practical Tribology Problems. Senior Technical Associate Herguth Laboratories, Inc.
22. R.F. Egerton. Electron Energy-Loss Spectroscopy in the Electron Microscope.
23. M.Von Heimendahl, W.Bell, G.Thomas. Applications of Kikuchi line Analyses in Electron Microscopy. Journal of Applied Physics 35 (1964) 3614.
24. C. Richard Brundle, Charles A. Evans Jr, Shaun Wilson. Encyclopedia of materials characterization, Butterworth-Heinemann Publications, 1992.
25. Joachim Mayer, Lucille A. Giannuzzi, Takeo Kamino, and Joseph Michael. TEM Sample Preparation and FIB-Induced Damage. Mrs Bulletin, Vol. 32, May 2007.
26. Binnig, G.; Rohrer, H.; Gerber, Ch.; Weibel, E. Phys. Rev. Lett. **1982**, 49, 57.
27. Binnig, G.; Quate, C. F.; Gerber, Ch. Phys. Rev. Lett. **1986**, 56, 930.
28. Magonov, S. N.; Whangbo, M.-H. Surface Analysis with STM and AFM; VCH: New York, 1996.
29. Strausser, Y. E.; Heaton, M. G. "Scanning Probe Microscopy Technology and Recent Innovations," Amer. Lab. **1991**, May, .
30. Magonov, S. N. "Surface Characterization of Materials at Ambient Conditions by Scanning Tunneling Microscopy and Atomic Force Microscopy" Appl. Spect. Rev. **1993**, 28, 1.

Glossary

Absorbance: Units used to measure the amount of IR radiation absorbed by a sample. Absorbance is commonly used as the Y-axis units in IR spectra. Absorbance is defined by Beer's Law, and is linearly proportional to concentration.

Absorption: Absorption of light at a defined wavelength from a compound.

Absorptive: Also known as the molar extinction coefficient in molecular spectroscopy, it is the wavelength-dependent absorption of an analyte as a function of concentration and pathlength and is expressed in units of concentration $-1 * cm^{-1}$.

Activate: Modifying the stationary phase into a chemical reactive form by introducing functional groups.

Adsorption: Interaction between the solute and the surface of an adsorbent.

Adsorption Chromatography: Basic LC mode that relies upon adsorption to the surface of an active solid to affect the separation.

Affinity Chromatography: Chromatography of analytes with stationary phases which contain specific ligands (i.e, enzyme, antigene, hormone).

Aliasing: If frequencies above the Nyquist Frequency are not filtered out, energy in these will appear as spectral artefacts below the Nyquist Frequency. Optical and electronic anti aliasing can be used to prevent this. Sometimes the higher frequencies are said to be "folded" back so the term "folding" is used.

Analyte: A sample component whose concentration is being measured (i.e. analyzed). In atomic absorption spectroscopy (AA), this is an element.

Angular Divergence: The spreading out of an infrared beam as it travels through the FTIR. Angular divergence contributes to noise in high resolution spectra, and can be a limit to achievable resolution.

Anion Exchanger: Ion exchanger with positive loaded solid ions and negative loaded counter ions which are exchangeable against other negative charged ions.

Apodization Functions: Functions used to multiply an interferogram to reduce the amount of sidelobes in a spectrum. Different types of apodization functions include boxcar, triangle, Beer-Norton, Hanning, and Bessel. The use of apodization functions unavoidably reduces the resolution of a spectrum.

Arm: The part of the microscope that connects the tube to the base. When carrying a microscope, grab the arm with one hand and place your other hand under the base.

Articulated Arm: A type of stand that holds a microscope body. The stand clamps to a table and has a variety of motion in three dimensions.

Artifact: A feature which appears in an NMR spectrum of a molecule which should not be present based on the chemical structure and pulse sequence used.

Ashing: Also referred to as charring, this is the step in a graphite furnace AA program that is designed to remove matrix constituents that might interfere with the measurement of the analyte. Ashing temperatures vary from 200 to 1800°C, depending on the matrix and the analyte element.

Atomization: The process of producing atoms for the atomic absorption measurement. The atom-forming process usually requires a high temperature (except for cold-vapor Hg methods) which is produced by a flame or by electrical current flowing through a resistive medium.

Background Spectrum: A single-beam spectrum acquired with no sample in the infrared beam. The purpose of a background spectrum is to measure the contribution of the instrument and environment to the spectrum. These effects are removed from a sample spectrum by ratioing the sample single-beam spectrum to the background spectrum.

Band Broadening: The process of increasing width and concomitant diluting of the chromatographic band as it moves down the column.

Base: The bottom support of the microscope (see arm above).

Baseline: Baseline is the line drawn by the data system when the only signal from the detector is from the mobile phase.

Baseline Correction: A spectral manipulation technique used to correct spectra with sloped or varying baselines. The user must draw a function parallel to the baseline, then this function is subtracted from the spectrum.

Beer-Lambert Law: the law that defines a linear relationship between concentration and absorbance. It is often written as absorbance = absorptivity * pathlength * concentration or $A = a^*b^*c$.

Biopolymer: Biological active macromolecules with molecular weight > MW 5000 (proteins, enzymes, nucleic acids, polysaccharides).

Biosensor: Device incorporating a biological sensing element connected to a transducer.

Body: This term is used mostly with the low power stereo microscopes and it is the basic heart of the microscope without any type of stand (base) or illuminators. It usually includes the eyepiece and objective lenses but not the focusing block.

Boxcar Truncation: With no apodization, all points in an interferogram are given equal weight, up to the edges of the interferogram. If the resolution is less than the smallest linewidth in the spectrum, oscillations appear on the baseline on both sides of the peaks.

Buffer: Aqueous solution of an acid/base pair at defined pH value. pH of a buffer will change a little by adding acids or bases.

Calibration: A quantitative procedure performed in order to relate the known concentration of standard solutions of the analyte element to the detector signal which is generated from the analyte in the unknown solutions. An extensive discussion of various methods of calibration including bracketing and standard addition can be found on this site in HTML, PDF, or MathCad formats

Calibration curve: Also known as a working curve, the relationship of instrument response (absorbance) as a function of concentration. Ideally, this should be a linear relationship in AA, under conditions that obey Beer's Law, where absorbance = (slope × concentration) + intercept. Minor curvature can be corrected by a curve-fitting algorithms.

Capillary Gas Chromatography (Capillary GC): Gas chromatographic separation in a capillary column.

Carrier Gas: Inert gas that constantly flows through the chromatography column and detector. Carrier gas mobilizes the sample by pushing it through the column. Inert refers to non-reactivity of the carrier gas with the sample and a low response of the detector to the carrier gas

Cation Exchanger: Resins or packings with functional groups that can separate cations.

Cell Density: Measurement of the Absorbance (also known as the optical density or OD) of a suspension of cells at 600 nm (OD 600 nm) can give a researcher an indication of when cells are ready for harvesting, e.g., *E coli.* It is important to note that for turbid samples such as cell cultures, the absorbance (OD) measured is due to light scattering, and not due to molecular absorption. All Biochrom spectrophotometers can measure OD at 600 nm.

Center burst: The sharp, intense part of an interferogram. The size of the center burst is directly proportional to the amount of infrared radiation striking the detector.

Chemical bonded Stationary Phases: Stationary phases which are created by a chemical reaction with a reagent and a packing material.

Chemical Screening: The screening of an applied magnetic field experienced by a nucleus due to the electron cloud around an atom or molecule.

Chemical Shift: A variation in the resonance frequency of a nuclear spin due to the chemical environment around the nucleus. Chemical shift is reported in ppm.

Chromatographic System: Instrument consisting of a pump, injector, guard column and column, detector and integrator.

Chromatography: Separation method

Chromatography column: Length of metal or fused silica tubing filled or coated with one or more packing materials that separates the compounds of interest as the sample mixture passes through.

Chromophor: Functional groups in a molecule which are absorbing in UV/VIS light.

Coadding: The process of adding interferograms together to achieve an improvement in signal-to-noise ratio.

Coarse Focus: This is the rough focus knob on the microscope. You use it to move the objective lenses toward or away from the specimen (see fine focus).

Coaxial Focus: A focusing system that has both the coarse and fine focusing knobs mounted on the same axis. Usually, the coarse knob is larger and on the outside and the fine knob is smaller and on the inside. On some coaxial systems, the fine adjustment is calibrated, allowing differential measurements to be recorded.

Coil: One or more loops of a conductor used to create a magnetic field. In NMR, the term generally refers to the radio-frequency coil.

Cold-Vapor: The method by which a cloud of atoms is produced from a solution containing Hg ions. Hg in the +2 state is reduced by the addition of stannous chloride and then swept by a flow of inert gas into a quartz-ended absorption cell kept at ~ 200°C.

Collimation: The ideal input beam is a cylinder of light. No beam of finite dimensions can be perfectly collimated; at best there is a diffraction limit. In practice the input beam is a cone that is determined by the source size or aperture used. The degree of collimation can affect the S/N and the resolution.

Column: Tube with end fittings. Tube contains the packing material.

Column Switching: The use of multiple columns connected by switching valves to effect better chromatographic separations or for sample cleanup.

Columns: Glass capillaries filled with stationary phase diameter: 0.2 – 1.0 mm, length: 5-50 m.

Complex Data: Numerical data with a real and an imaginary component.

Concentration using a Factor: The concentration of a sample is determined by multiplying the absorbance value by a specific factor. Concentration using a factor utilizes Beer Lambert Law where concentration is proportional to absorbance.

Concentration using a Standard Curve: Where a specific factor is not known, the concentration of a sample can be determined through constructing a standard curve by measuring the absorbance of standards that contain known concentrations of the analyte. Biochrom spectrophotometers offer a variety of curve fitting algorithms.

Condenser Lens: A lens mounted in or below the stage whose purpose is to focus or condense the light onto the specimen. The higher power objective lenses have very tiny diameters and require concentrated light to work properly. By using a condenser lens you will increase the illumination and resolution. Condenser lenses are not required on low power microscopes.

Constructive Interference: A phenomenon that occurs when two waves occupy the same space and are in phase with each other. Since the amplitudes of waves are additive, the two waves will add together to give a resultant wave which is more intense than either of the individual waves.

Continuous Wave (CW): A form of spectroscopy in which a constant amplitude electromagnetic wave is applied.

Contrast Plate: A circular opaque plate placed on the stage of a low power microscope. One side is white, the other is black. It can be flipped around depending on the coloration of your specimen.

Convolution: A mathematical operation between two functions.

Coordinate Transformation: A change in the axes used to represent some spatial quantity.

Coupling: Chemical reaction in which a biospecific adsorbent is prepared.

Cover Slip: A very thin square piece of glass or plastic placed over the specimen on a microscope slide. When used with liquid samples, it flattens out the liquid and assists with single plane focusing.

Cryopumping: The condensation of air onto a surface cooled by a cryogenic liquid.

Dead Volume: The volume outside of the column packing itself. The interstitial volume (intraparticle volume + interparticle volume) plus extracolumn volume (contributed by injector, detector, connecting tubing and end fittings) all combine to create the dead volume.

Dephasing Gradient: A magnetic field gradient used to dephase transverse magnetization.

Desorption: Removing of an adsorbed compound from the surface of the particles of a packing material.

Destructive Interference: A phenomenon that occurs when two waves occupy the same space. Since the amplitudes of waves are additive, if the two waves are out of phase with each other, The resultant wave will be less intense than either of the individual waves.

Detection: Is the method to monitor the eluting compounds at the column outlet.

Detection Limit: The minimum amount of an analyte that can be detected reliably. This is most often defined as three times the standard deviation of the blank measurement.

Detector: The part of the instrument that converts radiant energy from the light source to electricity. Typically a photomultiplier tube, but may also be a solid-state detector in more modern instrumentation

Detectors in GC:

- ECD - electron capture detector
- ELCD – electrolytic-conductivity detector
- FID - flame ionization detector
- FPD - flame photometric detector
- MSD - mass selective detector
- NPD - nitrogen phosphorous detector
- TCD - thermal conductivity detector

Detergents: Surface active substances, which are showing lipophilic as well as hydrophilic behavior.

Deuterium: Deuterium arc lamps measure in the UV region 190 - 370 nm. As Deuterium lamps operate at high temperatures, normal glass housings cannot be used for the casing. Instead, a fused quartz, UV glass, or magnesium fluoride envelope is used. When run continuously typical lamp life for a Deuterium lamp is approximately 1000 hours, however this can be extended by up to a factor of three using PTR technology. Deuterium lamps are always used with a tungsten halogen lamp to allow measurements to be performed in both the UV and visible regions.

Deuterium Arc Background Correction: The first successful method to correct for background absorption in furnace AA, this method employs a continuum radiation source (the deuterium arc, a *white* light source) that is passed through the atomic vapor cell along with the HCL radiation. While the deuterium arc is not significantly absorbed by atoms of the analyte, it behaves similar to the HCL radiation with respect to molecular absorption and scatter, thus allowing an accurate background correction. However, it is not as accurate as Zeeman or Smith-Hieftje methods at high background absorbances that approach or exceed 2.0 absorbance units.

Diaphragm: Generally, a five hole disc placed under the stage on a high power microscope. Each hole is of a different diameter. By turning it, you can vary the amount of light passing through the stage opening. This will help to properly illuminate the specimen and increase contrast and resolution. The diaphragm is most useful at the higher powers.

Diffuse Reflectance: The phenomenon that takes place when infrared radiation reflects off a rough surface. The light is transmitted, absorbed, scattered, and reflected by the surface. The light approaches the surface from one direction, but the diffusely reflected light leaves the surface in all directions. A reflectance sampling technique known as DRIFTS is based on this phenomenon.

Diffusion: Dilution of a compound in the direction of the column axis depending of differences in concentration.

Digital Filtering: A feature found on many news spectrometers which eliminates wrap-around artifacts by filtering out the higher frequency components in the time domain spectrum.

Direct Injection: Occurs when sample enters an inlet and is swept into a column by carrier-gas flow. No sample splitting or venting occurs during or after the injection.

Dispersion: Peak broadening of a compound during chromatographic transport through a column.

Dispersive Instruments: Infrared spectrometers that use a grating or prism to disperse infrared radiation into its component wave numbers before detecting the radiation. This type of instrument was dominant before the development of FTIR.

Double Beam: In a double-beam spectrophotometer the beam from the light source is split in two. One beam illuminates the reference cell holder and the other illuminates the sample. This configuration enables reference correction to be applied continually throughout a measurement and is commonly used when the sample and reference change over time, e.g., kinetics measurements. Although advances in electronics have greatly improved the reliability of single and split beam instruments, double beam instruments still provide the greatest accuracy.

Double-beam optics: The optical design whereby a percentage of the radiation from the light source of an AA is diverted before it reaches the atomization cell and monitored to compensate for drift in light source intensity.

Doubly balanced mixer: An electrical device, often referred to as a product detector, which is used in NMR to convert signals from the laboratory frame of reference to the rotating frame of reference.

Drain trap: A hole at the bottom of the mixing chamber that leads though a plastic tube to a water filled trap, allowing waste sample solution to drain from the mixing chamber but not allowing combustion gases to escape.

DTGS: Deuterated tri-glycine sulfate pyroelectric detectors are the most common detectors used in FT-IR instruments. They are chosen for their ease of use, good sensitivity, wide spectral responsivity and excellent linearity

Duplicate Range: For an interferogram, it is the ratio of the large centerburst signal at ZOPD to the smallest recorded signal (which must be greater than the noise for any benefit from signal averaging). The A/D used must have sufficient precision to measure the entire range as any clipping or distortion of the largest signal affects the whole spectrum.

Dynamic Range: Sometimes known as linear dynamic range or linear range, the analyte concentration range over which response is a well-defined (usually linear) function of the analyte concentration. The dynamic range can be increased by varying instrumental parameters, such as choice of analyte absorption line or decrease of absorption pathlength and sample volume.

Echo: A form of magnetic resonance signal from the refocusing of transverse magnetization.

Echo Time (TE): The time between the 90 degree pulse and the maximum in the echo in a spin-echo sequence.

Eddy Diffusion: Flow-heterogenity induced broadening and the multipath term.

Efficiency: A measure determined by the number of theoretical plates.

Electrodeless Discharge Lamp (EDL): A more intense radiation source for AA than the HCL, it consists of a sealed quartz tube containing a small amount of the element of interest and an inert gas. The lamp is placed in a radio frequency field, which excites the atoms to emit intense line radiation. EDL sources are less stable than HCL sources, but are far more intense and thus produce much better detection limits for elements such as As, Se, Hg, Sb, and Te.

Electrolytic-conductivity detector: In ELCD, the detector catalytically reacts halogen-containing solutes with hydrogen (reductive mode) to produce strong acid by-products that are dissolved in a working fluid. The acids dissociate and the detector measures increased electrolytic conductivity. Other operating modes modify the chemistry for response to nitrogen- or sulfur-containing substances

Electron-capture detector (ECD): Detector that is especially sensitive to halogen-containing molecules. Thermal electrons generated by the radioactive decay of 63Ni create a small, steady current within the detector. This current is measurably reduced when certain molecules capture electrons as they pass through the detector.

Eluens: Solvent or solvent mixture which is used to transport the analytes down the column.

Eluent: The mobile phase used to perform a separation.

Elution: The process of passing mobile phase through the column to transport solutes down a column.

Elution Profile: Dependence of concentration of a compound during the wash down.

Elutropic Series: A series of solvents (eluents) with an increasing degree of solvent strength (valid for a defined adsorbents).

Emission: Emitting light of a given wavelength after illuminating a compound.

Endcapping: Modification of an already silanized silica gel with a reactive short chain silane to cover still existing hydroxyl groups.

Exchange, Chemical: The interchange of chemically equivalent components on a molecule.

Exchange, Spin: The interchange of spin state between two nuclei.

Exclusion Chromatography: Method of HPLC to separate by size of the molecules (SEC = size exclusion chromatography).

Exclusion Limit: The upper limit of molecular weight (or size) beyond which molecules will be eluted at the same retention volume. For example, a 105 column of porous silica gel will exclude any compounds with a molecular weight greater than 100000 (based on a polystyrene calibration standard).

Eyepiece Lens: The lens at the top of the microscope that you look into. They are usually 10 × but also are available in 5 ×, 15 × and 20 ×. Widefield lenses have a large diameter and show a wide area of the field of view.

Felgett (multiplex) Advantage: An advantage of FT-IR instrument compared to scanning/single channel dispersive instruments. It is based on the fact that in an FTIR all the wave numbers of light are detected at once.

Field of View: Sometimes abbreviated "FOV", it is the diameter of the circle of light that you see when looking into a microscope. As the power gets greater, the field of view gets smaller. You can measure this by placing a clear metric ruler on the stage and counting the millimeters from one side to the other. Typically you will see about 4.5 mm at 40 ×, 1.8 mm at 100 ×, 0.45 mm at 400 × and 0.18 mm at 1000 ×. (See micrometer).

Fine Focus: This is the knob used to fine tune the focus on the specimen. It is also used to focus on various parts of the specimen. Generally, one uses the coarse focus first to get close then moves to the fine focus knob for fine-tuning.

Fixed Arm: A type of stand used with low power microscopes. The arm and body are integral parts of the microscope and connected solidly to the base.

Flame AA: The atomic absorption method that uses a flame as an atomization cell. Typical flames are air-acetylene (2400°C) and nitrous oxide-acetylene (3000°C).

Flame ionization detector (FID): Most common GC detector – detects all solutes containing at least 1 CHn unit.

Flame photometric detector (FPD): Highly selective for solutes containing S- or Sn-atoms.

Flow Spoiler: This is a plastic, fan-shaped device placed in the mixing chamber of a flame AA to improve the mixing of combustion gases and analyte solution droplets and facilitate the removal of large droplets down the drain trap at the bottom of the mixing chamber.

Flowrate (gas): The volumetric flowrate of combustible gases (L/min) into the mixing chamber of a flame AA instrument, or of inert gas used for graphite furnace and gas generation methods.

Flowrate (solution): The volumetric flowrate (mL/min) of solution uptake into the nebulizer of a flame AA instrument. This is typically from 5 to 10 mL/min.

Focus: A means of moving the specimen closer or further away from the objective lens to render a sharp image. On some microscopes, the stage moves and on others, the tube moves. Rack and pinion focusing is the most popular and durable type.

Fourier transform (FT): A mathematical technique capable of converting a time domain signal to a frequency domain signal and vice versa.

Fourier Transform: Calculation performed on an interferogram to turn it into an infrared spectrum.

Free induction decay (FID): A form of magnetic resonance signal from the decay of transverse magnetization.

Function: To respond to a particular analyte in a selective way through a biochemical reaction. To be used for the qualitative or quantitative determination of the analyte.

Gas chromatograph (GC): Type of instrument based on the injection of component mixtures onto a chromatography column, separation of the components as the mixture passes through the column, and sensing of the components of interest as they elute into a detector.

Gas sampling valve (GSV): A multiple-port, two-position valve that changes the direction of gas flows through chromatography columns and other parts of an instrument. Used to fill a sample loop with sample then inject the loop contents onto the column.

Gradient (G): A variation in some quantity with respect to another. In the context of NMR, a magnetic field gradient is a variation in the magnetic field with respect to distance.

Gradient Elution: Technique for decreasing separation time by increasing the mobile-phase strength over time during the chromatographic separation.

Graphite tube: The atomization cell used in an electrothermal atomizer for AA. Typically made of pyrolytic graphite and bathed in an inert gas such as Ar to prevent decomposition. Can be heated up to 3000°C

Gyromagnetic Ratio: The ratio of the resonance frequency to the magnetic field strength for a given nucleus.

Headspace Sampling: Gas-phase sampling technique in which the solute is removed from an enclosed space above a solid or liquid sample.

Hollow Cathode Lamp (HCL): The most common radiation source for AA, consisting of a low-pressure inert-gas-filled tube containing an anode and a hollow cathode made from the element for which the lamp is to produce atomic line radiation. Current flowing through the lamp (3 to 30 mA) is carried by the inert gas and sputters atoms of the analyte element from the cathode, which are subsequently collisionally excited to produce radiation characteristic of the analyte element.

Hydride-generation: The method by which hydride-forming elements, such as As, Se, Sb, and Te are released from solutions of their ions using sodium borohydride. The released hydrides are swept from the solution by an inert gas and decomposed to atoms in an absorption cell at a temperature of approximately 1000°C.

Hydrophilic Stationary Phase: Stationary phase with polar ligand.

Illuminator: A light source mounted under the stage. Three types of light are commonly used: Tungsten, Fluorescent and Halogen. Tungsten is the least expensive and most common. Fluorescent is bright, white and runs cool and Halogen is very bright and white but gives off heat like tungsten.

Imaginary Component: The component of a signal perpendicular to the real signal.

Imaging Sequence: A specific set of RF pulses and magnetic field gradients used to produce an image

Inert Gases: H_2, He, H_2

Integration: A process for identifying and calculating the amount of a component by measuring the area greater than the baseline defined by the instrument blank over a specific time period. In flame AA, integration times of three to ten seconds are most commonly used, since continuous signals are measured. In furnace and gas-generation methods that produce transient signals, the "peak" produced by the analyte is integrated from baseline to baseline.

Interferogram: A plot of infrared detector response versus optical path difference. The fundamental measurement obtained by an FTIR is an interferogram. Interferograms are Fourier transformed to give infrared spectra.

Interstitial Volume: The volume between the particles. It does not include the volume in the pores of the particles.

Inversion Recovery Sequence: A pulse sequence producing signals which represent the longitudinal magnetization present after the application of a 180° inversion RF pulse.

Inversion Time (TI): The time between the inversion pulse and the sampling pulse(s) in an inversion recovery sequence.

Ion Exchange Chromatography: Mode of chromatography in which ionic substances are separated on cationic or anionic sites of the packing in aqueous mobile phases.

Ion Exchanger: Stationary phase with bonded ions at its inner surface with counter ions.

Ion Pair: Anion (A^-) and cation (B^+) which can be paired or neutralized and separated as an ion pair on a reversed phase column.

Ion-pair Chromatography: Form of chromatography in which ions in solution can be paired by using an ion-pair reagent as modifier of the mobile phase.

Ion-trap detector: A mass spectrometric (MS) detector that uses an ion-trap device to generate mass spectra.

Isocratic Chromatography: Using a time invariant-eluent composition in LC.

Lipophilic Stationary Phase: Stationary phase with alkyl bonding.

Loading: Amount of mass of liquid stationary phase per mass of loaded packing material (used in partition chromatography).

Longitudinal Magnetization: The Z component of magnetization.

Lorentzian Lineshape: A function obtained from the Fourier transform of an exponential function.

Magnetic Resonance Imaging (MRI): An imaging technique based on the principles of NMR.

Magnitude: The length of a magnetization vector. In NMR, the square root of the sum of the squares of the Mx and My components, i.e., the magnitude of the transverse magnetization.

Matrix: Medium where the analyte is distributed in.

Matrix Modifier: An element or compound that is added to the sample in a graphite furnace AA measurement in order to increase the volatility of the matrix (and thus remove it during the ashing stage of the temperature program), or to decrease the volatility of the analyte element so that it can be atomized at high temperatures. Some typical matrix modifiers are palladium, nickel and ammonium phosphate.

Micrometer: Also called a micron it is the metric linear measurement used in microscopy. There are 1000 microns in a millimeter. If something is 1.8 mm long then it can also be expressed as 1800 microns (or micrometers) long.

Microwave Digestion: the preferred method for dissolving most samples in acid for analysis by AA. The method uses a closed Teflon container into which 5 to 10 mL of acid and approximately 0.5 gram of sample are subjected to an increasing microwave field for periods up to an hour. The high pressure and temperature inside the container rapidly dissolves most samples and no volatile analytes are lost since the container is sealed.

Mirror: Allows you to direct ambient light up through the hole in the stage and illuminate the specimen.

Mirror Displacement: The distance that the mirror in an interferometer has moved from zero path difference.

Mixing chamber: The heart of the sample introduction system for a flame AA, this is a plastic chamber in which combustible gases are mixed with the solution droplets from the nebulizer and then transported to the flame. Larger droplets (approximately 95% of the sample) are removed from the mixing chamber through the drain trap.

Mobile Phase: Solvent to transport the compounds to separate down the column.

Modulation: The periodic variation of the radiation from the light source, either electronically or mechanically with a chopper, at frequencies between 30 and 200 Hz. Modulation of the light source allows the instrument to discriminate against other sources of radiation that might reach the detector and bias the absorbance measurement.

Monochromator: A wavelength selection device used in AA spectrometers to isolate the absorbable radiation from the light source from other extraneous radiation, both from the source (non or weakly absorbing lines) and the atomizer (flame or furnace emission).

Monocular Head: A microscope head with a single eyepiece lens.

Multi Wavelength Measurement: This is an extension of single wavelength measurements where sample absorbance is recorded at multiple wavelengths. Mathematical manipulations of this data can often reveal details about the sample's composition or purity. Nucleic acid measurement at 260 nm and 280 nm is an example of a multiwavelength application to check sample purity. Absorbance at multiple wavelength used in combination with a mathematical formula is often used to define color intensity in, e.g., wine.

Multi-column Operation: Special technique in HPLC to enhance the separation.

Nebulizer: The component of a flame AA sample introduction system that draws aqueous solution into the mixing chamber and converts it to a fine mist of small droplets that are swept into the flame. It is typically a "pneumatic" nebulizer that operates on the principle of the Bernoulli effect, where the low pressure produced by air flowing rapidly into the mixing chamber through the nebulizer pulls the analyte-containing solution through a capillary tube.

Negative Frequency Artifact: The appearance of smaller in amplitude peaks in one half of the spectrum which are the mirror image of ones in the opposite half.

Net Magnetization Vector: A vector representing the sum of the magnetization from a spin system.

Nitrogen phosphorous detector (NPD): Nitrogen/phosphorus selective detector detects only solutes with N and/or P-atoms.

Noise: High frequency shifts of detector signal measured during base line measurement at highest sensitivity.

Normal Phase Chromatography: Chromatography with silica gel, aluminium oxide or polar bonded silica gels to be used with organic solvents as eluent.

Normalized: The process of dividing all the absorbance values in a spectrum by the largest absorbance value. This resets the Y-axis scale from 0 to 1.

Nosepiece: The part of the microscope that holds the objective lenses also called a revolving nosepiece or turret.

Nuclear Magnetic Resonance (NMR): A spectroscopic technique used by scientists to elucidate chemical structure and molecular dynamics.

Numerical Aperture (N.A.): This is a number that expresses the ability of a lens to resolve fine detail in an object being observed. It is derived by a complex mathematical formula and is related to the angular aperture of the lens and the index of refraction of the medium found between the lens and the specimen. To get the best possible image, you should have a condenser system that matches or exceeds the N.A. of the highest power objective lens on your microscope. (N.A. is only important with high power microscopes.)

Nyquist Frequency: A term widely used in information theory, but here applies to the highest frequency, shortest wavelength, that can be identified in an interferogram. It is the one for which there are exactly two points per cycle. The contribution of any higher frequency, signal or noise, can be represented by some lower frequency and so will appear aliased or folded into the spectrum.

Objective Lens: The lens closest to the object. In a stereo (low power) microscope there are objective pairs, one lens for each eyepiece lens. This gives the 3-D effect. On a high power binocular model there is still only one objective lens so no stereo vision.

On-column injection (OCI): In on-column injection, sample enters the column directly from the syringe and does not contact other surfaces. On-column injection usually signifies cold injection for capillary columns.

Optical Distance: Physical distance multiplied by the index of refraction of the medium.

Optical Path Difference: The difference in optical distance that two light beams travel in an interferometer.

Packing material: Granular, solid-phase powder, sometimes also coated with a liquid phase, that is used to fill a chromatography column. The packing material of a column is chosen for its ability to separate the molecules of interest from one another. Packing materials typically separate molecules by their size (molecular sieves), boiling points, or polarities.

Partition Chromatography: Chromatography using a liquid-liquid phase system.

PC Control: Full control of a spectrophotometer via a PC, including data export and analysis.

Peak: When the detector registers the presence of a compound, the normal baseline signal it sends to the data system changes, resulting in a deflection from the baseline called a peak. Well resolved peaks are symmetrical, touch the baseline, and do not interfere with other peaks.

Peak Area: The area enclosed between the peak and the peak base.

Peak capacity: The amount of solute that can be injected without a significant loss of column efficiency.

Peak Standard Deviation: Peak width in time or volume in 61% of peak height.

Phase Correction: A software procedure to compensate for not taking a data point exactly at ZOPD, and for frequency dependent variations caused by the beam splitter and signal amplification. The Mertz and Forman corrections are both used with the Mertz applied to double sided interferograms; this is considered in the most accurate approach.

Phase System: Mobile and stationary phase used for separation.

Photomultiplier: The most often used detector in an AA instrument. It consists of a vacuum tube containing an alkali-element photocathode that produces electrons when struck by photons of sufficient energy (the photoelectric effect). Each photoelectron is then multiplied by collisions with a series of dynodes so that the electrical signal produced by each photon is greatly amplified.

Pixel: Picture element.

Platform Atomization: Also known as the L'vov platform, it is a small platform onto which a sample is placed inside the graphite furnace tube rather than placing the sample on the tube wall. This delays sample atomization until the gas temperature inside the furnace is higher than it would be for wall atomization, which reduces some interference effects. A good discussion of graphite furnace history and the design of modern instruments can be found at this site at the University of Umea.

Polarity: Measure of lipophilic or hydrophylic property of a stationary or mobile phase.

Polarizibility: Measure for charge shifts in a molecule.

Pore Volume: Cavity volume of porous particles of stationary phase in the column.

Precess: A rotational motion of a vector about the axis of a coordinate system where the polar angle is fixed and the azmuthal angle changes steadily.

Press to Read (PTR): Biochrom press to read (PTR) technology helps to prolong lamp life and reduce the cost of ownership by ensuring that the lamp is only on when a measurement is being taken.

Print: Print directly from the instrument. Options offered by Biochrom include built-in thermal printers, and connection to external devices such as chart recorders, or serial printers.

Proportionality Constant: A constant used to convert one set of units to another.

Pulse Sequence: A series of RF pulses and/or magnetic field gradients applied to a spin system to produce a signal whose behavior gives information about some property of the spin system.

Purge- and-trap sampling: A concentration technique for volatile solutes. Sample is purged with an inert gas that entrains volatile components onto an adsorptive trap. The trap is then heated to desorb trapped components into a GC column.

Quadrature Detection: Detection of Mx and My simultaneously as a function of time.

Quantitative Analysis: The determination of the amount or concentration of the components in the sample.

Radio Frequency: A frequency band in the electromagnetic spectrum with frequencies in the millons of cycles per second.

Raw data: The Mx and My data as a function of time and/or other parameters in an NMR pulse sequence. This is also called k-space data.

Real: The component of a signal perpendicular to the imaginary signal.

Redox Potential: Measure for reducing and oxidizing power of a redox pair.

Regeneration: Returning the packing in the column to its initial state after gradient elution

Repetition Time: The time between repetitions of the basic sequence in a pulse sequence.

Resolution: A measure of how well an IR spectrometer can distinguish spectral features that are close together. For instance, if two features are 4 cm^{-1} apart and can be discerned easily, the spectrum is said to be at least 4 cm^{-1} resolution. Resolution in an FTIR is mainly determined by the optical path difference.

Resonance: An exchange of energy between two systems at a specific frequency.

Retention: Delay time of separated compounds due to their interaction with the stationary phase.

Retention Factor: The period of time that the sample component resides in the stationary phase relative to the time it resides in the mobile phase.

Retention Time: The time between injection and the appearance of the peak maximum.

Retention Volume: The volume of mobile phase required to elute a substance from the column.

Reticle: A very tiny grid pattern inserted in an eyepiece lens. It is used to make actual measurements of the size of objects seen through the microscope.

Reversed-phase Chromatography: Chromatography on lipophilic supports or stationary phases.

Reversed-phase Supports: Supports or stationary phases with lipophilic properties (Styrene divinyl benzene co-polymeres, silanized silica gels).

Revolving Nosepiece: See nosepiece

RF Coil: An inductor-capacitor resonant circuit used to set up B_1 magnetic fields in the sample and to detect the signal from the sample.

RF Pulse: A short burst of RF energy which has a specific shape.

Ring Light: An independent light that usually connects to the microscope body and gives off a ring of light.

Rotation Matrix: A matrix used to describe the rotation of a vector.

Saddle Coil: A coil geometry which has two loops of a conductor wrapped around opposite sides of a cylinder.

Sample Look: Length of hollow metal tubing whose fixed volume is filled with sample to a repeatable pressure and its contents injected onto a chromatography column. Consistent sample loop volume, pressure and temperature ensure that the same amount of sample is injected for every analysis.

Sample Preparation: All steps to prepare a sample before separation.

Sample Probe: That portion of the NMR spectrometer containing the RF coils and into which the sample is placed.

Sensitivity: The relationship of analyte concentration to instrument response. Mathematically, this is the slope of the linear plot of "instrument response vs. analyte concentration". Traditionally, in AA, the sensitivity is defined as the concentration of the analyte that produces an instrument response of 0.0044 absorbance units (1% absorption).

Sensor: Device that detects or measures a physical property and records, indicates or otherwise responds to it (OED).

Sensor Vs Detector: A sensor has generally some additional properties quantitative, repeatable, regenerable.

Sidelobes: Spectral features that appear to the sides of an absorbance band as undulations in the baseline. Sidelobes are caused by having to truncate an interferogram, as a result of finite scan distance, and can be removed from a spectrum by multiplying the spectrum's interferogram by an apodization function.

Silica with Polar Bonding: Silica gel with polar functional groups like Cyano, Diol, Amino, Nitro, etc.

Silylation: Reaction of an organochloro- or organoalkoxysilane with a compound containing a reactive group.

Sinc Pulse: An RF pulse shaped like Sin(x)/x.

Single Beam: These are the simplest and most economical instruments. In a single beam spectrophotometer all of the light passes through the sample holder. Measurements are made by placing a reference in the sample holder, which is measured to standardize the instrument. This reference value is subtracted from subsequent sample measurements to remove effects from the solvent and the cell.

Single Beam Spectrum: The spectrum that is obtained after Fourier transforming an interferogram. Single beam spectra contain features due to the instrument, the environment, and the sample.

Single Wavelength Measurement: Also known as fixed wavelength. This is the simplest application of a spectrophotometer and allows for measurement of the Absorbance or transmittance (percentage of transmission %T) at a specified wavelength. All Biochrom spectrophotometers and colorimeters can perform single wavelength measurements.

Sipper: Used with a flow cell this enables measurement of controlled sample volumes. Sippers are ideal for speeding up sample analysis, e.g., for monitoring continuous production processes but they also reduce sample handling when measuring toxic materials.

Slide: A flat glass or plastic rectangular plate that the specimen is placed on. It may have a depression or well to hold a few drops of liquid.

Slip Clutch: When students bring the focus all the way up or down and continue to try turning the knob, damage to the focusing system can occur if there wasn't a slip clutch. It is a mechanical device that protects the gears of the microscope.

Smith-Hieftje Background Correction: A method to correct for background absorption in furnace AA that pulses the HCL at low and then at high current. During the high current pulse, a large cloud of atoms is formed in front of the lamp cathode. This cloud essentially prevents absorbable radiation from reaching the analyte in the atomic vapor cell and thus allows discrimination of atomic absorption from other sources of absorption.

Spacer: Spacer via which a functional group is bonded to the surface of the packing material.

Spin: A fundamental property of matter responsible for NMR and MRI.

Spin Density: The concentration of spins.

Spin Packet: A group of spins experiencing the same magnetic field.

Spin-Echo: An NMR sequence whose signal is an echo resulting from the refocusing of magnetization after the application of 90° and 180° RF pulses.

Spin-Lattice Relaxation: The return of the longitudinal magnitization to its equilibrium value along the +Z axis.

Spin-Lattice Relaxation Time (T_1): The time to reduce the difference between the longitudinal magnitization and its equilibrium value by a factor of *e*.

Spin-Spin Relaxation: The return of the transverse magnitization to its equilibrium value (zero).

Spin-Spin Relaxation Time: The time to reduce the transverse magnetization by a factor of *e*.

Split injection: The sample size is adjusted to suit capillary column requirements by splitting off a major fraction of sample vapors in the inlet so that as little as 0.1% enters the column. The rest is vented.

Splitless injection: A derivative of split injection. During the first 0.5–4 min of sampling, the sample is not split and enters only the column. Splitting is restored afterward to purge the sample remaining in the inlet. As much as 99% of the sample enters the column.

Stage: The flat plate where the slides are placed for observation.

Stage Clips: Clips on the stage used to hold the slide in place.

Stage Plate: On a low power microscope, there is a frosted circular glass plate that fits in over the lower illuminator. This is called the stage plate. (See also contrast plate.)

Stand: On a low power microscope, the type of connection between the microscope body and the base. There are three main types: the post, the fixed arm and the universal boom stand.

Standard Addition: a method of calibration that compensates for matrix-induced enhancement or supression of analyte signals. A known concentration of analyte element is added to the sample and the instrument response of the known concentration of added element is used to calibrate the instrument response for the sample.

Stationary Phase: The immobile phase involved in the chromatographic process.

Stereo: Related to microscopes, seeing with both eyes through separate eyepiece and objective lenses. With two objectives, the image looks 3-D, we see it in "stereo". (See also Binocular head.)

Store: Stores data on the instrument's internal memory, allowing download or export to removable media, e.g., SD card, Bluetooth® or USB mass storage devices for storage or data manipulation on a PC.

Superconduct: To have no resistance. A perfect superconductor can carry an electrical current without losses.

Support: Refers to solid particles carrying the stationary phase (= packing material).

Surface Modification : Specific modification of chemical and physical properties of a packing material.

T_2*: Pronounced T-2-star. The spin-spin relaxation time composed of contributions from molecular interactions and inhomogeneities in the magnetic field.

Temperature Control: Applications that require sample measurements at non-ambient temperatures need some form of temperature control, e.g., life science methods at 37°C. Biochrom offers two methods for temperature control; electrical control (also known as Peltier control), which gives precise temperature control over a wide range of temperatures and water circulated temperature control (water jacket) an economical alternative for precise temperature control.

Theoretical Plate Count: Measure of column efficiency.

Theoretical Plate Height: Height equivalent to a theoretical plate. A measure of a column`s efficiency.

Thermal Conductivity detector (TCD): Unselective, detects all types of solutes.

Timing Diagram: A multiaxis plot of some aspects of a pulse sequence as a function of time.

Transducer: Device that converts an observed change (physical or chemical) in a measurable signal (usually electronic signal).

Transmission Sampling: A sampling method where the infrared beam passes through the sample before it is detected. Samples are typically diluted or flattened to adjust the absorbance values to a measurable range.

Transverse magnetization: The XY component of the net magnetization.

Tungsten Halogen: Also known as quartz Iodine lamps, these measure most effectively in the visible region from 320 - 1100 nm. Instruments that only use Tungsten halogen lamps as the light source will only measure in the visible region. Using Press to Read technology the lamp life of Tungsten halogen lamps can be extended to approximately 3000 hours.

UV/Visible: Often abbreviated to UV/VIS or UV-VIS. UV-Visible spectroscopy offers the maximum flexibility and is suitable for applications in the wavelength range 190 to 1100 nm. In UV/Visible spectroscopy the UV region is considered to be any wavelength less than 340 nm. Nucleic acid, purified proteins and other organic molecules are often measured in the UV region. Another name for a UV/VIS spectrophotometer is UV/VIS spectrometer. Special cells and cuvettes have to be used in UV/VIS spectrophotometers.

Van Deemter Graph: Dependence of theoretical plate height from linear velocity.

Visible: Often abbreviated to Vis. A visible spectrophotometer or colorimeter typically measures in the visible region of the electromagnetic spectrum (340 - 750 nm), although Biochrom spectrophotometers can measure up to 1100 nm (the near infrared region). If your samples are coloured this is an indication that they will absorb visible light. If you do not need to measure in the UV region (<340 nm) a visible spectrophotometer or colorimeter may be a more economical alternative.

Void Time: The time for elution of an unretained peak.

Wavelength Scanning: Also known as spectral scanning or wavescan measurements. The absorbance or % transmittance of a sample is measured over a specified wavelength range.

The position and magnitude of these peaks and valleys will give an indication of sample composition and purity.

Xenon: Xenon flash lamps provide a high energy light source with a short warm up time and long lamp life. They measure in both the UV and visible regions of the electromagnetic spectrum from 190 - 1100 nm. Xenon lamps typically provide 80 flashes per second giving them their distinctive bird-like noise.

Zeeman Background Correction: a method to correct for background absorption in furnace AA that uses a magnetic field around the atomizer. The field splits the energy levels of the absorbing atoms and allows discrimination of atomic absorption from other sources of absorption.

Index